The Tectonics of Structural Systems

The Tectonics of Structural Systems provides an architectural approach to the theory of structural systems. This book combines:

- structural recommendations to follow during the architectural design of various structural systems;
- the tectonic treatment of structural recommendations in architecture.

Written expressly for students, this book makes structures understandable and useful, providing:

- practical and useful knowledge about structures;
- a design-based approach to the subject of structures;
- a bridge in the gap between structures and the theory of design.

Successful architectural examples for each structural system are given in order to demonstrate that tectonics can be achieved with sound technical knowledge. Over 300 illustrations visually unpack the topics being explained, making the book ideal for the visual learner.

Yonca Hurol is a Professor at the Department of Architecture, Eastern Mediterranean University, North Cyprus. She has been teaching structure to students of architecture for more than 30 years. She has worked at the Middle East Technical University and Gazi University, Turkey, and is currently working at Eastern Mediterranean University, North Cyprus. She has published many articles in the areas of her research interests: structures in architecture, earthquake architecture, and ethics and architecture.

'This beautifully illustrated book provides a useful insight, from an architect's perspective, into the often problematic interface between structural requirements and the tectonics of architecture. A particular strength is the large number of high quality graphic depictions of the functioning of a wide range of structural types and case-study buildings.'

Angus Macdonald, Professor, University of Edinburgh, UK

'*The Tectonics of Structural Systems* explains the principles of building structures in an architectural context. It introduces a comprehensive range of structural systems and imparts a wealth of practical structural knowledge that will find application in the design studio. Architectural students, especially, will appreciate the large number of explanatory diagrams.'

Andrew Charleson, Associate Professor, School of Architecture,
Victoria University of Wellington, New Zealand

The Tectonics of Structural Systems

An architectural approach

Yonca Hurol

With contributions from guest author Baydu Can Al

Routledge
Taylor & Francis Group
LONDON AND NEW YORK

First published 2016
by Routledge
2 Park Square, Milton Park, Abingdon, Oxon OX14 4RN

and by Routledge
711 Third Avenue, New York, NY 10017

Routledge is an imprint of the Taylor & Francis Group, an informa business

British Library Cataloguing-in-Publication Data
A catalogue record for this book is available from the British Library

Library of Congress Cataloging-in-Publication Data
Hurol, Yonca.
The tectonics of structural systems : an architectural approach / Yonca Hurol with contributions from guest author Baydu Can Al.
pages cm
Includes bibliographical references and index.
1. Structural engineering. 2. Buildings. I. Baydu, Can Al. II. Title.
TA633.H87 2015
624.1—dc23 2015003897

ISBN: 978-1-138-85548-9 (hbk)
ISBN: 978-1-138-85553-3 (pbk)
ISBN: 978-1-315-72030-2 (ebk)

Typeset in Avenir
by Keystroke, Station Road, Codsall, Wolverhampton

To the memory of Selçuk Sait

CONTENTS

List of figures xv
List of tables xxiii
Acknowledgements xxv
List of abbreviations xxvii

1 Introduction 1
Explanation of key concepts 1
Research problem 4
Objectives of this book 4
Methodology for determining structural guidelines 5
Methodology for providing analytical tools 5
Methodology for discussing the relationship between structural guidelines and tectonics 6
Contents of this book 6
References 8

2 Tectonics and Structural Guidelines 10
History of the concept of tectonics 10
Evaluation of tectonic value 13
An analytical approach to tectonics 14
An analytical approach to aesthetic value of structures 14
An analytical approach to tectonic value of structures 15
Notes 15
References 15

PART 1 GENERAL STRUCTURAL GUIDELINES 17

3 General Structural Guidelines in Relation to Economy 19
Structural efficiency and the optimum 19
Possible approaches to the economy of structures 19
Designing for evolutionary structural optimisation 20
Designing for the optimum structure 21
Optimisation of the designed structure 21
Percentages of different approaches to economy 23
Structural guidelines for different approaches to economy 23
Note 24
References 24

4 General Structural Guidelines Originating from the Requirements of Strength, Stability, Equilibrium and Deformation Limit 25
Strength 25
Types of forces 25
Forces affecting building structures 26
Types of stress 27
Stability 33
Overturning 33
Buckling 34
Sliding 34
Uneven settlement 35
Wind instability 36
Equilibrium 36
Deformation limit 39
Structural guidelines originating from these requirements 41
Problems to solve 42
References 43

5 Structural Guidelines in Relation to Form and Size 44
Classification of modern structural systems 44
Exceptions not fitting into this classification 47
Size and proportions of structural systems 48
Structural guidelines and building form 49
References 49

PART 2 THE TECTONICS OF MASONRY STRUCTURES 51

6 The Tectonics of Traditional Approaches to Masonry Structures 53
The tectonics of a traditional approach to stone masonry 53
Structural guidelines for a stone wall 53
Structural guidelines for a stone building 55
Structural guidelines for stone masonry 59
Case study 1: Cologne Cathedral, Germany 60
Case study 2: The stone tower at Norman Castle, UK 61
The tectonics of a traditional approach to brick masonry 62
Types of bricks 62
Thickness of brick walls 63
Structural guidelines for brick masonry 63
Case study 3: Summer House, Finland 64

The tectonics of a traditional approach to adobe masonry 65
Types of adobe masonry walls 65
Differences between adobe and stone masonry structures 66
Structural guidelines for adobe masonry 67
Case study 4: Harran Houses, Turkey 68
Case study 5: Great Mosque of Djenne, Mali 69
The tectonics of a traditional approach to timber masonry 70
Comparing traditional timber masonry with timber balloon frames 71
The role of structural elements in traditional timber masonry 72
Structural guidelines for timber masonry 72
Case study 6: İsmail Hacı Çakır House, Turkey 73
Conclusions 75
Note 75
References 75

7 The Tectonics of Masonry Roof Structures 77
Masonry arch 77
Form of masonry arches 77
Structural behaviour of masonry arches 77
Construction process of masonry arches 79
Span of masonry arches 80
Structural guidelines for masonry arches 80
Case study 7: The arches at Cordoba Mosque, Spain 80
Masonry vault 81
Form of masonry vaults 81
Structural behaviour of masonry vaults 81
Construction process of masonry vaults 82
Span of masonry vaults 82
Structural guidelines for masonry vaults 83
Case study 8: The vaults at Cologne Cathedral, Germany 83
Masonry dome 84
Form of masonry domes 84
Structural behaviour of masonry domes 84
Construction process of masonry domes 86
Span of masonry domes 86
Structural guidelines for masonry domes 87
Case study 9: The dome at Pantheon, Italy 87
Conclusions 88
References 89

8 The Tectonics of Contemporary Approaches to Masonry Structures 90
Types of reinforced masonry 90
Flexibility provided by reinforced masonry 91
Structural guidelines for reinforced masonry structures 92
Hybrids of reinforced masonry and reinforced concrete frame systems 92
Hybrids of reinforced masonry and steel frame systems 94
Case study 10: Villa Müller, Czech Republic 95
Case study 11: Falling Water, USA 97
Conclusions 98
References 98

PART 3 THE TECTONICS OF FLEXURAL STRUCTURES 101

9 The Tectonics of Frame and Shear Wall Systems 103
Elements of frame systems 104
Beams 104
Columns 105
Slabs 106
Partition walls 109
Stairs 110
Foundations 114
Expansion joints 116
Shear wall systems and use of shear walls with frame systems 118
Finding internal forces in frames 120
Drawing N, V and M diagrams and the deflected shape of determinate systems 120
The Portal Method 126
When to use frame systems 129
Organisation of elements 129
Earthquake-resistant design of frame systems 131
Dimensions of columns in relation to dimensions of beams 132
Plan irregularities 132
Vertical irregularities 134
Structural guidelines for frame and shear wall systems 136
Case study 12: Villa Savoye, France 138
Case study 13: National Assembly in Dacca, Bangladesh 139
Case study 14: Church of Light and Sunday School, Japan 140
Case study 15: Barcelona Pavilion, Spain 142
Discussion on the case studies 142

Conclusions 143
Problems to solve 144
References 145

10 The Tectonics of Frame Systems in Interior Architecture 147
Making subtractions from frame systems 147
Subtraction of columns, beams and shear walls 147
Subtraction of slabs 148
Subtraction of stairs 148
Subtraction of rigid infill walls 148
Subtraction of lightweight infill walls 148
Making additions to frame systems 148
Connection with the existing structure 149
Order 149
Lightweight structures 149
Structural engineering input 149
Structural guidelines for subtractions and additions to existing frame systems 149
Case study 16: Photographer's Penthouse, USA 150
Case study 17: Suspended Bedroom, France 151
Conclusions 152
References 152

11 The Tectonics of High-Rise Building Structures 154
The Cantilever Method 155
Strategies to increase height 157
Use of aerodynamic form 157
Selecting the appropriate structural system 159
Use of damping systems 163
Structural guidelines for high-rise building structures 163
Case study 18: World Trade Center, USA 164
Case study 19: Shanghai World Financial Center, Republic of China 165
Conclusions 167
Problems to solve 167
References 168

PART 4 THE TECTONICS OF FORM-RESISTANT STRUCTURES 171

12 The Tectonics of Tensile Structures 173
General characteristics and problems of tensile structures 173
Cables and suspension structures 176
Analysis of Akashi Kaikyo Bridge 176
Analysis of the former Federal Reserve Bank Building (now Marquette Plaza) 177
Analysis of Dorton Arena 178
Analysis of Yale Hockey Rink 179
Structural guidelines for cables and suspension structures 179
Case Study 20: Zagreb Arena, Croatia 179
Cable trusses, bicycle-wheel structures and suspended glass systems 181
Structural behaviour of cable trusses 181
Construction process of cable trusses 182
Bicycle-wheel structures 182
Suspended glass systems 183
Structural guidelines for bicycle-wheel structures and suspended glass systems 184
Case study 21: Science and Technology Museum, France 185
Membrane structures 186
Structural behaviour and construction of membranes 186
Organisation of membrane units 187
Structural guidelines for membrane structures 188
Case study 22: Olympic Stadium in Munich, Germany 188
Pneumatic structures 189
Types of pneumatic structures 190
Structural guidelines for pneumatic structures 190
Case study 23: Swarovski Pavilion, Switzerland 190
Negative curvature shells 191
Conclusions 192
Problems to solve 192
References 192

13 The Tectonics of Compression Structures 195
General characteristics and problems of compression structures 195
Steel vaults and domes 197
Steel vaults 197
Steel domes 197
Structural guidelines for steel vaults and domes 198
Case study 24: The dome at Parliament Building in Berlin, Germany 198

Geodesic domes 200
- Geometry and the structural behaviour of geodesic domes 200
- Examples of geodesic domes 200
- Construction methods of geodesic domes 201
- Structural guidelines for steel geodesic domes 202
- Case study 25: US Pavilion at Expo '67, Canada 202

Shell structures 202
- Thin and thick shells 203
- Form of shell structures 204
- Ways to increase the span of shell structures 206
- Construction methods of shell structures 207
- Structural guidelines for shell structures 207
- Case study 26: Kimbell Art Museum, USA 207
- Case study 27: Small Sports Palace, Italy 208
- Case study 28: TWA Airport Building, USA 209

Grid shells 211
- Form and organisation of grid shells 211
- Details and construction of grid shells 211
- Structural guidelines for grid shells 212
- Case study 29: Suan Lien Center, Republic of China 213

Conclusions 213

Problems to solve 214

References 214

14 The Tectonics of Tension and Compression Structures 216

2D trusses 216
- Analysis of internal forces in trusses 217
- Span and depth of trusses 219
- Use and organisation of trusses in buildings 220
- Structural guidelines for trusses 221
- Case study 30: Cluj Arena, Romania 221

3D trusses 222
- Span and depth of 3D trusses 223
- Structural guidelines for 3D trusses 223
- Case study 31: Waterloo Terminal, UK 223

Space frames 224
- Span and depth of space frames 225
- Structural guidelines for space frames 226
- Case study 32: Water Cube, Republic of China 226

Conclusions 227
Problems to solve 227
References 228

15 The Tectonics of Folded Plates 229
Form of folded plates 229
Structural behaviour and span of folded plates 230
Structural guidelines for folded plates 231
Case study 33: Yokohama International Port Terminal, Japan 231
Conclusions 232
References 232

PART 5 THE TECTONICS OF OTHER STRUCTURES 233

16 The Tectonics of Hybrid Structures 235
Addition of the same structural units in a different way 235
Addition of different structural units in an uncommon way 235
Integration of different structural units to form another system 236
Addition of unique structural units to form a hybrid structure 237
On hybridity of structures 238
Case study 34: Sydney Opera House, Australia 238
Conclusions 240
References 240

17 Evaluation of Case Studies and Conclusions 241

Index 248

FIGURES

1.1 A sketch of Santiago Calatrava's State Hermitage Museum, St Petersburg, 2012 3
1.2 A sketch of Richard Murphy's Circus Lane House, Edinburgh, 2006 3
1.3 Plan of this book 7
2.1 A Doric temple 11
2.2 A Caribbean hut: earthwork, hearth, framework/roof and enclosing membrane 12
2.3 Masjid-i Jami, Isfahan, 771 12
3.1 Finding the optimum solution 19
3.2 A sketch of Felix Candela's Los Manantiales Restaurant, Xochilmico, Mexico, 1958 20
3.3 A sketch of Pier Luigi Nervi`s Lanificio Gatti, Rome, 1953 20
3.4 A sketch of La Cité de la Muette, Marcel Lods, Drancy, France, 1932–1934 21
3.5 A sketch of John Burgee Architects with Philip Johnson`s Lipstick Building, New York, 1986 22
3.6 A sketch of Frank Gehry`s Guggenheim Museum, Bilbao, Spain, 1997 22
4.1 Types of forces 25
4.2 Ways of creating a moment around a table 26
4.3 Variation of wind load due to building height 27
4.4 Tensile and compressive stress 28
4.5 Stress versus strain diagram of steel and reinforced concrete 29
4.6 Form of effective area in axially loaded members 29
4.7 Buckling in slender compression members 30
4.8 Simple shear in a bolt joining two pieces of timber 30
4.9 A structural element in bending 30
4.10 Bending stress and its distribution in the cross-section 30
4.11 Successful and unsuccessful shapes against bending 31
4.12 Bending a ruler to examine the role of depth against bending 31
4.13 Horizontal and vertical shear stress in bending structural elements 32
4.14 Torsion in a circular structural element 32
4.15 Elimination of torsion in a beam 33
4.16 Finding centre of gravity in complex shapes 33
4.17 Overturning due to gravity loads 34
4.18 Shapes that are weak against overturning 34
4.19 A sketch of Stephen Svetko, Stephen Durkovic and Barnabas Kissling`s Slovak Radio Building, Bratislava, Slovakia, 1967–1983 34
4.20 Reducing column slenderness against buckling 35
4.21 Sliding 35
4.22 Uneven settlement 35
4.23 Wind instability of a tensile structure 36
4.24 Support types 36
4.25 A problem of finding reactions 37
4.26 Example 4.1: finding reactions 39

4.27 Example 4.2: finding reactions 39
4.28 Deflection limit of a beam 40
4.29 Pre-stressing of a beam 40
4.30 Deflection limit of a skyscraper 40
5.1 Three large families of modern building structural systems 44
5.2 Tensile structures 45
5.3 Compression structures 45
5.4 Bending structures 46
5.5 Structures with beams 46
5.6 Structures with trusses 47
5.7 A sketch of Richard Rogers' Millennium Dome, London, 1999 47
5.8 Flat arch 47
6.1 Rubble stone wall and cut stone wall 53
6.2 Reinforced concrete and stone continuous foundations 54
6.3 Organisation of stone pieces in a section of stone wall 54
6.4 Horizontal layers in a rubble stone wall 55
6.5 Determination of the dimensions of a room according to slab structure 56
6.6 Use of trusses 56
6.7 Continuous foundations under a building 57
6.8 Use of cut stone corners 58
6.9 Arrangement of openings on a masonry wall 58
6.10 Lintel/arch over an opening 58
6.11 Use of horizontal tie-beams 59
6.12 Cross-walls and buttresses 59
6.13 A sketch of case study 1: Cologne Cathedral, Cologne, Germany, 1248–1880 60
6.14 Plan and section of Cologne Cathedral 61
6.15 A sketch of case study 2: the stone tower at Norman Castle, Rochester, UK, twelfth century 62
6.16 Plans and section of the stone tower at Norman Castle 62
6.17 A sketch of case study 3: Summer House, Muuratsalo, Finland, 1953 64
6.18 Plan and section of Summer House 65
6.19 A sketch of case study 4: Harran Houses, Urfa, Turkey, 2500–600 BC 68
6.20 Plan and section of Harran Houses 69
6.21 A sketch of case study 5: Great Mosque of Djenne, Mali, 1907 69
6.22 Plan of Great Mosque of Djenne 70
6.23 Types of timber masonry 70
6.24 Structural elements of traditional timber masonry and timber balloon frames 71
6.25 Plans and elevation of case study 6: İsmail Hacı Çakır House, Kütahya, Turkey, 1781 74
7.1 Forms of arches 77
7.2 Forces affecting the key-stone of stone arches 78

7.3 Flat stone arch 78
7.4 Example of a stone arch in earthquake regions 78
7.5 Transfer of arch weight 78
7.6 Horizontal force problems in arches and associated solutions 79
7.7 Construction of arches without centring 79
7.8 Construction of arches with centring 80
7.9 A sketch of case study 7: the arches at Cordoba Mosque, Cordoba, Spain, eighth century 81
7.10 Plan and section of the arches at Cordoba Mosque 81
7.11 Forms of vaults 82
7.12 Raising the vault above ground level with the help of beams 82
7.13 Solutions to horizontal force problems in masonry vaults 82
7.14 Construction of vaults with centering 83
7.15 Vaults with and without ribs 83
7.16 A sketch of case study 8: the vaults at Cologne Cathedral, Cologne, Germany, 1248–1880 84
7.17 Forms of domes 84
7.18 Understanding shallow and high domes with the help of an orange 85
7.19 Stress types in high and shallow domes 85
7.20 Horizontal force problems of domes and associated solutions 86
7.21 Construction of domes without centering 86
7.22 A dome with ribs: the dome of San Lorenzo Church in Turin, Italy 86
7.23 A sketch of case study 9: the dome at Pantheon, Rome, Italy, AD 126 87
7.24 Plan and section of the dome at Pantheon 88
8.1 Types of reinforced masonry 90
8.2 Integration of frame elements with a reinforced masonry structure 94
8.3 An example of balanced distribution in reinforced masonry walls within a hybrid system 94
8.4 A sketch of case study 10: Villa Müller, Prague, Czech Republic, 1928 95
8.5 Plans and section of Villa Müller 96
8.6 A sketch of case study 11: Falling Water, Pennsylvania, USA, 1935 97
8.7 Plans and section of Falling Water 98
9.1 Frame system and post and lintel system 103
9.2 Deformation of frames under vertical and horizontal loads 103
9.3 Axial force and shear force in the elements of frames 104
9.4 Two-dimensional frames within a three-dimensional frame 105
9.5 Directions of columns in plan 106
9.6 Reinforced concrete slab types 107
9.7 Structural plan (with reinforcement) and section of one-way and two-way slabs 107
9.8 Different applications of a flat slab 108
9.9 Structural plan and section of a ribbed slab 108
9.10 Dimensions of ribs 108

9.11 Structural plan and section of a waffled slab 109
9.12 An example of the arrangement of ribs and beams in irregularly formed slabs 109
9.13 Use of secondary steel beams and trusses in steel slabs 110
9.14 Structure of a flight with stringer beams 111
9.15 Transfer of flight weight to a frame using beams 111
9.16 Use of beams to carry flights and landings in different arrangements 112
9.17 Cantilevering steps 112
9.18 Whole staircase as a cantilever 113
9.19 A problematic and a correct solution for slab structures adjacent to staircases 113
9.20 Plan and section of a structure with individual footings 114
9.21 Connecting individual footings to each other 114
9.22 Section of a slab-on-ground foundation 115
9.23 Various applications of a raft foundation 115
9.24 Pile foundations and the use of friction piles 116
9.25 Places to use expansion joints 117
9.26 Different applications of expansion joints 117
9.27 (a) Minimum dimensions of reinforced concrete shear walls; (b) Transfer of horizontal load by reinforced concrete shear walls 118
9.28 Bracing in steel shear walls 119
9.29 Internal forces in a steel shear wall 119
9.30 Distribution of shear walls within a structure 119
9.31 Steps to draw an N diagram 121
9.32 Steps to draw a V diagram 121
9.33 Steps to draw an M diagram 122
9.34 Steps to draw deflected shape 122
9.35 Drawing a parabolic curve in the M diagram 122
9.36 A system to draw N, V and M diagrams and deflected shape 123
9.37 N, V and M diagrams and deflected shape of the system in Figure 9.36 123
9.38 An alternative system to draw N, V and M diagrams and deflected shape 124
9.39 N, V and M diagrams and deflected shape of the system in Figure 9.38 124
9.40 Types of columns in determinate systems 124
9.41 A system with columns to draw N, V and M diagrams and deflected shape 125
9.42 N, V and M diagrams and deflected shape of the system in Figure 9.41 125
9.43 An alternative system with columns to draw N, V and M diagrams and deflected shape 126
9.44 N, V and M diagrams and deflected shape of the system in Figure 9.43 126
9.45 Finding shear in columns 127
9.46 Finding moments in columns and beams 127
9.47 Finding shear in beams 127
9.48 Finding axial force in columns 127

9.49 Finding axial force in beams 128
9.50 N, V and M diagrams and deflected shape of a frame 128
9.51 A frame to draw N, V and M diagrams and deflected shape 128
9.52 Finding internal forces in the elements of a frame 128
9.53 N, V and M diagrams and deflected shape of the frame in Figure 9.51 129
9.54 Economic height limits for various frame applications 130
9.55 Transfer of load to foundations 130
9.56 Two-dimensional frame systems taking place within a three-dimensional frame 130
9.57 Types of column axes 131
9.58 Column axes starting at one end of the structure and ending at the other, with intersecting beams and unconnected frame pieces 131
9.59 Earthquake map of the world 132
9.60 Separating deep recesses with the help of expansion joints 133
9.61 Examples of problematic arrangements for galleries 133
9.62 Earthquake force (F) and resistance of structure (R) 134
9.63 Reducing eccentricity 134
9.64 Acceptable and unacceptable arrangements of rigid partition walls and windows within frame systems 134
9.65 Examples to solve soft-storey problems 135
9.66 Form of openings and short-column problems 136
9.67 Examples of causes of short-column problems 136
9.68 A sketch of case study 12: Villa Savoye, Poissy, France, 1928 138
9.69 Plans and section of Villa Savoye 139
9.70 A sketch of case study 13: National Assembly in Dacca, Bangladesh, 1962–1974 140
9.71 Plan and section of Dacca National Assembly 140
9.72 A sketch of case study 14: Church of Light and Sunday School, Osaka, Japan, 1999 141
9.73 Plan and section of Church of Light and Sunday School 141
9.74 A sketch of case study 15: Barcelona Pavilion, Barcelona, Spain, 1928–1929 142
9.75 Plan and section of Barcelona Pavilion 142
10.1 A sketch of case study 16: Photographer`s Penthouse, New York, USA, 1992 150
10.2 Plan and section of Photographer`s Penthouse 151
10.3 A sketch of case study 17: Suspended Bedroom, Paris, France, 2004 152
10.4 Plan and section of Suspended Bedroom 152
11.1 High-rise building structure and cantilevering beam 154
11.2 Finding the center of gravity for a frame 155
11.3 Finding the axial force in columns and the shear in beams for the top floor of a frame 156
11.4 Finding the moment in beams and columns and finding shear in columns and axial force in beams for the top floor of a frame 156
11.5 N, V and M diagrams of the top floor of the given system 156

11.6 Wind effects on a building depending on the wind direction 157
11.7 Wind movement towards the top of a building 158
11.8 Wind movement around a building 158
11.9 The moment of inertia for plans of frames, shear wall systems and tubes 160
11.10 Shear lag in tubes 160
11.11 Structural configurations of various tubes 161
11.12 Height ranges of high-rise building structures 161
11.13 Use of outrigger systems and belt trusses to connect inner and outer structural systems 162
11.14 A sketch of case study 18: World Trade Center, New York, USA, 1972 164
11.15 Plan and partial section of World Trade Center 164
11.16 Facade of the World Trade Center towers 165
11.17 A sketch of case study 19: Shanghai World Financial Center, Shanghai, Republic of China, 2008 165
11.18 Plans and sections of Shanghai World Financial Center 166
12.1 Change of form depending on load 173
12.2 Examples of strategies to avoid wind instability 174
12.3 Reactions at the supports 174
12.4 Drawing the best form for tensile structures 175
12.5 A schematic sketch of Akashi Kaikyo Bridge, Kobe, Japan, 1998 177
12.6 A schematic sketch of the former Federal Reserve Bank Building, Minneapolis, USA, 1972 178
12.7 A schematic sketch of Dorton Arena, North Carolina, USA, 1952 178
12.8 A schematic sketch of Yale Hockey Rink, New Haven, USA, 1958 179
12.9 A sketch of case study 20: Zagreb Arena, Zagreb, Croatia, 2009 180
12.10 Plans and sections of Zagreb Arena 180
12.11 Types of cable truss 181
12.12 Structural behaviour of a cable truss 182
12.13 A bicycle-wheel structure 183
12.14 Use of cable trusses between floors 183
12.15 Use of cable trusses in Banque Populaire de l'Ouest et de l'Armorique, Montgermont, France, 1990 184
12.16 A sketch of case study 21: Science and Technology Museum, Paris, France, 1983–1998 185
12.17 A schematic sketch of the suspended glass system in the Science and Technology Museum 185
12.18 Structural elements of membranes 186
12.19 Different applications of membranes 187
12.20 Columbus'92 'Bigo', Genoa, Italy, 1992 188
12.21 An example of order of form in membranes 188
12.22 A sketch of case study 22: Olympic Stadium in Munich, Germany, 1972 189
12.23 Drawings of Olympic Stadium in Munich 189
12.24 Types of pneumatic structures 190

12.25 A sketch of case study 23: Swarovski Pavilion, Basel, Switzerland, 2008 191
12.26 Plan and sections of Swarovski Pavilion 191
12.27 Negative curvature shell with cables 191
13.1 Drawing best form for compression structures using a moment diagram 195
13.2 Curvature 196
13.3 Horizontal forces in compression structures 196
13.4 Structural elements in steel vaults 197
13.5 Structural elements in simple steel domes 198
13.6 A sketch of case study 24: the dome at Parliament Building in Berlin, Germany, 1992–1999 199
13.7 Plan and section of the dome at Parliament Building in Berlin 199
13.8 Different possibilities for geodesic domes 200
13.9 Construction of small geodesic domes made out of panels 201
13.10 Construction of large geodesic domes using organisation of the elements 201
13.11 A sketch of case study 25: US Pavilion at Expo '67, Montreal, Canada, 1967 202
13.12 Plan and section of US Pavilion at Expo '67 203
13.13 Curvature type of shell structures 203
13.14 Simple forms for shell structures 204
13.15 Pieces of simple forms 204
13.16 Addition of pieces of simple forms 205
13.17 Hyperbolic paraboloid form 205
13.18 Pieces of hyperbolic paraboloid forms 206
13.19 Addition of pieces of hyperbolic paraboloid forms 206
13.20 An example of a complex form for a shell structure 206
13.21 Corrugation of the surface of shell structures 207
13.22 Folding the edges of a shell form 207
13.23 A sketch of case study 26: Kimbell Art Museum, Fort Worth, USA, 1967–1972 208
13.24 Plans and section of Kimbell Art Museum 208
13.25 A sketch of case study 27: Small Sports Palace, Rome, Italy, 1958 209
13.26 Plan and partial section of Small Sports Palace 209
13.27 A sketch of case study 28: TWA Airport Building, New York, USA, 1956–1962 210
13.28 Plans and elevation of TWA Airport Building 210
13.29 An example of grid shells 211
13.30 Examples of construction details for timber grid shell joints 212
13.31 A sketch of case study 29: Suan Lien Center, Taipei, Republic of China, 2009 212
13.32 Plans and section of Suan Lien Center 213
14.1 Deformation by shear forces: a triangle in comparison to a square 216
14.2 A simple truss 216
14.3 Types of trussed systems 217
14.4 Similarity between beams and trusses 217

14.5	Type of internal forces in trusses	217
14.6	Finding internal forces in trusses 1	218
14.7	Finding internal forces in trusses 2	219
14.8	Creation of counter moment using the force couple in the top and bottom chords of a truss	219
14.9	Economic depth of trusses in relation to their span	220
14.10	Loading on trusses	220
14.11	A joint detail from a steel truss	220
14.12	Support and truss connection	220
14.13	Organisation of trusses	221
14.14	A sketch of case study 30: Cluj Arena, Cluj-Napoca, Romania, 2011	222
14.15	Plan and section of Cluj Arena	222
14.16	Different forms of 3D trusses	223
14.17	A sketch of case study 31: Waterloo Terminal, London, UK, 1993	224
14.18	Schematic plans and section of Waterloo Terminal	224
14.19	Joints in space frames	224
14.20	Arrangement of columns in a space frame structure	225
14.21	Connection of space frame to columns	225
14.22	A sketch of case study 32: Water Cube, Beijing, Republic of China, 2008	226
14.23	Plans and section of Water Cube	227
15.1	Folding a piece of paper	229
15.2	Examples of prismatic folds	229
15.3	Examples of non-prismatic folds	229
15.4	Examples of faceted folds	230
15.5	A slice of a folded plate used as a beam	230
15.6	Sketches of case study 33: Yokohama International Port Terminal, Yokohama, Japan, 2002	231
15.7	Plans and sections of Yokohama International Port Terminal	232
16.1	A schematic drawing of the 'Bird's Nest'	235
16.2	Schematic drawings of Lyon Satolas Airport Railway Station	236
16.3	Roof of Oklahoma State Fair Arena under gravity effect and wind suction	236
16.4	Ribs of Sydney Opera House	237
16.5	The unique structural unit in Stansted Airport	237
16.6	Plans and section of Stansted Airport	238
16.7	A sketch of case study 34: Sydney Opera House, Sydney, Australia, 1957–1973	239
16.8	Plan and section of Sydney Opera House	239

TABLES

4.1 Finding moment around B 38
4.2 Finding moment around A for Example 4.1 38
4.3 Finding moment around B for Example 4.2 39
4.4 Structural guidelines originating from the requirements of strength, stability, equilibrium and deformation limit and the associated value system 41
5.1 Relationship between structure type and the spans commonly used 48
5.2 Relationship between type of structure and economic building height 49
6.1 Structural guidelines for stone masonry structures and the associated value system 59
6.2 Structural guidelines for brick masonry structures and the associated value system 63
6.3 Structural guidelines for adobe masonry structures and the associated value system 67
6.4 Structural guidelines for timber masonry structures and the associated value system 73
7.1 Structural guidelines for masonry arches and the associated value system 80
7.2 Structural guidelines for masonry vaults and the associated value system 83
7.3 Structural guidelines for masonry domes and the associated value system 87
8.1 Conservative structural guidelines for reinforced stone masonry structures and the associated value system 92
8.2 Conservative structural guidelines for reinforced brick masonry structures and the associated value system 93
8.3 Conservative structural guidelines for reinforced adobe masonry structures and the associated value system 93
8.4 Structural guidelines for reinforced masonry and reinforced concrete frame hybrid systems and the associated value system 95
9.1 Limits of reinforced concrete and steel beams and slabs 129
9.2 General structural guidelines for frame (and shear wall) systems with any structural material and the associated value system 137
9.3 Structural guidelines for reinforced concrete frame (and shear wall) systems and the associated value system 138
9.4 Structural guidelines for steel frame (and shear wall) systems and the associated value system 138
10.1 Structural guidelines for making subtractions from existing frame systems and the associated value system 150
10.2 Structural guidelines for making additions to existing frame systems and the associated value system 150
11.1 Guidelines for structural design of high-rise buildings and the associated value system 163
12.1 Structural guidelines for cables and suspension structures and the associated value system 179
12.2 Structural guidelines for bicycle-wheel structures and the associated value system 184
12.3 Structural guidelines for suspended glass systems and the associated value system 184
12.4 Structural guidelines for membrane structures and the associated value system 188
12.5 Structural guidelines for pneumatic structures and the associated value system 190

13.1	Structural guidelines for steel vaults and the associated value system	198
13.2	Structural guidelines for steel domes and the associated value system	198
13.3	Structural guidelines for steel geodesic domes and the associated value system	202
13.4	Structural guidelines for shell structures and the associated value system	207
13.5	Structural guidelines for grid shells and the associated value system	212
14.1	Structural guidelines for trusses and the associated value system	221
14.2	Structural guidelines for 3D trusses and the associated value system	223
14.3	Structural guidelines for space frames and the associated value system	226
15.1	Structural guidelines for folded plates and the associated value system	231
17.1	Attitude towards structural guidelines and its relation to architectural concept in case studies	241
17.2	Attitude towards structural guidelines, earthquake risk, dominant physical entities and innovation in case studies	243

ACKNOWLEDGEMENTS

I would like to thank my teachers Mustafa Pultar and Mehmet Emin Tuna, who contributed much to my development in the area of structural design in architecture. Without their efforts, generous teaching and pedagogic approach, this book could not have been possible.

I would like to thank my son Baydu Can Al, who contributed as a guest author to this book, for reading and criticising all chapters, re-writing some parts and introducing me to different research material. His valued support as an engineer encouraged me to write this book.

I would like to thank Nicholas Wilkinson and Emmanuel Chengi for long discussions about the idea of the book and discussions about the use of various concepts within the book. Thanks to Hugh Clarke for the dedicated proofreading he provided: his contribution certainly raised the quality of the book. I would like to thank my colleague Netice Yıldız for her help in finding some masonry case studies. Finally, my thanks to Taylor & Francis for a very positive publishing experience.

Cover image: Drawn by Yonca Hurol with the help of http://en.wikipedia.org/wiki/Dico_si_Tiganas.

ABBREVIATIONS

a	acceleration (m/s^2)
A	area (m^2)
d	distance (m)
e	eccentricity (m)
E	elastic modulus (kN/m^2)
F	force (kN)
I	moment of inertia (m^4)
m	mass (kg)
M	moment (kNm)
N	axial force (kN)
V	shear force (kN)
W	total weight (ton)
σ	stress (kN/m^2)
σ all	allowable stress (kN/m^2)
σ ult	ultimate stress (kN/m^2)
ε	strain
Δ	deflection (m)

1 Introduction

System,
Structural system,
Tectonics,
Tectonics of structural systems.

It is first necessary to explain what the author understands from these concepts.

EXPLANATION OF KEY CONCEPTS

The concept of 'system' implies the presence of some parts within a whole and the presence of an organisational relationship between these parts (Checkland, 1981). Modern buildings can be analysed as combinations of various systems from a technical point of view. The best way of understanding these parts within buildings is to overview the professions that take place within a building team. Even a small building team contains an architect, a structural engineer or a civil engineer (a structural engineer specialises only in structures whereas a civil engineer has a wider area of knowledge including transportation and water structures), a mechanical engineer and an electrical engineer. Sometimes the team becomes larger with the inclusion of an interior architect, a landscape designer or an acoustic engineer. The structural engineer designs the building structure together with the architect. The mechanical engineer designs the clean water system, the dirty water system, the sewage system, the ventilation/heating/air-conditioning system and elevator/escalator systems together with the architect. The electrical engineer designs the electrical systems within the building together with the architect and the mechanical engineer. The architect is involved within all systems' design processes and s/he is the coordinator of the building team. The architect is responsible for the holistic design of the building, including the functional and artistic issues of architecture and design, and coordination of the building systems.

The structure system is the part of the building that carries all the loads affecting the building. A structural element carries its own weight, the weight of other building parts and its share from all other loads affecting the building. A non-structural element carries only its own weight and loads. Thus, the structural system is specialised to carry loads in the building including dead weight, wind load and earthquake load.

Most probably you are reading this book sitting in a building. If you look at the building elements surrounding you, you might differentiate at least some of these building systems and the structural system. You can see some pipes or shafts. You can see some devices to control the heat. You can see some plugs, switches and lighting fixtures. There might be some columns or walls belonging to the structural system. There might also be some beams and slabs. Even if they are plastered and painted, you can differentiate the presence of different materials by hitting these elements with a metal ring on one of your fingers.

This book is about structural systems in architecture. It does not cover mechanical and electrical systems. However, this book does not handle structural systems as an isolated entity. It covers structural systems as they are needed in architectural design. This means that this book has to cover knowledge of structures, which is a quantitative type of knowledge determining what is right and what is wrong; and the systematic knowledge of design, which helps in transforming the quantitative knowledge of structures into qualitative knowledge to judge functionality and aesthetics of the building elements. The individual parts that take place within a system are important, but the whole is also important. This book bridges the gap between analytical knowledge about structures and artistic knowledge about architecture.

In order to handle the artistic knowledge of architecture in a systematic way, this book introduces the concept of tectonics. Tectonics in architecture (or architectonics) is a holistic and qualitative approach to artistic use of technology in architecture. Thus, the tectonics of structural systems means the role of the structure system within the artistic use of all systems and other physical entities that take place in architectural design. Tectonics covers the artistic use of the following physical entities (Şahali, 2009):

- Structure
- Building materials
- Details
- Mechanical systems

- Electrical systems
- Information technology
- Construction methods
- Topography/context (physical and social)
- Climatic response of the building
- Technology and culture relationship
- Light.

This list can be enlarged by adding other issues, such as acoustic precautions.

In a good architectural project, all of these physical entities are in the service of tectonics. Each might have equal roles or one/two/three of them might be more dominant in achieving the tectonic quality of the building. For example, Santiago Calatrava's Auditorio de Tenerife in Spain exaggerates the structural system in order to achieve the tectonic qualities of the building. All other systems in the building are in the service of the tectonics achieved through the structure. (You can search the internet to see images of this building). Louis Kahn's Indian Institute of Management Ahmedabad in Gujarat, India, exaggerates the use of bricks as the main structural material to achieve tectonic qualities. Thus, the materials and the structure are dominant in achieving this building's tectonic value. Carlo Scarpa's Museo di Castelvecchio in Verona, Italy, is famous for the details of reinforced concrete elements and its stairs. Renzo Piano and Richard Rogers' Centre Pompidou in Paris, France, uses all mechanical systems in the building to represent the tectonic idea. Every colour, especially at the back facade of the building, shows the presence of a different system. On the other hand, all the buildings in New York Times Square, USA, are covered with electronic screens that are used for advertising and artistic purposes. (You can search the internet for the night view of New York Times Square in order to see the screens' contribution to the quality of space for the whole square.) Type of formwork used in reinforced concrete buildings affects the tectonic qualities of those buildings. Thus, the construction method might affect the tectonics of the building. The integration of topography and Frank Lloyd Wright's Falling Water in Pennsylvania, USA, can be seen as one of the dominant tectonic ideas behind this building. Renzo Piano's Tjibaou Cultural Center in Noumea, New Caledonia, responds to the climate in a tectonic way by interpreting the vernacular technology. This building is also an example of interpretation of vernacular technology to achieve a cultural relationship with the existing context. Finally, Tadao Ando's Church of Light and Sunday School in Osaka, Japan, is one of the best examples of the use of openings/light in order to achieve the tectonic qualities of the building.

The concept of 'tectonics of structural systems' often makes people think about buildings which were designed by Santiago Calatrava and Zaha Hadid because structure is the dominant contributor to the tectonic qualities of their projects. However, all good examples of architecture can be analysed in terms of their structures' contribution to tectonics whether or not structure is the dominant contributor. If another building system and/or environmental characteristic is dominant within the tectonics of the building, structure is still contributing to this tectonic quality. It can be in harmony with the tectonic quality or make a contrast with it. Thus, all good examples of architecture can be analysed in terms of tectonics of their structural systems. This means that the contribution of structure to the buildings' tectonics is equally discussible for Santiago Calatrava's State Hermitage Museum in Russia (see Figure 1.1) and Richard Murphy's Circus Lane House in the UK (see Figure 1.2). The first building uses structure in order to achieve a sculptural form, whilst the other building integrates masonry and frame systems in order to achieve larger openings without losing the contextual value of the building that takes place in a historical context of stone buildings. Determining the role of structure within the tectonic qualities of buildings is one of the interest areas of this book.

How does structure affect the design concept (or the image) of the building?
How does structure affect the architectural space?
How does structure affect the relationship between the building and its context?

To start bridging the gap between the structural system (analytical) and tectonics (balance between the artistic and technology), it can be stated that each structural system has a potential (or

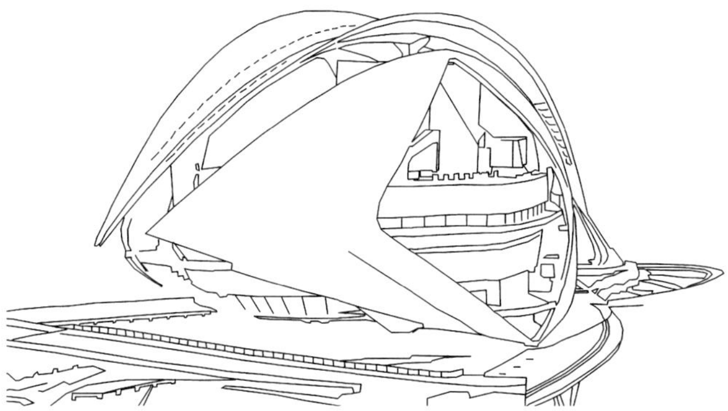

1.1 A sketch of Santiago Calatrava's State Hermitage Museum, St Petersburg, 2012 (drawn with the help of URL1, 2012)

1.2 A sketch of Richard Murphy's Circus Lane House, Edinburgh, 2006 (drawn with the help of URL2, 2006)

tendency) to make certain contributions to the tectonic qualities of buildings. This potential/tendency can be studied with the help of the concept of 'structural guidelines'. Structural guidelines are whatever structural engineers recommend for the design of a structural system from the perspective of values such as safety and/or economy, and they determine the potential of that structural system to achieve tectonic qualities. For example, having the beam and column dimensions close to each other in reinforced concrete frame systems in earthquake regions is a structural recommendation that takes place within structural guidelines. Not having steel frame systems in a building over 30 storeys is another structural recommendation.

Although all structural guidelines have a quantitative basis, they can also be related to the qualitative value system of tectonics. The quantitative and quantitative value systems start to be related when the sanction power of recommendations that take place within structural guidelines is questioned. It is known that designers do not follow all structural guidelines strictly. There are many successful buildings that do not follow some structural guidelines. For example, the 102-storey Empire State Building in New York, USA, which was completed in 1931, has a steel frame system. Yet, from an economy standpoint, buildings with rigid steel frame systems should not be used for more than around 30 storeys.

On the other hand, there are many successful buildings for which structural guidelines have been followed consistently. This means that it is somehow optional whether designers choose to follow structural guidelines. This is the point at which the value system of structural engineers and architects differ from each other. Structural engineers try to guarantee safety and economy, whilst architects try to achieve safety (*firmitas*), functionality (*utilitas*), aesthetics (*venustas*) and expected level of economy simultaneously. What architects and structural engineers understand from 'economy' can also differ. Structural engineers' value system is hierarchical: safety is more important than economy. However, all architectural values are equally important and none of them can be sacrificed. It may seem as if there is a contradiction between architects not following some structural recommendations and still needing to achieve safety, but actually there is no contradiction. Either safety is achieved in another way, or the recommendation in the structural guideline is about economy (which has less sanction power in comparison to other structural recommendations about safety).

Architects need to classify recommendations within structural guidelines according to their sanction power. This shift from the quantitative value system (of right and wrong) into the qualitative value system as regards the structural recommendations (to follow them, leave them, replace them, compensate for them or transform them) forms the interface between quantitative and qualitative in building design. It is urgent for architects and structural engineers to explain the nature of this shift from quantitative to qualitative that is very common within modern architecture.

There are structural guidelines in one hand and there are tectonics of architecture in the other hand. The nature of each is different and also the nature of their combination. Following structural guidelines forms the structural engineering point of view. However, tectonics in architecture might combine contradictory characteristics. It is even possible to say that such contradictory characteristics might increase the tectonic value. This book combines 34 case studies, each corresponding to successful architectural examples of different structural systems. As concrete examples, the case studies present the relationship between structural guidelines and tectonics.

RESEARCH PROBLEM

The way structure courses are given in departments of architecture and interior architecture does not help students to understand the structural guidelines and the variety in the sanction power of the recommendations within these guidelines. The courses are usually mathematical courses of statics, strength and structural analysis, through which students are expected to develop their own values about structures by solving as many structural problems as possible. For architects, this is similar to digging a hole with a needle. Thus, students of architecture usually find these courses very abstract and not practical. Many of them complain about not remembering anything after the last exam.

When a student of architecture starts his/her education, s/he is desperate to learn about building technology and structural systems. However, soon after beginning these mathematical courses, the student sees that structure courses do not give the practical knowledge needed in design studios. A student also learns that structural engineering recommendations have varying sanction powers for application during the design process. For these reasons, some students lose interest in structure courses.

Existing literature about structures is also very limited for architects. Engineering books are too abstract and not practical enough. Another problem is that building codes use the same language as structural engineering. There are some books in which structural engineering knowledge is simplified for architects, by either simplifying the mathematics or by avoiding the mathematics altogether (see Salvadori & Heller, 1963; Macdonald, 1997; Place, 2007; Ambrose & Tripeny, 2010; Dabby & Bedi, 2012). There are some books that translate technical knowledge with the help of drawings (see Engel, 1997; Ching, 2008). These books are very useful for architecture and interior architecture design and construction studios. On the other hand, the books about tectonics in architecture (see Hartoonian, 1994; Frampton, 2001; Leatherbarrow & Mostafavi, 2005) are not related to the knowledge of structures: instead, they are theory books of architecture. Similarly, the books about the history of structural systems (see Billington, 1985; Mainstone, 2001; Littlefield & Jones, 2013) explain the developments in the history of structures step by step. There are four books in the literature concerning the use of structural systems in architecture (see Macdonald, 2001; Charleson, 2005; Sandaker, 2007; Sandaker et al., 2011). Sandaker, Eggen and Cruvellier's *The Structural Basis of Architecture* can be used as a textbook in structure courses for departments of architecture. This book does not use tectonics, but it adapts another approach from the philosophy of aesthetics. One of the problems in the interdisciplinary subject of 'structural design in architecture' is the absence of textbooks that handle structural engineering and tectonic dimensions equally and in a systematic way.

OBJECTIVES OF THIS BOOK

The objectives behind this book are to:

- Prepare 'structural guidelines' that are useful design tools.
- Discuss the sanction power of the recommendations within these guidelines according to the value system behind them.
- Develop the concept of 'tectonics of structural systems' in order to discuss the relationship between structural guidelines and tectonics in architecture in a systematic way.

- Integrate the knowledge of 'structural guidelines' about all structural systems and the theory about 'tectonics of structural systems' within a textbook that can be used in the structure courses for departments of architecture and interior architecture.
- Interpret the nature of the relationship between structural guidelines and tectonics in architecture.

These five objectives can be seen to be the originality of this book and its contribution to the area of structure in architecture.

METHODOLOGY FOR DETERMINING STRUCTURAL GUIDELINES

The author is an architect and has been teaching structure and design courses in departments of architecture and interior architecture for approximately 30 years (which means being a mentor for about 300 students' projects and evaluating a minimum of 1,000 student projects), researching structural systems that are useful for architects and analysing pertinent literature. Using structure books for architects and building codes such as: *International Building Code* from the International Code Council, 2011; ACI 318-08 for reinforced concrete structures; ACI 530-02/ASCE 5-02/TMS 402-02 for masonry structures; AISI S100-07 and AISI S200 for steel structures; and Paz's *International Handbook of Earthquake Engineering*, 1995, for other building codes; the concept and types of 'structural guidelines' were developed as follows:

- General structural guidelines:
    - General structural guidelines in relation to economy.
    - General structural guidelines originating from the requirements of strength, stability, equilibrium and deformation limit.
    - Structural guidelines in relation to form.
- Structural guidelines in order to solve common structural problems of families of structures.
- Specific structural guidelines for each structural system:
    - Structural guidelines in relation to definition (configuration) of the structure.
    - Structural guidelines in relation to limits of the structure.
    - Structural guidelines in relation to building codes of each system.
    - Structural guidelines in relation to economy of each system.

METHODOLOGY FOR PROVIDING ANALYTICAL TOOLS

Using experience from teaching design, the author analysed the literature concerning tectonics in architecture and developed the following steps for the analysis of tectonics of structural systems in particular cases of architecture:

- Identification of the design concept/image of design as the artistic dimension of the project.
- Identification of the role of various physical entities in architecture (such as structure, selected materials, details, mechanical systems, electrical systems, information technology, building techniques, topography/context, climatic response of the building, technology and culture, and light) on the design concept/image of design.
- Effect of 'structural recommendations within the guidelines' on the formation of the design concept/image of design.

It is accepted that the structural system of any building has an ontological value if it provides safety, the expected level of economy and aesthetics simultaneously. The 34 case studies, which take place between chapters 6 and 16 of this book, are successful examples that have ontological value.

The concepts of 'design concept' and 'image of design' represent the aesthetic value of a building. Both these concepts and the concept of 'ontological value' will be explained in chapter 2, which introduces the way in which theory of tectonics is interpreted within this book.

METHODOLOGY FOR DISCUSSING THE RELATIONSHIP BETWEEN STRUCTURAL GUIDELINES AND TECTONICS

The knowledge about 'structural guidelines' and the theory of 'tectonics of structural systems' meet within this book for the analysis of case studies which takes place after the explanation of structural guidelines for each structural system. These case studies are selected as well-known examples of architecture and their design concepts/images of design are found in the associated architectural literature. The effect of each physical entity, including structural system, on design is discussed by analysing the architectural literature about each building.

The 34 case studies were selected to be:

- Well-known examples of architecture.
- Representatives of particular structural systems.

Structural systems' level of dominance within these case studies is also questioned.

CONTENTS OF THIS BOOK

This book contains 5 parts covering 17 chapters including an introduction and a conclusion.

Chapter 2 starts with the history of the theory of tectonics in architecture. A particular interpretation of the theory of tectonics is made in order to achieve an analytical approach to tectonics, which bridges the gap between the quantitative and the qualitative. The concepts of 'design concept', 'image of design' and 'ontological value' are explained in this chapter. The narrative concerning the relationship between structural guidelines and tectonics is introduced with the help of these concepts.

The organisation of the rest of this book follows the types of 'structural guidelines'. Chapters 3, 4 and 5, which form the first part of this book, concern general structural guidelines. Chapter 3 covers general structural guidelines in relation to economy. This chapter introduces structural guidelines concerning three approaches to economy, including designing for evolutionary structural optimisation, design of the optimum structure and optimisation of the designed structure. Chapter 4 covers structural guidelines originating from the requirements of strength, stability, equilibrium and deformation limit. Chapter 5 contains structural guidelines relating to form and structure size. For this purpose, structural systems are classified according to their form and the stress type they generate. The relationship between form and stress type forms a bridge between architecture and structural engineering.

The second part of this book concerns the tectonics of masonry structures. Chapter 6 covers traditional approaches to masonry structures. This chapter contains structural guidelines and case studies about stone, brick, adobe and timber masonry. Chapter 7 covers structural guidelines and masonry roof structures with case studies about masonry arch, vault and dome. Chapter 8 covers contemporary approaches to masonry structures. Rather than a traditional approach to masonry, contemporary architects often integrate masonry and modern structures in order to achieve larger openings in modern masonry structures. This chapter introduces structural guidelines about the techniques of making larger openings and analyses two case studies in relation to the subject.

The third part of this book concerns the tectonics of flexural structures, such as frame and shear wall systems. Chapter 9 covers the tectonics of frame and shear wall systems, which present an interplay between structural guidelines and creativity in modern buildings. Chapter 10 covers the interior changes made to existing buildings with frame systems. This chapter discusses structural guidelines concerning adding and subtracting parts from frame systems. Chapter 11 covers structural guidelines relating to high-rise building structures. Recommendations within this chapter are categorised under three strategies of achieving higher buildings: the aerodynamic design of the form of the building in order to reduce the wind load; the selection of the appropriate structural system (such as types of tubes); and the use of details (such as details containing damping material).

The fourth part of this book concerns the tectonics of form-resistant structures, which are categorised as tension, compression, tension and compression structures, and folded plates. These are the structures that span long distances. Chapter 12 covers the tectonics of tension structures with structural guidelines and case studies about cable, membrane, pneumatic and suspended glass systems. Chapter 13 covers compression structures including shells in various forms (such as vaults, domes and shells designed with modern materials). Chapter 14 discusses the structures that generate tension and compression simultaneously. This chapter covers structural guidelines and case studies about truss, 3D truss and space-frame structures. Chapter 15 covers the tectonics of folded plates, which are flexural structures. These are studied within the fourth part of this book, together with form-resistant structures, due to the important role their form plays in resisting loads affecting them.

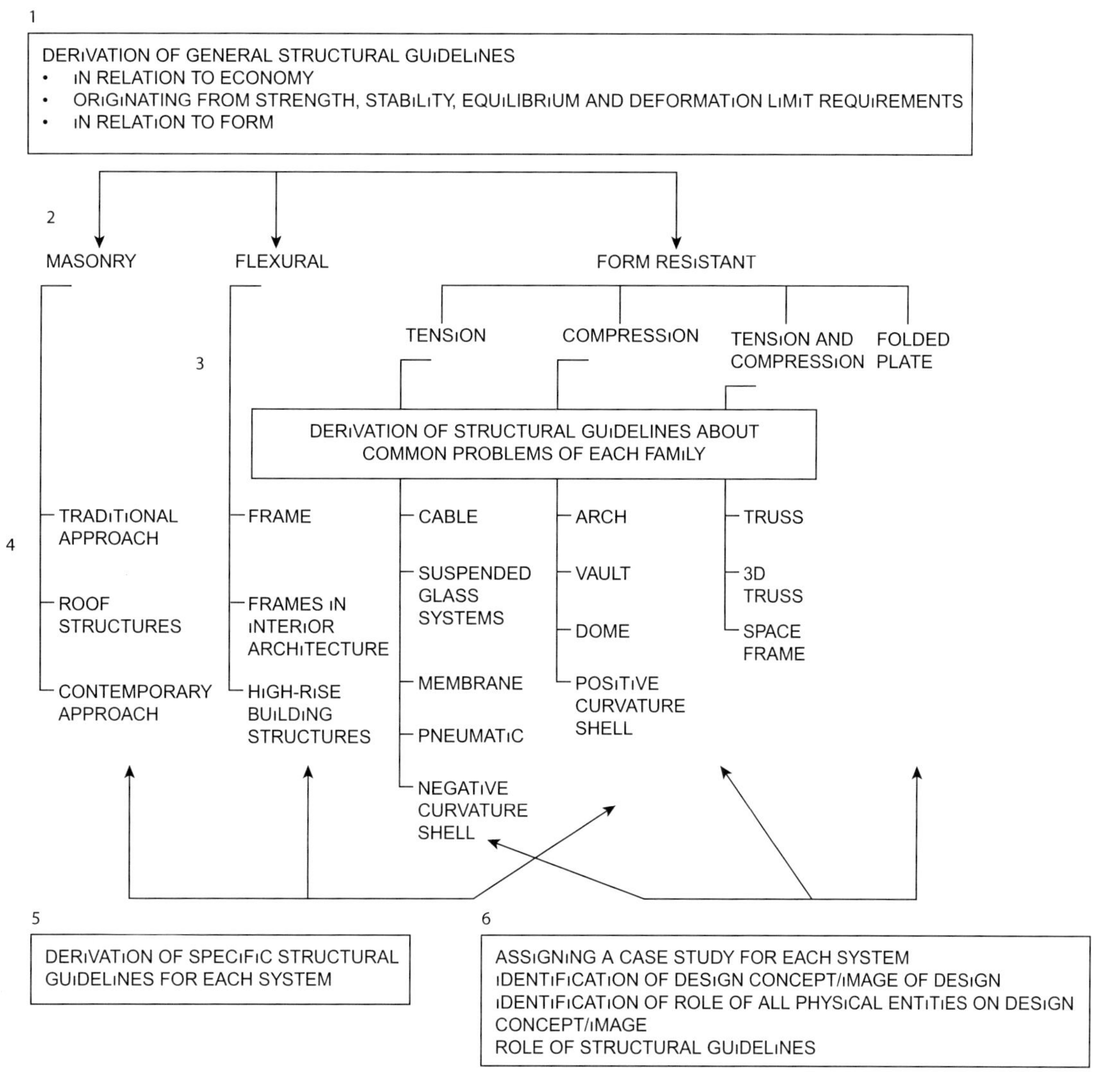

1.3 Plan of this book

The fifth part of this book concerns hybrid structural systems that integrate different types of structures in order to achieve the tectonic qualities demanded by architecture. This part shows that architects are not limited by defined types of structures in order to achieve specific tectonic qualities. Acting with structural engineers, architects can encourage innovative design.

The conclusion discusses the usefulness of the concepts of structural guidelines and tectonics in order to achieve a more practical and artistic approach to structures in architecture. This chapter also contains the analysis of case studies, which integrates structural guidelines and tectonics, in order to make an interpretation about the role played by structural systems in architectural design.

The plan of this book is shown in Figure 1.3. The first step is the derivation of general structural guidelines, which apply to all structural systems. It should be remembered that whenever a structural guideline is derived, the sanction power of the recommendations in it is discussed through determining the value system(s) behind it. The second step is the categorisation of structural systems: masonry, flexural and form resistant (including tensile, compression, tensile and compression structures, and folded plates). The third step is the derivation of structural guidelines concerning common structural problems of each structure family. The fourth step further categorises the families of structural systems. The fifth step is the derivation of particular structural guidelines for each structural system. The sixth step assigns a minimum of one case study for each structural system, to understand the design concept/image of design of each case study, to discuss the role of physical entities on the tectonics of the building, and to assess the role of structural systems amongst all physical entities.

REFERENCES

ACI 318-08 Building Code (2009) *Requirements for Structural Concrete* (viewed 5 September 2013: www.concrete.org/tempComDocs/-28807/r._stehly.pdf)

ACI 530-02/ASCE 5-02/TMS 402-02 (2002) *Building Code Requirements for Masonry Structures*, Masonry Standards Joint Committee (viewed 5 September 2013: https://engineering.purdue.edu/~ramirez/CE479/FALL05/MasonryBuildingCode1-3-02.pdf)

AISI S100-07 (2007) *North American Specification for the Design of Cold Formed Steel Structural Members*, American Iron and Steel Institute (viewed 5 September 2013: www.ce.jhu.edu/cfs/cfslibrary/AISI-S100-07%20Specification.pdf)

AISI S200 (2007) *North American Standard for Cold Formed Steel Framing – Truss Design*, Supplement 2, American Iron and Steel Institute (viewed 5 September 2013: http://steelframing.org/PDF/design_manuals/AISIS214-07_S2-08_Final_Version9-19-08.pdf)

Ambrose, J., Tripeny, P. (2010) *Simplified Engineering for Architects and Builders*, John Wiley and Sons: New York.

Billington, D.P. (1985) *The Tower and the Bridge: The New Art of Structural Engineering*, Princeton University Press: New Jersey.

Charleson, A. (2005) *Structure as Architecture: A Source Book for Architects and Structural Engineers*, Elsevier: Amsterdam.

Checkland, P. (1981) *Systems Thinking, Systems Practice*, John Wiley and Sons: New York.

Ching, F.D.K. (2008) *Building Construction Illustrated*, 4th edition, John Wiley and Sons: New York.

Dabby, R., Bedi, A. (2012) *Structure for Architects, A Primer*, John Wiley and Sons: New York.

Engel, H. (1997) *Structure Systems*, 3rd edition, Gerd Hatje Publishers: Ostfildern, Germany.

Frampton, K. (2001) *Studies in Tectonic Culture: The Poetics of Construction in Nineteenth and Twentieth Century Architecture*, ed. J. Cava, The MIT Press: Cambridge, Massachusetts.

Hartoonian, G. (1994) *Ontology of Construction*, Cambridge University Press: New York.

International Code Council (2011) *2012 International Building Code*, International Code Council: Illinois.

Leatherbarrow, D., Mostafavi, M. (2005) *Surface Architecture*, The MIT Press: Cambridge, Massachusetts.

Littlefield, D., Jones, W. (2013) *Great Modern Structures: 100 Years of Engineering Genius*, Carlton Books: London.

Macdonald, A.J. (1997) *Structural Design for Architecture*, Architectural Press: New York.

Macdonald, A.J. (2001) *Structure and Architecture*, 2nd edition, Architectural Press: New York.

Mainstone, R. (2001) *Developments in Structural Form*, 2nd edition, Taylor and Francis: London.

Paz, M. (1995) *International Handbook of Earthquake Engineering: Codes, Programs and Examples*, Springer-Science: Berlin.

Place, J.W. (2007) *Architectural Structures*, John Wiley and Sons: New York.

Şahali, O. (2009) *The Issues of Ontology and Scenography in Tectonics of Buildings with Frame Systems in Architecture*, unpublished Master thesis, Eastern Mediterranean University: North Cyprus.

Salvadori, M., Heller, R.A. (1963) *Structure in Architecture: The Building of Buildings*, Prentice Hall: New Jersey.

Sandaker, B.N. (2007) *On Span and Space – Exploring Structures in Architecture*, Routledge: London.

Sandaker, B.N., Eggen, A.P., Cruvellier, M.R. (2011) *The Structural Basis of Architecture*, 2nd edition, Routledge: London.

URL1 (2012) *Russia's State Hermitage Museum Exhibits Contemporary Architect Santiago Calatrava* (viewed 26 September 2014: www.justluxe.com/lifestyle/arts/feature-1789950.php)

URL2 (2006) *Circus Lane House* (viewed 26 September 2014: www.edinburgharchitecture.co.uk/circus-lane-houses)

2
Tectonics and Structural Guidelines

The concept of tectonics implies a nostalgic approach to technology that replaces the contemporary concept of technology with its ancient versions. Many architectural historians, such as Kennett Frampton (2001) and Gevork Hartoonian (1994), believe that the relationship between technology and architecture cannot be explained with the help of the contemporary concept of technology mainly because this is quantitative, and is usually evaluated as right or wrong. Although the author acknowledges the dominant role of technology on modern architecture, a more qualitative approach is preferred. At this point the old concepts of technology become more useful.

There are some philosophers such as Martin Heidegger (1977) and social scientists such as Richard Sennett (2008, 2012) who think that this transformation of the concept of technology is flawed. According to Heidegger (1977), the existing concept of technology leads to an inappropriate approach towards man-made and natural environments by conceptualising everything as 'standing reserve', by which he means that everything in nature is seen as ready to be used for production. Richard Sennett in his book *The Craftsman* (2008) defends 'the desire to do a job well done for its own sake', which is increasingly uncommon. Sennett's book can also be seen as a book about the history of work ethics. Later, in his book *Together* (2012) Sennett says that 'doing a job well done', which describes a particular approach to technology, could produce a more reasonable (dialogical and emphatic[1]) dialogue type between productive people. Sennett believes that production relationships can be improved if people are more connected to what they produce. In this sense, his approach to work ethics has similarities to Martin Heidegger's (1995) concept of authenticity which is based on making step-by-step changes (improvements) on the daily routine of work. These authors, like many others, address the concept of tectonics, a term which has ancient Greek origins.

Today, it is beneficial for architects to go through the old concepts of technology in order to identify their viewpoints. For this purpose, this chapter begins with an analysis of the historical origins of technology by studying the evolution of the concepts of 'techne' and tectonics. Next, an analytical approach to the relationship between the concepts of tectonics and structural guidelines is developed.

HISTORY OF THE CONCEPT OF TECTONICS

The concept of tectonics originates from the ancient Greek concepts of 'tekton' meaning carpenter, and 'techne' meaning craft, art or technical knowledge (not scientific). Both concepts indicate a skill and a method for producing something. The concept of 'archi-tekton' meaning master-builder ('archi' meaning the principle), appeared later than the concept of techne (*Stanford Encyclopedia of Philosophy*, 2007).

The concept of techne changed considerably between the fourth century BC and the third century AD (*Stanford Encyclopedia of Philosophy*, 2007). According to Xenophon, who lived during the fourth century BC, the word meant the knowledge of doing something in an organised way. It was linked with the concept of 'episteme', which meant scientific knowledge (knowing something for the sake of knowledge). Plato, who lived between 428 and 347 BC, considered techne to mean a more practical knowledge and episteme a more theoretical knowledge. Although techne related to physical products, it was also affected by episteme.

Aristotle, who lived between 384 and 322 BC related techne to the knowledge of particular things, and episteme to the general knowledge of things. For him, the concept of techne included the process of creation and it was related to poetics (Aristotle, 330 BC, 1988). The concept still had a connection with the tragic. The Stoics, who became influential during the third century BC, believed that the main problem of human beings in this world is their emotions. According to them, techne was not as reliable as episteme because it was not stable. Conversely, episteme was unshakable by reason. For Alexander of Aphrodisias, who lived during the second century AD, the pure knowledge of episteme was superior to human action, which was determined by techne. These thoughts ran parallel to the value system, which gave more credit to knowledge (the education of noble people) than to physical work/production (which was carried out by slaves) (Sennett, 2008). Plotinus, who lived during the third century AD, did not use the concept of techne at all. For him only episteme existed. Gradually the concepts of technique and technology appeared but were dependent upon scientific knowledge (*Stanford Encyclopedia of Philosophy*, 2007).

The old concept of tectonics started to be reused as an alternative to the concept of technology after the Industrial Revolution affected the world of production. It could be said that the Industrial Revolution affected crafts negatively and professional design positively. According to Gilles Deleuze and Felix Guattari (1993: pp.362–8), this was a slow process, causing the development of new needs that were satisfied with the production of new types of products (and buildings). Deleuze and Guattari give the first iron bridges as examples of this process. The design and production of these bridges were out of the state norm for the production of stone bridges. It was necessary to increase the speed of construction. This need caused a change in the materials used, the design/production processes and the division of labour. The process of modern bridge construction started with Thomas Telford (1757–1834) and continued with the competition between the early engineers such as French bridge designer Francois Hennebique (1843–1921), Swiss bridge designers Wilhelm Ritter (1847–1906) and Robert Maillart (1872–1940) and German designers Franz Dischinger (1887–1953) and Walter Bauersfeld (1879–1959). French and Swiss designers were integrating aesthetics (of lightness) and engineering in differing ways, whilst German designers ignored aesthetics for the sake of scientific methods (Billington, 1983: pp.27–44, 147–70). The success of the German scientific approach was the contemporary victory of episteme over techne. Specific formulas corresponding to specific forms were dominating the design of buildings. People criticised the buildings designed by the German designers as: 'Form follows formula'. Later, the use of scientific methods in structural design became the state norm of the modern world. However, new methods and new structural systems were developed and gradually it became possible to design almost any form.

Developments in civil and structural engineering profoundly affected architecture. The use of new structural materials and new structural systems determined the tectonic qualities of modern architecture. However, architecture always supported techne against episteme by transforming the discussions about the particularity and practicality of techne into discussions about its relation to aesthetics. The discussions around Crystal Palace (UK: 1851), The Chicago School of Architecture (USA: 1880s and 1890s) and the Eiffel Tower (France: 1889) were concerned with the effect of new technologies on aesthetic taste. The modern theories of tectonics in architecture began with these early examples of modern architecture. As early as 1844, Karl Botticher (1852) started writing his three volumes *The Tectonics of the Hellenes* in which he argued that the 'outer cladding' (in other words the aesthetic dimension) of the new iron buildings should be supporting the characteristics of its 'inner structure' as it was done in the ancient Doric temples (see Figure 2.1). Gottfried Semper (1802–1879) developed the concept of tectonics, to denote the light and linear elements in a building, and stereotomics, to denote the heavyweight elements by dividing a building into four elements (see Figure 2.2. for earthwork, hearth, framework and roof, and enclosing membrane) on the basis of method of construction. Unlike Botticher, according to Semper (1851), it is mainly the cladding and the joints that make style possible.

The twentieth-century theoreticians of architecture focused on the role played by structure and construction in achieving tectonic qualities. Eduard Sekler (1965) distinguished the concepts of structure (abstract knowledge and structural systems) and construction (building methods) from tectonics, and discussed the role played by each on tectonics. According to him, in the Masjid-i Jami in Isfahan, Iran, (see Figure 2.3) both structure and

2.1 A Doric temple (drawn with the help of URL1, 2014)

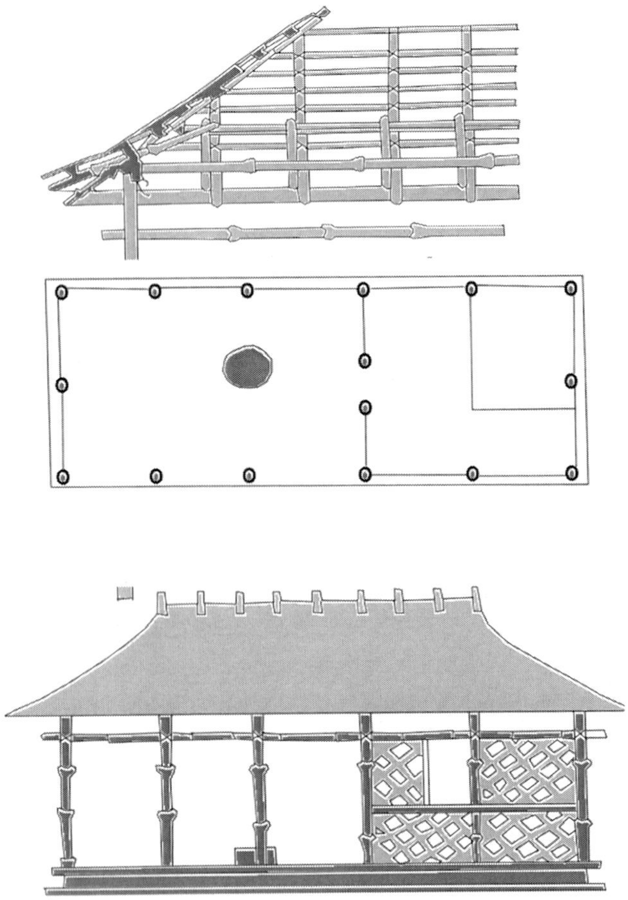

2.2 A Caribbean hut: earthwork, hearth, framework/roof and enclosing membrane (drawn with the help of Chakroff, 2006)

2.3 Masjid-i Jami, Isfahan, 771 (drawn with the help of URL2, n.d.)

construction affected tectonics positively. The structural system determined the form of the building, whilst the blue ceramic tiles covering it create a dematerialised effect, making the building (and especially the dome) look much lighter than it really is.

Following Sekler, some other architectural theoreticians also saw structure/construction and tectonics as separate issues. Carles Vallhonrat (1988) studied the impact of tectonics on techniques (structure and construction technique). Vittorio Gregotti (1996) believed that details form a relationship between tectonics and techniques. Marco Frascari (1996) explained that the tectonic significance of modern architecture is due to the developments in structural systems. The postmodern philosopher Fredric Jameson (1994) also agreed that technology determines modern architecture by believing that modern architecture is more about structure/construction than it is about space and form. The theoreticians of digital tectonics also follow Sekler's point of view, believing that digital tectonics balance the modern separation between structure/construction and space/form (Liu & Lim, 2006; Oxman, 2012).

Similar to Semper, Kennett Frampton (2001) did not separate structure/construction and tectonics. He believed that tectonics is the poetics of construction and that the joint is the smallest unit to affect tectonics. Frampton explained that the structures, by affecting the form of the building, also determine the tectonics of modern buildings. David Leatherbarrow and Mohsen Mostafavi (2005) studied the tension between representation and technology in modern architecture in their book *Surface Architecture*. According to Gevork Hartoonian (1994) secularisation is the main characteristic of modern techne. Richard Sennett's *Flesh and Stone* (1994) also supports the similarity between tectonics and architectural technology by relating aesthetic sensitivity in architecture to aesthetic sensitivity towards the human body.

This book supports the idea that the tectonics of any particular building cannot be a separate issue from its structure and construction technique. If they are separate, it would not be possible for architects to play with structural (or constructional) recommendations from structural engineers. By highlighting practical knowledge of structures and particularity of tectonics, this book is more about techne than episteme. However, the presence and powers of episteme is acknowledged.

EVALUATION OF TECTONIC VALUE

Since techne (practical knowledge – the knowledge about a particular product) is affected by episteme (scientific knowledge – knowledge for its own sake), an evaluation of tectonic value requires the elimination of the contradictions between the value system of techne and the value system of episteme. Scientific knowledge is based on being right/wrong (usually in natural and engineering sciences) or acceptable/unacceptable (usually in social sciences). On the other hand, practical knowledge about a particular product (such as a building) is based on being right/wrong, good/bad and beautiful/ugly. The research methods used to determine right/wrong, good/bad and beautiful/ugly are all different. One can use quantitative engineering methods to determine whether the technologies used in a building were used appropriately. Determining if a building is useful or not (good/bad) requires the use of statistical evaluation of data collected by using social science techniques such as questionnaires. It is also possible to make an interpretation on the usefulness of a building through the analytical evaluation of interviews with users. Finally, an evaluation of the aesthetic value of an object requires interpretation of data collected through the methods of arts and humanities, which can be based on knowledge about human psychology, phenomenology or conceptual analysis. There is a tension between these three value systems because there is only one correct design, whilst there can be many good and beautiful designs. This tension increases further when the preferences of the designers are taken into account.

The evaluation of tectonic value requires a method that can bridge the gap between the quantitative and the qualitative: a method that can evaluate the artistic value (the qualitative) and the engineering value (the quantitative) simultaneously. This is the reason why the philosophical method of ontology is preferred by the theoreticians of tectonics.

One object is ontologically different to others if its characteristics have been supported by a reason or reasons. In architecture, these reasons can be safety, economy, functionality and/or aesthetics. According to Manuel Delanda (2007), the form of a structure is ontological if it is supported by a structural reason. Delanda's concept of the ontological is close to an engineering approach because it suggests the right form for structures. According to this approach, the form with a reason is the natural and aesthetical form. Kenneth Frampton's concept of ontological architecture is the opposite. According to Frampton (2001), the aesthetic characteristics of a building can be ontological if they are achieved through the materials and systems which are used in the building. If aesthetic characteristics are achieved through make up, then they are scenographic. Frampton does consider the concepts of right or wrong. If materials and systems used in a building are acceptable (by the building's client and building codes) and if they contribute to the aesthetic quality of the building, then that building has ontological characteristics. Frampton's value system also includes climatic comfort, cultural background and topology of the physical environment. Thus, he uses multiple values to evaluate tectonic quality. According to Frampton (2001) tectonics is mainly an aesthetic category.

The sections about the use of plaster in architecture in Richard Sennett's book *Together* (2012) force us to imagine another concept of the ontological. According to Sennett, modern building techniques and styles usually do not support skilful work, such as the aesthetic use of stucco. However, from the point of view of craftsmen, the aesthetic use of stucco is very appropriate for improving routines of workmanship, which is an ontological approach to work. According to this approach, the quality of the production process determines the ontological quality of the product. Thus, there can be various definitions of the ontological in architecture.

This book is about the ontological value of products (buildings), and not production processes. The book combines the approaches of Delanda and Frampton by considering all architectural values simultaneously. According to this approach, a designer should know what is right whether or not s/he uses it in order to achieve the safe, good and the aesthetic. However, this means that some compromise is needed from the engineering value system, which is based on finding out the only correct possibility, in order to have a multiple value system.

AN ANALYTICAL APPROACH TO TECTONICS

It is difficult to bridge the gap between the engineering value system and other value systems by using a method such as phenomenology, which depends on extensive observation of the object or understanding the conceptions of other people about an object. This gap is between the knowledge about a particular thing and the knowledge about the general. However, an analytical approach can produce interpretations according to all value systems. Thus, the methodology used in this book is to make interpretations about the aesthetic and engineering value of buildings through logical analysis. The value system about functionality (the good) is not considered in this book unless it directly affects structural design.[2]

Both the methods of evaluating aesthetic value (usually phenomenology) and engineering value (usually quantitative) are replaced with the analytical. The analytical approach makes it possible to discuss the role of each physical entity (such as structure, materials, details, mechanical systems, electrical systems, topography, climate and light, etc.) on the aesthetic quality of the building. It also becomes possible to consider less appropriate engineering alternatives.

AN ANALYTICAL APPROACH TO AESTHETIC VALUE OF STRUCTURES

The aesthetic value of specific buildings can be analysed with the help of two different approaches:

- By evaluation of the design concept of the building.
- By evaluation of the image of the building within its context.

The first approach is more applicable if the building was designed according to a distinctive idea (design concept) within its context. The second approach is preferable if the image of the building is integrated with its context. The first type is less continuous with its context whilst the second type is more continuous with its context.

Design concepts are 'ideas which integrate various elements into a whole' (McGinty, 1979: p.208). They can be useful at various stages of design (White, 1975). There can be three types of design concepts. The first group of design concepts are based on making analogies or metaphors. The designers seek a similarity with the poetic meaning of something (McGinty, 1979: p.223) or they seek for an abstract or fictive relationship between two or more things (McGinty, 1979: p.228). Designs within this category include those by Santiago Calatrava (such as the idea of a man throwing a ball) and Frank Gehry (such as the abstract idea of a fish). The second group of design concepts are based on certain ideals, ideologies or philosophies which represent society. Here the designers are inspired from various types of thought. Examples of ecological architecture and zero carbon architecture can be seen within this group. The third group of design concepts are the rational responses to the design problem. Here the designer defines what is needed rationally and realises these needs autonomously. In other words, s/he identifies some major problems in the design of a specific type of buildings and concentrates on the solution of these specific problems in an autonomous manner.

Although the context is important for all types of architecture, design concepts determine what is distinctive (discontinuous) about specific buildings. On the other hand contextual design is based on continuities with the physical, social and natural context. Here, the designers develop images concerning these continuities (see Bachelard (1994) for the role of continuities and discontinuities on the poetics of space).

Thus the aesthetic value of architecture can be analysed through questioning:

- What makes that architecture distinct?
- What provides the continuity with the context?

The answers to these questions can be found in literature covering the specific architecture.

After understanding the design concept/image of the design of a building, it is necessary to analyse which physical entities (such as structure, materials, details, mechanical systems, climate and light, etc.) are supporting the design concept/image of design. Does the use of materials support the design concept/ image of design? Does the response of the building to the climate support the design concept/image of design? The main question to answer within this book is: Does the structural system of the building support its design concept/image of design by providing discontinuities or continuities? Answering this question for specific buildings is the main step of evaluating the aesthetic value of the structure of the buildings. Then one can ask whether that particular structure is dominant in the determination of the design concept/image of design in comparison to other physical entities.

AN ANALYTICAL APPROACH TO TECTONIC VALUE OF STRUCTURES

After determining the aesthetic value of the structure of a building, it is necessary to consider the engineering value in order to make an interpretation about that structure's tectonic value.

A building's structure has an engineering value if it is safe and economical. Structures that are safe and economical form a larger set than that formed by right/correct structures. Structural engineers can make some structural recommendations to designers in order to use specific structural systems in a safe and economical way. However, these recommendations are usually aimed towards achieving the right/correct solution. (Chapter 3 covers engineering recommendations aimed towards achieving economy.) However, the response of architects to these recommendations is never a direct application. These recommendations are applied in either a poetic way (the traditional approach, which is like preparing a banquet out of a strict diet), or they are transformed, replaced or even ignored (the modern approach, which is like achieving the same weight loss without applying a diet, but with sports). This is the play dimension in design. A comparison of the contribution of the structure of a building to its design concept/idea of design, with the possible structural recommendations of engineers about that type of structure, makes the play dimension in the architectural design of structures explicit.

Gaston Bachelard (1987) believed that the desire for playing with the right/wrong can be explained by the Prometheus Complex, which is a desire for playing with fire in a skilful way. Families do not let their children play with fire because they do not want their homes burnt. Yet due to this restriction, the children (and adults) like to make fires without causing any trouble and this desire is related to their love for life. Thus, playing with the structural guidelines in architecture can be related to the Prometheus Complex and it should be supported if it is done carefully and skilfully in order not to cause problems.[3]

Following engineering recommendations in an architectural way (applying the recommendations in a poetic way) is an affirmative approach to architectural design. On the other hand, playing with engineering recommendations (transforming, replacing or ignoring these recommendations) is a contravening approach to architectural design. It is a serious business to play with fire. One should know that both the recommendations and the play dimension are equally important for the satisfaction of the intellectual needs of architects.

NOTES

1 Richard Sennett's use of the word 'emphatic' has a different meaning than that implied by the common usage of the word. Sennett means a kind of objective attitude when people negotiate with each other in a reasonable way.

2 See sources about International Style for more information.

3 The Prometheus Complex shows itself in a different way in engineers. They force the limits of systems or try to achieve impossible targets through innovation.

REFERENCES

Aristotle (1988) *On the Art of Poetry*, trans. I. Bywater, Oxford University Press: Oxford.

Bachelard, G. (1987 [1938]) *The Psychoanalysis of Fire*, 2nd edition, trans. A.C.M. Ross, Beacon Press: Boston.

Bachelard, G. (1994 [1958]) *The Poetics of Space*, trans. M. Jolas, Beacon Press: Boston.

Billington, D.P. (1983) *The Tower and Bridge*, Basic Books: New York.

Botticher, K. (1852) *The Tectonics of the Hellenes*, Postdam: Germany.

Chakroff, E. (2006) *Convergence* (viewed 26 September 2014: http://evanchakroff.com/page/6/?attachment_id)

Delanda, M. (2007) *Intensive Science and Virtual Philosophy*, 4th edition, Continuum: New York.

Deleuze, G., Guattari, F. (1993 [1980]) *A Thousand Plateaus – Capitalism and Schizophrenia*, 4th edition, University of Minnesota Press: Minneapolis.

Frampton, K. (2001) *Studies in Tectonic Culture: The Poetics of Construction in Nineteenth and Twentieth Century Architecture*, ed. J. Cava, The MIT Press: Cambridge, Massachusetts.

Frascari, M. (1996) 'The tell-the-tail detail' in ed. K. Nesbitt *Theorizing a New Agenda for Architecture: An Anthology of Architectural Theory 1965–1995*, Princeton Architectural Press: New York: pp.498–515.

Gregotti, V. (1996) 'The exercise of detailing' in ed. K. Nesbitt *Theorizing a New Agenda for Architecture: An Anthology of Architecture Theory 1965–1995*, Princeton Architectural Press: New York: pp.494–497.

Hartoonian, G. (1994) *Ontology of Construction*, Cambridge University Press: New York.

Heidegger, M. (1977 [1927]) *The Question Concerning Technology and Other Essays*, trans. W. Lovitt, Harper & Row: New York.

Heidegger, M. (1995[1927]) *Being and Time* (Sein un Zeit), 13th edition, trans. J. Macquarrie & E. Robinson, Blackwell Publishers: Oxford.

Jameson, F. (1994) 'The constraints of postmodernism' in *The Seeds of Time*, Colombia University Press: New York.

Leatherbarrow, D., Mostafavi, M. (2005) *Surface Architecture*, The MIT Press: Cambridge, Massachusetts.

Liu, Y.T., Lim, C.K. (2006) 'New Tectonics: A Preliminary Framework Involving Classic and Digital Thinking' *Design Studies*, Vol. 27, No. 3: pp.267–307.

McGinty, T. (1979) 'Concepts in architecture' in ed. J.C. Snyder & A.J. Catanese *Introduction to Architecture*, McGraw Hill Book Company: New York.

Oxman, R. (2012) 'Informed Tectonics in Material Based Design' *Design Studies*, Vol. 33, No. 5: pp.427–455.

Sekler, E.F. (1965) 'Structure, construction, tectonics' in ed. G. Kepes *Structure in Art and in Science*, George Braziller: New York.

Semper, G. (1851) *The Four Elements of Architecture and Other Writings*, Cambridge University Press: New York.

Sennett, R. (1994) *Flesh and Stone: The Body and the City in Western Civilization*, W.W. Norton & Company: New York.

Sennett, R. (2008) *The Craftsman*, Yale University Press: New Haven.

Sennett, R. (2012) *Together – The Rituals, Pleasures and Politics of Cooperation*, Yale University Press: New Haven.

Stanford Encyclopedia of Philosophy (2007) 'Episteme and techne' (viewed 30 September 2013: http://plato.stanford.edu/entries/episteme-techne/)

Vallhonrat, C. (1988) 'Tectonics Considered: Between the Presence and the Absence of the Artifice' *Perspecta*, Vol. 24: pp.122–135.

White, E.T. (1975) *Concept Sourcebook, A Vocabulary of Architectural Forms*, Architectural Media Ltd: Arizona.

URL1 (2014) *Three Doric Temples* (viewed 26 September 2014: http://benedante.blogspot.com.tr/2014/01/three-doric-temples.html)

URL2 (n.d.) *Masjid-i Jami'-i Isfahan* (viewed 26 September 2014: www.pinterest.com/pin/128000814382835310/)

PART 1
GENERAL STRUCTURAL GUIDELINES

Specific structural recommendations can be made for each structure to form structural guidelines. However, there can also be common structural recommendations that are valid for all structural systems. These can be collected under a general set of structural guidelines.

General structural guidelines can be divided into three:

- General structural guidelines in relation to economy.
- General structural guidelines originating from the requirements of strength, stability, equilibrium and deformation limit.
- Structural guidelines in relation to form and size of the structure.

Chapter 3 discusses structural guidelines in relation to economy. Chapter 4 contains guidelines originating from strength, stability, equilibrium and deformation limit requirements. Chapter 5 covers structural guidelines in relation to form and size of structures.

3

General Structural Guidelines in Relation to Economy

BY YONCA HUROL AND BAYDU CAN AL

Buildings are designed within various economic constraints. The budget can be very limited for one building and limitless for another. Whatever the economic conditions, engineering ethics define a certain approach to the subject of economy. The first step in understanding this approach is to understand the concepts of 'efficiency' and 'optimum'. Then the different approaches to building economy can be discussed and structural guidelines in relation to economy can be derived.

STRUCTURAL EFFICIENCY AND THE OPTIMUM

Structural engineering methods, which can be analytical or computational, help designers to find ways of improving the performance of structures whilst decreasing their cost. If the performance is high and the cost is low, then the structure is efficient. Performance of a structure can be related to its load-bearing capacity, amount of deflection, functionality and aesthetics. On the other hand, cost can cover the amount of structural materials, labour, maintenance and construction time (Xiaoye, 2011). The most important contemporary aesthetic concept that has affected both architecture and structural engineering is lightness. Lightness is both an aesthetic and an economic category (Billington, 1983).

There are different approaches for achieving structural efficiency. However, all of these approaches are based on the concept of optimisation. If we consider the amount of structural material per unit area and the amount of specialised labour as two factors that affect the cost of the building, it could be said that these two factors are contradictory. Decreasing the amount of structural materials might result in an increase in specialised labour, which in turn increases the cost. Here, we have more than one factor which determines the economy of the structure/building:

- The amount of structural materials.
- The type of labour for construction.

Similarly, the use of cheap labour might bring an increase in the amount of structural material. Thus, choosing the most appropriate (both light and economic) solution in a situation requires determination of the optimum possibility.

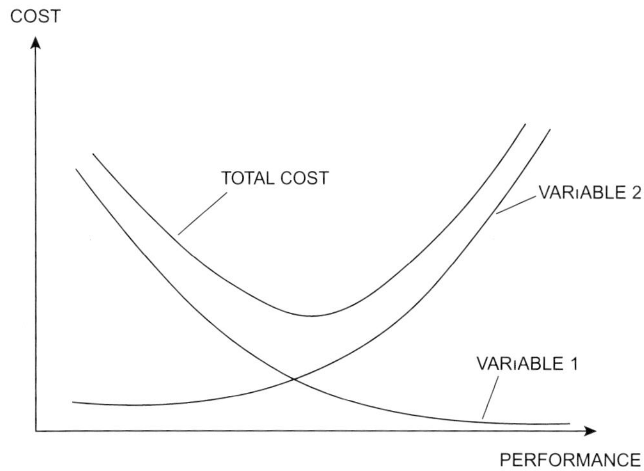

3.1 Finding the optimum solution (after Macdonald, A. J. (2001) *Structure and Architecture*, Architectural Press: Oxford, p. 65, Figure 6.3)

The optimum is the best: either the maximum or the minimum. Figure 3.1 shows two criteria that contradict each other in their relation to cost. The performance of variable 1 increases as its cost decreases. In contrast, the performance of variable 2 changes parallel to its cost. If these measurements are taken from the same body, such as a structure, then the best solution for both variables can be achieved at the point where the cost is minimised. (See Vignaux (2004) for more information about multi-attribute decision problems.)

The concepts of efficiency and optimum present quantitative approaches to design.

POSSIBLE APPROACHES TO THE ECONOMY OF STRUCTURES

Possible approaches to the economy of structures can be explained by using the concepts of structural efficiency and optimisation as follows:

- Designing for evolutionary structural optimisation.
- Designing for the optimum structure.
- Optimisation of the designed structure.

Designing for evolutionary structural optimisation

The amount of structural material per unit area increases considerably if the span or the height of the structure increases. This is the point at which the structure starts to determine the economy of the building. It is therefore necessary to minimise the amount of structural material needed for long-span and high-rise structures. The history of long-span and high-rise structures is related to developments in increased structural efficiency through evolutionary structural optimisation (Xiaoye, 2011) by inventing new structural systems. (See Mainstone (2001) for developments in structural systems.) The structural efficiency of skyscrapers has been measured and compared as the weight of the structural material per unit area (kg/m^2).

There are some engineer–architects who designed whole structures to minimise structural material. For example: Felix Candela, Heinz Isler and Pier Luigi Nervi. Candela and Isler designed using thin shell structures, whilst Nervi used various structural systems. Figure 3.2 shows Candela's Los Manantiales Restaurant in Mexico. Manantiales Restaurant consists of the intersection of four half hyperbolic paraboloids. The diameter of the circular plan of the building is 42m. The depth of the shell is 4cm and there are no edge ribs (Burger & Billington, 2006; URL6, 2008). The lightness of this structure was important for Candela: it is known that he tried to get rid of the edge beams to achieve the thinnest possible surface.

In Lanificio Gatti, Italy, (see Figure 3.3), Pier Luigi Nervi put the reinforced concrete ribs of the waffled slab parallel to the principal stress lines.[1] He thought that the use of mathematics for minimising the structural material results in a natural form that is also beautiful. Since this building is not a long-span or high-rise

3.2 A sketch of Felix Candela's Los Manantiales Restaurant, Xochilmico, Mexico, 1958 (drawn with the help of URL1, 2012)

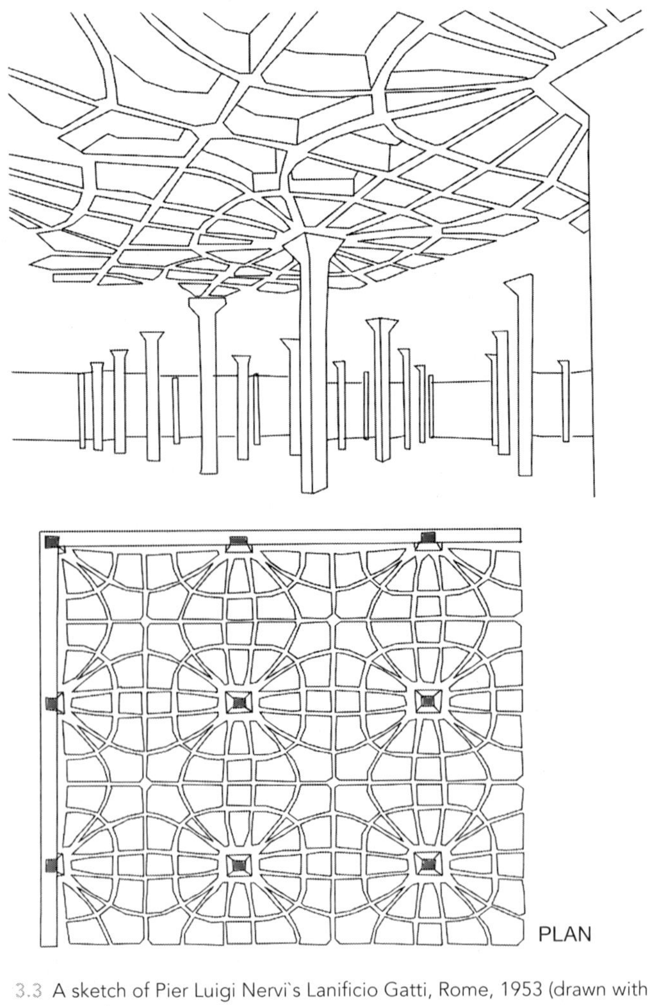

3.3 A sketch of Pier Luigi Nervi's Lanificio Gatti, Rome, 1953 (drawn with the help of URL2, n.d.)

building, the minimisation of structural material did not result in an economic solution because the labour and special formwork required to build these ribs were expensive.

Suspended glass systems are contemporary examples of designing for evolutionary structural optimisation. The cable trusses of suspended glass systems are designed in such a way that the amount of structural material is minimised in order to achieve the dematerialised effect of these glass surfaces. Examples of suspended glass systems can be seen with an internet search. These systems are studied in chapter 12 of this book.

Designing for evolutionary structural optimisation requires engineering knowledge and it is usually applied by talented structural engineers. The need to use mathematics to achieve the most efficient form usually prevents architects from directly using this approach in design. However, there are some buildings that are specifically designed for lightness by the architects. In such cases the architects seek the help of structural engineers in order to achieve their target. For example, the design of the Science and Technology Museum in Paris, France. The architect Bernard Tschumi designed the glass surfaces on the front facade of the building to be as transparent as possible and asked for the guidance of structural engineer Peter Rice, who invented the suspended glass systems for this building. This building is studied as one of the case studies in chapter 12.

Designing for the optimum structure

Between and after the two World Wars there was a great and urgent need to build many houses with very limited budgets. Many people in Europe were homeless and many people were migrating from villages to cities all over the world. The method of optimisation in order to achieve the maximum number of houses with the least cost became very important during the 1960s.

Buildings built during this period were called 'matchbox buildings' because it required less cost to build reinforced concrete rectangular buildings around 10 or 15 storeys with the effective use of foundations. The employment of cheap, non-specialised labour was possible because of the use of rectangular geometries

3.4 A sketch of La Cité de la Muette, Marcel Lods, Drancy, France, 1932–1934 (drawn with the help of URL3, 2012)

and the use of reinforced concrete as the structural material. All other characteristics of the building were optimised in order to achieve the least cost. The dimensions of the building, the span between two columns, the type of slab system and the dimensions of all structural elements, etc., were determined with the help of the engineering technique of optimisation. Marcel Lods' La Cité de la Muette in France (see Figure 3.4) is a good example of this type of building. Industrialised building techniques were used in this mass housing project.

The main characteristic of this period was the domination of engineering decisions during the design process. Architects were trying to achieve good quality and aesthetics within the boundaries of engineering decisions. This approach is still very useful for projects with limited budgets.

Optimisation of the designed structure

Starting from the 1960s, an increase in the wealth of Europe and the USA had an effect on people's approach to buildings. For many people, having more money helped them to better express their architectural taste and identity through special buildings. Clients started to support architects in rejecting matchbox solutions and designing more artistic and meaningful buildings. Architects were able to design their projects more freely, without being forced to use the optimum forms and dimensions. Architecture therefore became the main determinant in the design process. The structures of these buildings were optimised without changing the main decisions of the architects. The optimisation process is done to determine the optimum dimensions of structural members in order to ensure the safety of the structure.

John Burgee Architects with Philip Cortelyou Johnson's Lipstick Building in New York, USA, which is also known as 53rd at Third, (see Figure 3.5) is an example of this architectural approach to economy. This red building is 34 storeys. It uses less base area in comparison to other buildings surrounding it and most of its ground floor is used for the building's entrance hall. Neither the height of the building, nor the form, is at its optimum. These decisions were given by the architect and only the structural elements' dimensions were optimised by the structural engineer.

Another well-known example of this architectural approach to economy is Jorn Utzon's Sydney Opera House in Australia. This project won an architectural competition in 1957 but it was completed in 1973 after causing many economic problems for the Australian government. This building is presented as a case study in chapter 16.

This architectural approach to economy is also affected by the neo-liberal economy. The concept of symbolic capital, which was used by Pierre Bordieu (1984), explains this change. If someone gains money because of his/her recognition or prestige within a society, the source of his/her recognition/prestige is called symbolic capital. For example, the degree of PhD is symbolic capital for an academic. Architecture can also be a source of recognition and prestige. Having an office building designed by a famous architect in a prestigious district of a city could bring many valuable job opportunities to the owner of that building. In a situation like this the building becomes symbolic capital. The architectural

3.5 A sketch of John Burgee Architects with Philip Johnson's Lipstick Building, New York, 1986 (drawn with the help of URL4, 2005)

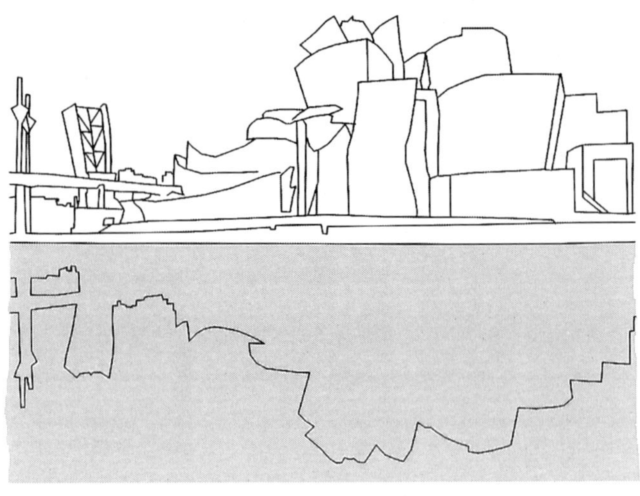

3.6 A sketch of Frank Gehry's Guggenheim Museum, Bilbao, Spain, 1997 (drawn with the help of URL5, 2010)

approach to economy for buildings that are expected to be symbolic capital is different than that for other buildings. Money spent on these buildings is seen to be an investment. A good example of symbolic capital in architecture is Frank Gehry's Guggenheim Museum in Spain (see Figure 3.6).

The Guggenheim Museum was originally suggested for the purpose of contributing to the economy and revitalisation of Bilbao. The architect Frank Gehry was encouraged to design a different and innovative building. He designed random and formless curves and covered them with titanium sheets reminiscent of fish scales. Titanium, which is an expensive material used in aerospace technology, was being used in a building for the first time. Many people from all over the world have visited the exhibitions in this building since it was opened to the public in 1997. Tourists visiting the building caused a considerable increase in the economic activity of the city.

Percentages of different approaches to economy

Designing for evolutionary structural optimisation, designing for the optimum structure, optimisation of the designed structure and designing for symbolic capital are all professional approaches to building economy. Although each originates from a specific time period, they all coexist within twenty-first-century professional architecture – and they are all needed.

The percentage of design for symbolic capital and design for evolutionary structural optimisation is very small in comparison with other architectural approaches to economy. Most of the buildings designed by professionally recognised architects still follow optimum forms which minimise the cost. City views seen from aeroplanes are evidence of this situation.

On the other hand, the majority of buildings in the world are traditional, vernacular architecture designed and built by master builders, or houses designed and built by squatters. Traditional architecture usually occurs in rural areas and squats usually appear in big cities. According to Mike Davis (2006), 99.4% of city populations live in illegal squats in Ethiopia and this figure is 42.6% for Turkey.

STRUCTURAL GUIDELINES FOR DIFFERENT APPROACHES TO ECONOMY

The content of structural guidelines should differ according to the economic approach. Thus, it is one of the first subjects to be discussed between the architect and the structural engineer in order to avoid unnecessary conflict. The client's economic expectations for the building should be clarified at the beginning of the design process.

If the design of a building aims for evolutionary structural optimisation (in order to have an extraordinary and light structure), then structural guidelines will cover minimising structural materials but not economy. All of the physical characteristics of the structure will be used to minimise the amount of structural materials. If one wishes to design a shell structure as such, then the form of that shell will be very important. If it is a high-rise tube structure, then the slenderness ratio, height and span of the beams should be considered carefully. Although these structures are usually designed by architect–engineers, there are some examples in which the idea of the light structure was developed by the architect (such as the dematerialised suspended glass systems of the Science and Technology Museum in France).

If the approach to economy is the design of the optimum, then structural guidelines should concentrate on minimising the cost of the structural system and the building. The professional team is expected to optimise the amount of structural materials and the cost of labour. Thus, all physical characteristics of the structure can be subject to structural guidelines in order to decrease the cost. For example, the type of structural material, selection of the structural system, use of rectangular or other geometries, span size, and number of basement floors can affect the cost of a 10-storey apartment building.

If the approach to economy is the optimisation of design, then structural guidelines will cover the safe dimensions of structural elements. The structural engineer should try to achieve the architectural characteristics given by the architect. Thus, most of the structural guidelines will be about the dimensions of structural elements. The structural engineer might ask some structural elements to be larger or smaller. Structural advice about achieving certain challenging forms might also be needed. Designing for symbolic capital also needs the same type of structural guidelines as optimisation of design, however an innovative engineering approach might also be needed.

NOTE

1 The abstract lines on which only compression or tension occur.

REFERENCES

Billington, D.P. (1983) *The Tower and Bridge*, Basic Books: New York.

Bordieu, P. (1984) *Distinction: A Social Critique of the Judgement of Taste*, Harvard University Press: Cambridge, Massachusetts.

Burger, N., Billington, D.P. (2006) 'Felix Candela, Elegance and Endurance: An Examination of the Xochimilco Shell' *Journal of the International Association for Shell and Spatial Structures*, Vol. 47, No. 3: pp.271–278.

Davis, M. (2006) *Planet of Slums*, Verso: London & New York.

Mainstone, R. (2001) *Developments in Structural Form*, Routledge: New York.

Vignaux, G.A. (2004) *Multi-Attribute Decision Problems* (viewed 27 September 2013: www.mcs.vuw.ac.nz/courses/OPRE251/2004T1/Lecture-Notes/multi.pdf)

Xiaoye, Y. (2011) *Improving the Efficiency of Structures Using Mechanics Concepts*, unpublished PhD thesis, University of Manchester: UK.

URL1 (2012) *Restaurante Los Manantiales* (viewed 26 September 2014: http://fresharquitectos.blogspot.com.tr/2012/12/restaurante-los-manantiales-mexico-df.html)

URL2 (n.d.) *Historia* (viewed 26 September 2014: http://webs.demasiado.com/forjados/historia/hormigon/nervi/gatti.htm)

URL3 (2012) *Les Graffiti de la Cité de la Muette* (viewed 26 September 2014: http://archives.seine-saint-denis.fr/Les-graffiti-de-la-Cite-de-la.html)

URL4 (2005) *Philip Courtelyou Johnson* (viewed 26 September 2014: http://archpaper.com/news/articles.asp?id=1596#.VCUhSE1xljo)

URL5 (2010) *Modern Creations in Spain Made by a Genius* (viewed 26 September 2014: http://www.spain-holiday.com/blog/modern-creations-in-spain-made-by-a-genius.php)

URL6 (2008) *Felix Candela and Restaurant Los Manantiales* (viewed 2 December 2013: http://anengineersaspect.blogspot.com/2009/06/felix-candela-and-restaurant-los.html)

4
General Structural Guidelines Originating from the Requirements of Strength, Stability, Equilibrium and Deformation Limit

BY YONCA HUROL AND BAYDU CAN AL

The major technical requirements for a structure can be listed as follows:

- Strength
- Stability
- Equilibrium
- Deformation limit.

Each requirement determines a series of structural guidelines. It can also be stated that the specific structural guidelines for different structural systems arise from these major requirements.

STRENGTH

Every structural element is made out of one or more structural materials and is subjected to external loads. Strength is a characteristic of structural materials. It denotes the capacity of the material to withstand various types of stress that occur due to loading. Homogeneous building materials have certain strength values against all types of stress. However, not all building materials are homogeneous. For example, concrete is strong against compression, but it is weak against tension because it is a composite material.

In order to study the subject of strength further, it is necessary to clarify the types of forces affecting building structures and causing stresses.

Types of forces

There are three basic types of forces which effect building structures. These are:

- Concentrated force
- Distributed force
- Moment.

A concentrated force is applied to a point and it can be represented by a vector, which can be defined with its magnitude, its direction and its application point. Concentrated forces, which take place on two-dimensional surfaces, can be in the vertical, horizontal or inclined directions. A two-dimensional distributed force is applied over a line (see Figure 4.1). On the other hand, within the same boundary conditions a moment can be created in the following different ways:

- Direct application of a moment to a point.
- With the help of one force, which is applied with a distance (lever arm) to the point around which the moment is created (see Figure 4.2).
- With the help of a force couple (equal and opposite forces), which causes a moment at the mid-point between the two forces (see Figure 4.2).

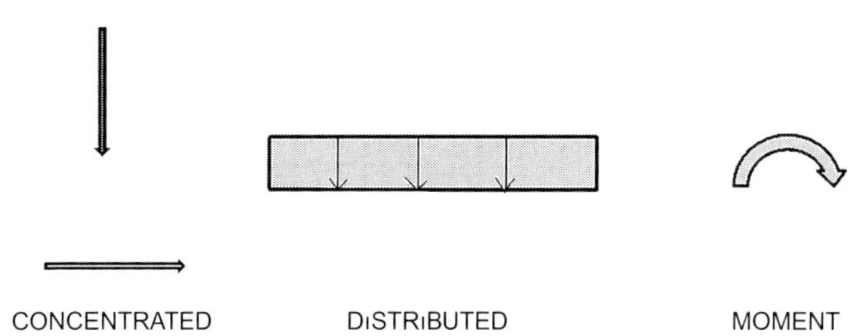

4.1 Types of forces

Magnitude of a moment can be found as:

$$M = F \times d$$

where:

M is moment (kNm),
F is force (kN),
d is distance between the force and the point of application of moment (m).

A practical example from everyday life can be given by applying a moment to a table with the help of three different force applications. A moment can be applied from one corner of the table. A force can be applied with a distance to the centre of gravity of the table. The centre of gravity is the point at which all the weight of the table can be concentrated and it is the mid-point of the table. This causes the table to turn around its centre of gravity. A force couple also causes a moment around the centre of gravity of the table (see Figure 4.2).

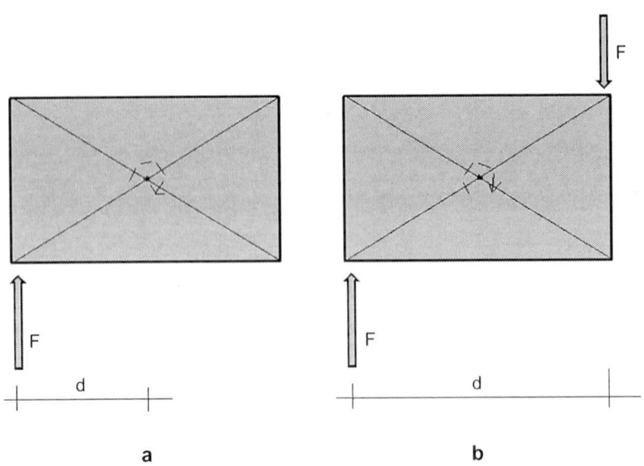

4.2 Ways of creating a moment around a table

Forces affecting building structures

The most common load types affecting building structures can be listed as follows:

- Dead load
- Live load
- Wind load
- Earthquake load
- Temperature load
- Construction load.

Dead load affecting a structure is the weight of the structure itself, which is in the direction of gravity. For example, a reinforced concrete slab (a horizontal structural surface) should carry its own weight. The weight of a slab of 3m × 4m and 15cm deep can be calculated by multiplying its volume with the unit weight (density) of reinforced concrete, which is 2.4 tons/m^3, as follows:

$$W = (3 \times 4 \times 0.15) \times 2.4 = 4.32 \text{ tons}$$

By using this approach total dead weight of a structure or a building can be calculated.

Live loads are the variable and mobile loads: for example, the weight of people, furniture and rainwater, etc. Live loads, which might have an effect on slabs of buildings, are defined in building codes according to the functions of specific buildings.

Wind load affects the outer facades of a building and it might create vortexes that affect the building structure. The building components that form the outer facades transfer this load to the structural elements, which transfer the load to the foundations and finally to the earth. Wind load changes according to wind velocity, building form and height, and presence of adjacent buildings. See Figure 4.3 for the change in wind load with respect to building height. The speed of wind increases with the increase in building height due to an atmospheric boundary layer that changes according to many factors such as obstacles surrounding the building. Keeping the building height lower is a structural recommendation to decrease the wind load. The subject of wind effect

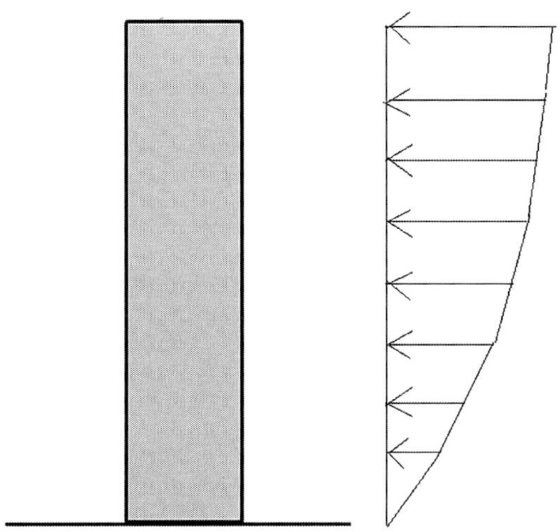

4.3 Variation of wind load due to building height

is studied further in chapters 11 and 12 where high-rise buildings and tensile structures are discussed.

Earthquake loads occur due to movements of the earth`s crust. These movements exert horizontal and sometimes vertical accelerations to the building structure. Earthquake load is proportional to the mass of the building ($F = m \times a$; where m is mass and a is acceleration). Thus, decreasing the weight of the building is a structural recommendation in order to decrease the earthquake load. Earthquake loads, like wind loads, are dynamic loads, causing buildings to sway. Earthquake resistance in frame structures is studied in chapter 9.

Temperature load occurs due to temperature variations. Colder temperature causes a shortening of linear elements whereas hotter temperature causes elongation. Since the movement of the structure is restricted by the supports (such as foundations), these changes in the dimensions of structural elements cause them to exert forces on each other. The magnitude of these forces increases with the size of the building. In order to eliminate temperature loads, buildings that have larger plan dimensions than 30m × 30m are divided into pieces with the help of expansion joints. Expansion joints are made by dividing a structure into independent pieces and by leaving a distance of approximately 3–5cm between them. Thus, it is a structural recommendation to divide buildings that are larger than 30m × 30m into pieces. However, this is not applicable for long-span and high-rise buildings because they cannot be divided into pieces. The high temperature loads created by the long elements of these large structures need to be considered during the structural design process. The subject of expansion joints is studied further in chapter 9.

Structural elements are subjected to loads when they are in their actual locations within the structure. However, they might be subjected to some other loads before they are put in their actual locations. For instance, they might be transported by trucks in a different position, or lifted by cranes. These types of applications might exert different loads from prefabricated structural elements. These loads are classified as construction loads.

Dead load is a static load. It does not change with time. However, live load, wind load and earthquake load are variable in time. These types of variable loads are known as dynamic loads. Dynamic loads can cause resonance, which is not desirable in structures. Resonance occurs if the rhythmic characteristic of a load causes movement in a structure which has the same rhythm. Fundamental time is the time required for a structure to complete one movement in one direction and then to turn back to its original position. If the rhythm of the load is equal to the fundamental time of the structure, then the structure will move more and more like a swing under the effect of this loading. Such an increase in the deflection or sway of the structure can finally cause its collapse. This is the reason for the collapse of bridges when soldiers march over them with the same tempo. Tacoma Narrows Bridge in Washington, USA, collapsed in 1940 as a result of resonance due to wind load.

The fundamental time of high-rise buildings is similar to the fundamental time of swampy earth. Thus, preventing the design of high-rise buildings on swampy earth is a structural requirement to avoid further complication due to resonance load.

Types of stress

Stress is defined as force per unit area. If the structural element is thicker, it will have less stress than thinner structural elements under the same loading conditions. There are five types of stress:

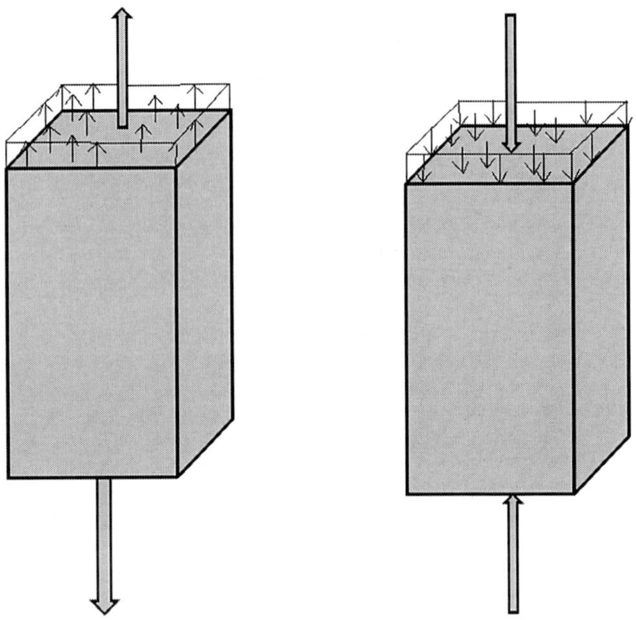

4.4 Tensile and compressive stress

- Tension
- Compression
- Shear
- Bending
- Torsion.

Tension and compression are grouped as axial types of stress, because they are parallel to the axis of the structural member. Tension causes elongation of the structural element whilst compression causes shortening. Both tensile and compressive stresses are uniformly distributed in the effective area of the structural element, as shown in Figure 4.4. This means that the force is shared equally within the area of the structural element.

If there is uniform stress distribution, stress can be calculated as:

$\sigma = F / A$
where σ is stress,
F is force,
A is the effective area.

According to the Allowable Stress Theory, the required area in relation to a known force can be found as follows:

$A = F / \sigma$ all
where σ all is the allowable stress that can be taken by the structural material.

Providing strength for all structural members by providing the sufficient cross-sectional area is a general structural requirement:

σ all = σ ult / safety factor
where σ ult is the ultimate stress that can be taken by the structural material, the safety factor is equal to 2 or 3 (depending on the country).

Depending on the ductility/brittleness of the materials used, σ ult can be replaced by the yield strength of the material, because σ ult and yield strength are very close to each other for brittle materials, whilst they are very different from each other for ductile materials.

Strain is deformation per unit length. If a 2m long column is shortened by 2cm, the strain in that column is 1cm/m (1%).

$\varepsilon = \Delta / L$
where ε is strain,
Δ is total deformation,
L is length of the element.

Figure 4.5 shows a stress versus strain diagram of steel and reinforced concrete. As can be seen, steel is more capable of undergoing plastic deformation before rupture. In other words, it is more ductile. On the other hand, reinforced concrete is more brittle and collapses without showing much deformation.

The angle of the straight part of the stress–strain diagram gives the elastic modulus (Young`s modulus) of the material. Elastic modulus is applicable to yield strength.

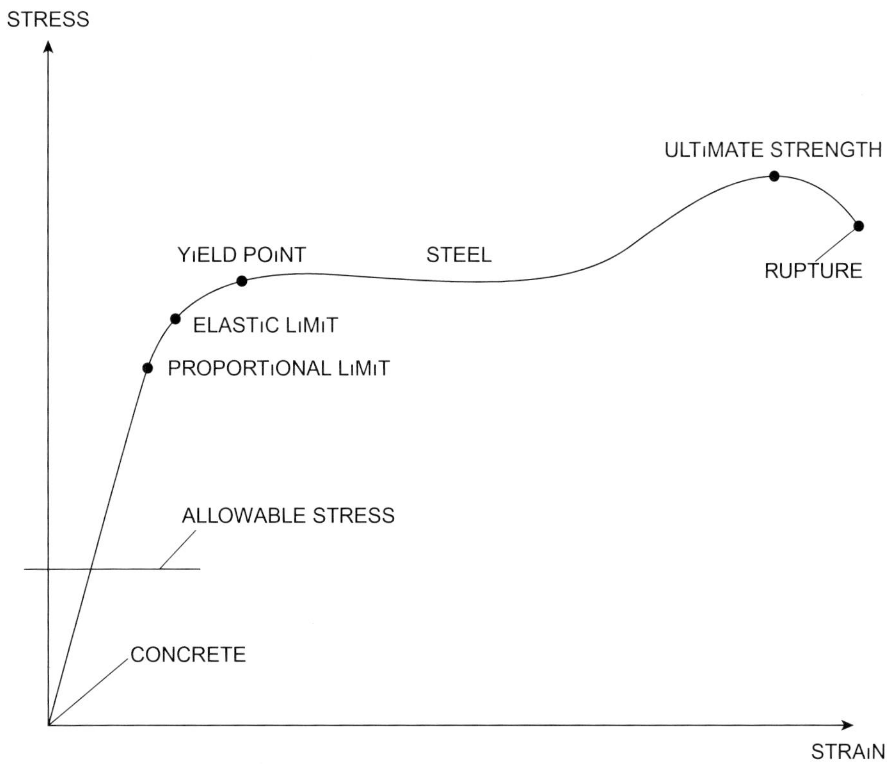

4.5 Stress versus strain diagram of steel and reinforced concrete

$E = \sigma / \varepsilon$
where E is the elastic modulus.

The elastic modulus of a material is an indication of its stiffness.

In order to carry a tensile or compressive force, it is necessary to have a cross-sectional area. However, the form of the effective area is not important. An area of any form can carry the same force. This means that for structural members, which have only an axial force, the effective area can be circular, rectangular or amorphous, etc., as seen in Figure 4.6. Thus, the possibility of having any cross-sectional shape is a structural recommendation for structural elements in pure tension or pure compression. However, if there is any risk of buckling due to compression, some shapes can be more advantageous in comparison to others.

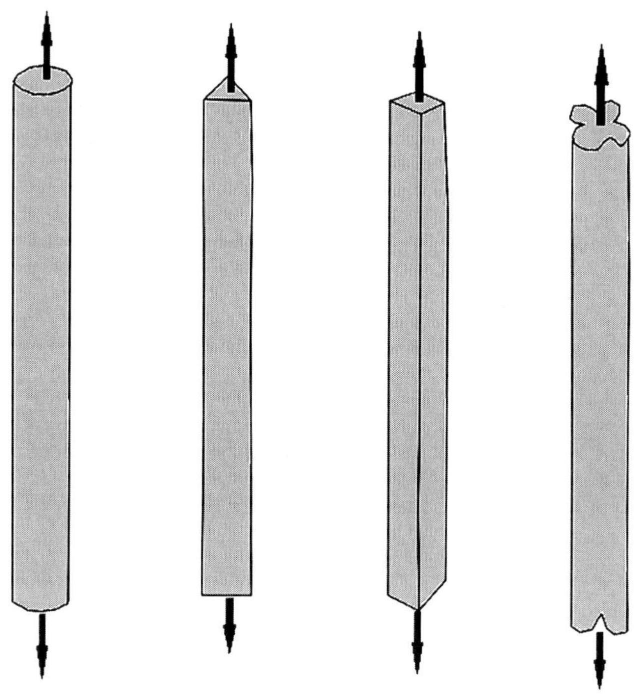

4.6 Form of effective area in axially loaded members

Although tension and compression seem to be equal and opposite to each other, they are not, because slender structural elements in compression have a danger of buckling, as shown in Figure 4.7. The famous comedian Charlie Chaplin is always remembered walking with the help of a buckled stick in one hand. Buckling is a sudden instability type. The structural requirement is to avoid buckling in building structures. Thus, increasing the cross-sectional area or decreasing the length of slender compression members are the structural recommendations to avoiding buckling.

Simple shear occurs due to the application of two equal and opposite forces perpendicular to the axis of the structural element. A shear force cuts. A scissor cutting a piece of paper is

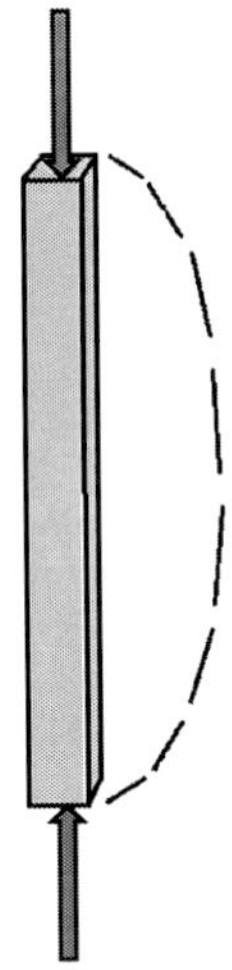

4.7 Buckling in slender compression members

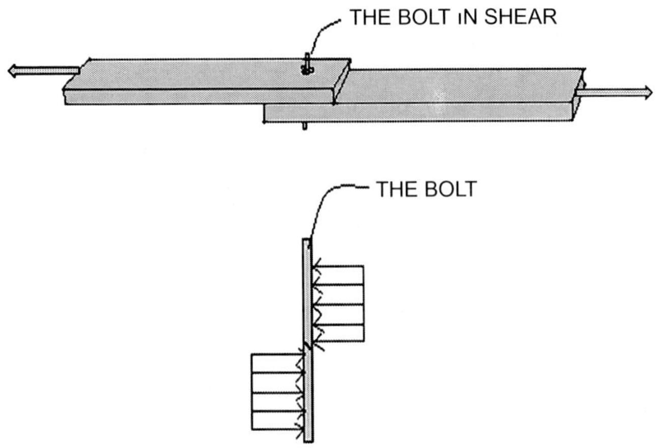

4.8 Simple shear in a bolt joining two pieces of timber

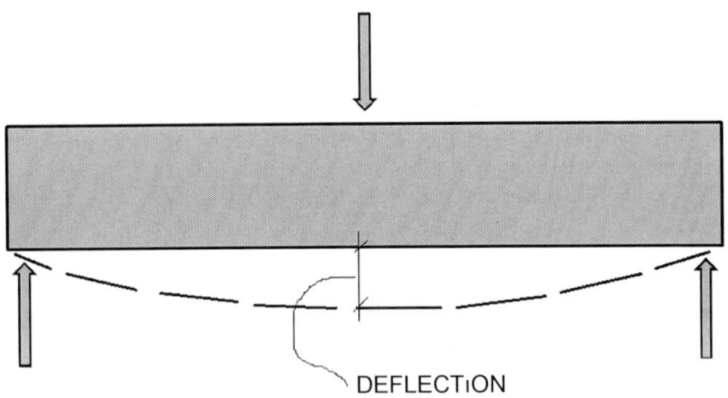

4.9 A structural element in bending

a good example of simple shear. Figure 4.8 shows two timber pieces connected to each other with the help of a bolt. When the two pieces are loaded with two equal and opposite forces, the bolt joining the two pieces will be in simple shear. Simple shear is also uniformly distributed in the effective area. Thus, it can be recommended to design structural elements in simple shear with any cross-sectional form.

Bending stress occurs when a linear structural element is loaded perpendicularly to its axis. As shown in Figure 4.9, any element in bending deflects. The type of deformation in bending is deflection.

To understand the nature of bending stress, one can draw two parallel lines on one side of a beam before applying any loading. After application of the load (as shown in Figure 4.10), these two lines will no longer be parallel to each other. If the new shape of the two lines is studied, it will be seen that the distance between the two lines has changed. This distance is decreased at the top and increased at the bottom. The only place that this distance has

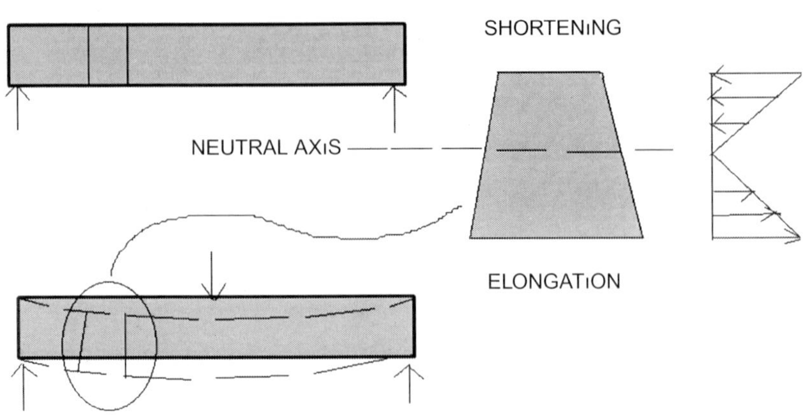

4.10 Bending stress and its distribution in the cross-section

not been changed is the mid-point, where the neutral axis of the element passes.

The decrease of the distance between the two lines at the top of the element shows that there is compression at the top fibres of the element. The increase of distance at the bottom shows that there is tension at the bottom fibres of the element. Similarly, there is no stress at the mid-point, because there is no change in the length of that fibre. The maximum compression is at the top. It decreases as it gets closer to the mid-point. The maximum tension is at the bottom and this also decreases as it gets closer to the mid-point. This type of stress distribution is known as triangular stress distribution.

Bending stress is calculated as:

$\sigma = (M \times y) / I$

where M is moment (kNm),

y is the distance between the top (or bottom) of the element and the neutral axis (m),

I is the moment of inertia of the structural element (m^4).

The moment of inertia is a tool to measure the strength of structural elements against bending. It depends on the shape of the cross-sectional area. Since the stress distribution is no longer uniform, the shape of the cross-sectional area affects the bending strength of the element. The cross-sectional area should be designed by placing more material in the places where the stress is at its maximum. Figure 4.11 shows some successful and unsuccessful shapes against bending. Since the box profile shape and the I profile shape places the maximum amount of material at the top and bottom of the element, their moment of inertia is higher than other elements even if they have the same cross-sectional area. This means that the elements with a higher moment of inertia can carry the same amount of load as smaller effective areas. Thus, the design of the cross-sectional shape by placing structural material at the places where stress is at its maximum is a structural recommendation for bending elements.

One way of testing the moment of inertia is to load a ruler in two different ways, as shown in Figure 4.12. This experiment shows that the use of a deeper cross-section is more efficient

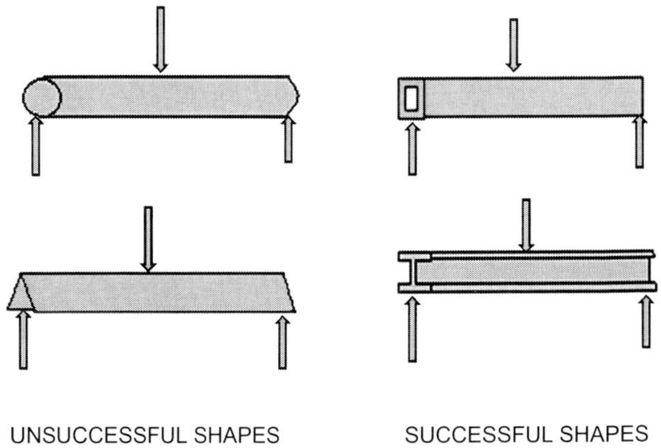

4.11 Successful and unsuccessful shapes against bending

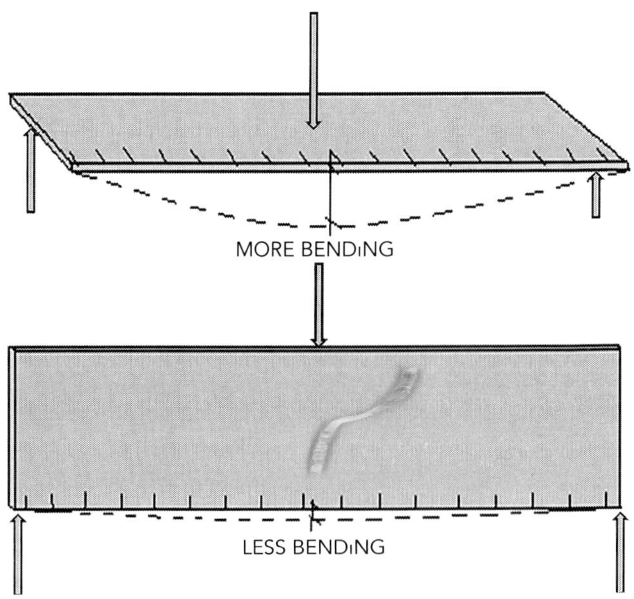

4.12 Bending a ruler to examine the role of depth against bending

against bending than the use of a shallow cross-section. It is very difficult to bend the ruler in the second case, whereas it is very easy in the first. The moment of inertia in the second case is higher than the first case.

The moment of inertia for a rectangular element can be calculated as follows:

$I = (b \times d^3) / 12$
where b is the width of the element,
d is the depth of the element.

Thus, depth affects the moment of inertia in proportion to its cube. Increasing the depth of cross-sectional shape and increasing the moment of inertia are structural recommendations regarding bending elements.

The variation in the amount of bending along a structural element also causes the occurrence of shear stress. Figure 4.13 shows how horizontal and vertical shear stress balance the difference between bending stress at two different locations.

Torsion occurs if a structural element is subject to twisting, as shown in Figure 4.14. After loading, the points a1 and b1 on the element change their locations to a2 and b2. It is clear that the line at the second location is longer than the first one. Thus there is tension parallel to that line. The tension stress in the element decreases closer to the mid-point of the cross-section. Thus, hollow, circular cross-sectional shapes are more effective against torsion.

It is better to eliminate torsion in building structures: thus, elimination of torsion is a structural recommendation. In the case of high-rise structures, wind loading may cause torsion and this will be studied further in chapter 11. Figure 4.15a shows an example of a structure in which one of the elements is subjected to torsion. The torsion in this beam should be balanced as shown in Figure 4.15b.

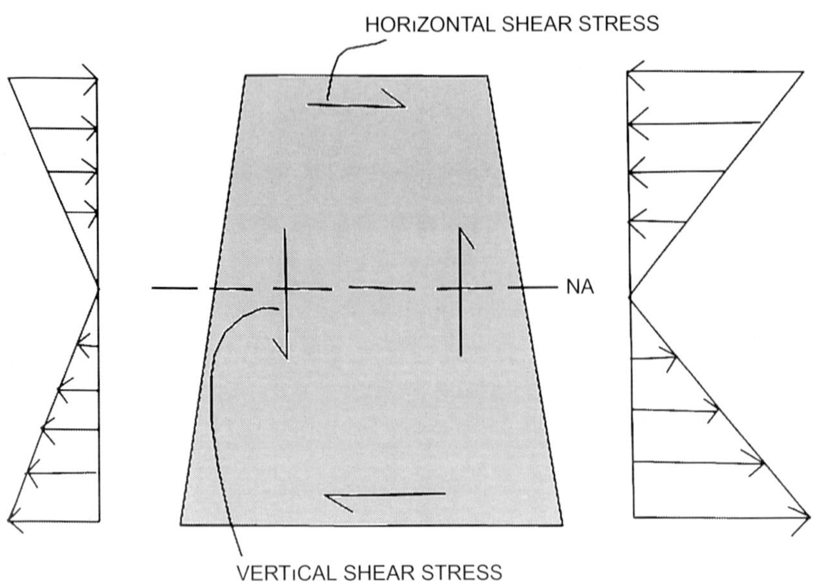

4.13 Horizontal and vertical shear stress in bending structural elements

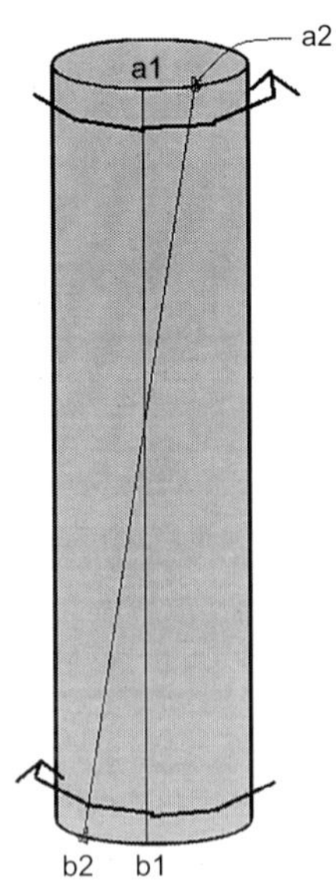

4.14 Torsion in a circular structural element

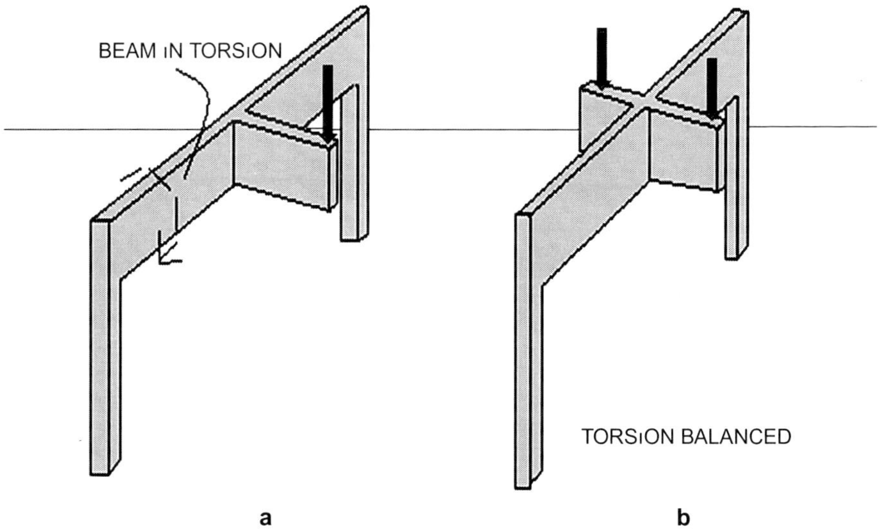

4.15 Elimination of torsion in a beam

STABILITY

The opposite of stability is instability, meaning the sudden and unpredictable movement of the structure as a whole. The avoidance of all instability types is a structural requirement. Types of instability are as follows:

- Overturning
- Buckling
- Sliding
- Uneven settlement
- Wind instability.

Overturning

Structures can overturn under the effect of gravity loads and horizontal (lateral) loads. A structure overturns due to gravity loads if its centre of gravity is outside its body. The centre of gravity of a simple shape (for example, the shape of a building`s elevation) is at the middle of the shape. The centre of gravity of a complex shape, which is a composition of two simple shapes (as shown in Figure 4.16.a) can be found by:

1 Dividing the shape into two simple shapes.
2 Finding the centre of gravity of each simple shape.
3 Connecting the centre of gravity of these shapes with a straight line. This is the centre of gravity of the whole shape.
4 Calculating the areas of the simple shapes.
5 If the total area can be represented with 1, finding the proportion of each area within the whole.
6 Finding the place of the final centre of gravity by dividing the line between the centre of gravities of the two simple shapes into two. This point should be closer to the larger area.

If the shape is a composition of three simple shapes, then the centre of gravity of the two simple shapes is found with the same method and then the centre of gravity of the whole is found between the two remaining centre of gravities, as shown in Figure 4.16b.

If the complex shape has been achieved through the subtraction of a smaller shape from a larger one, then the centre of gravity is found as shown in Figure 4.16c, by following the same steps until the last two steps. After connecting the two centre of gravities to each other with a line, this line is extended in the opposite direction to the empty shape in proportion to the area of empty shape.

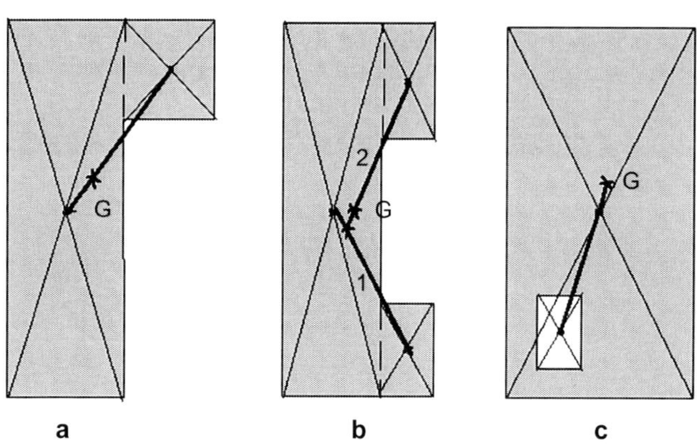

4.16 Finding centre of gravity in complex shapes

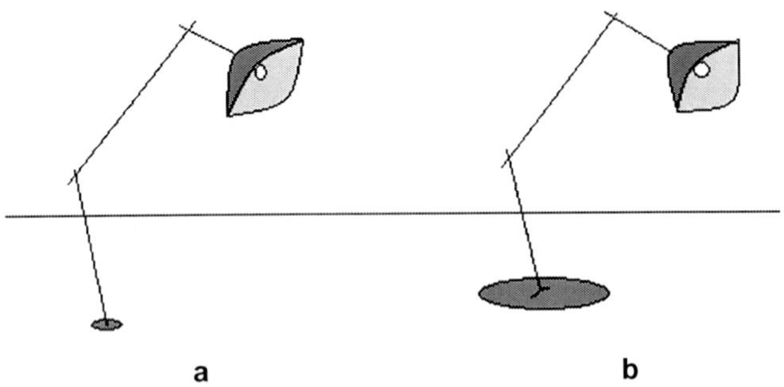

4.17 Overturning due to gravity loads

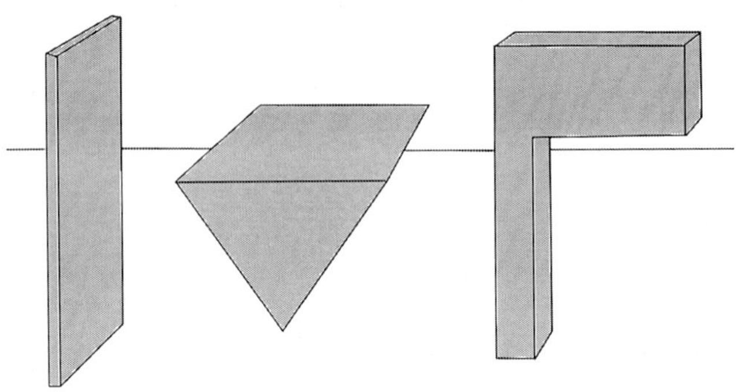

4.18 Shapes that are weak against overturning

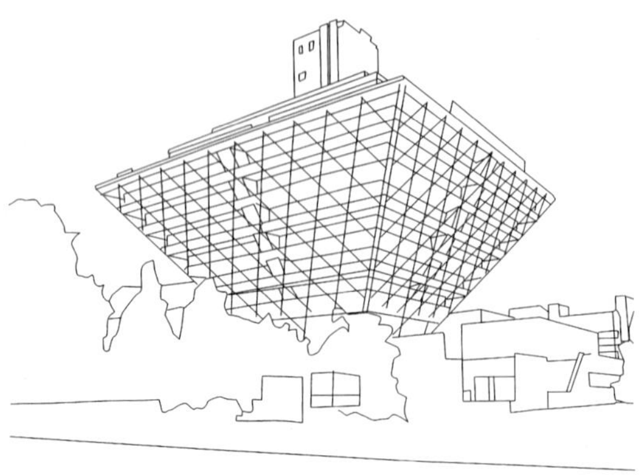

4.19 A sketch of Stephen Svetko, Stephen Durkovic and Barnabas Kissling's Slovak Radio Building, Bratislava, Slovakia, 1967–1983 (drawn with the help of URL1, n.d.)

If the centre of gravity is outside the shape of the structure as it is in Figure 4.17a, then the structure overturns due to gravity loads. However, the structure in Figure 4.17b does not overturn due to its own weight because its bottom plate is heavier than the first structure. To eliminate overturning of the desk lamp in Figure 4.17, it is possible to either increase the weight of its base or to connect the base to the table.

Structures might also overturn due to the overturning moment that is created by horizontal forces. If this overturning moment is larger than the counter moment that can be produced by the base of the structure, the structure overturns. Figure 4.18 shows some shapes that can easily overturn because they cannot produce the required amount of counter moment.

The basic structural recommendation is to avoid these shapes, which are weak against overturning. However, it may be possible to achieve these shapes and solve the overturning problem but this will cause an increase in the cost of the building. Figure 4.19 shows the Slovak Radio Building in Bratislava, Slovakia, which has an upside down pyramidal shape.

Buckling

The instability of buckling has been introduced earlier in this chapter, when discussing compressive stress (see p. 29). Structural elements or structures that are under the effect of compressive forces are subject to a potential buckling problem. Buckling should be considered when designing slender high-rise buildings. Keeping the slenderness ratio (height/width) of high-rise buildings low is a structural recommendation. Similarly, it is recommended to reduce the slenderness of columns either by increasing their thickness or by connecting them with a horizontal element, as seen in Figure 4.20.

Sliding

Sliding instability occurs if the structure is placed over an inclined topography as shown in Figure 4.21. To avoid this instability it

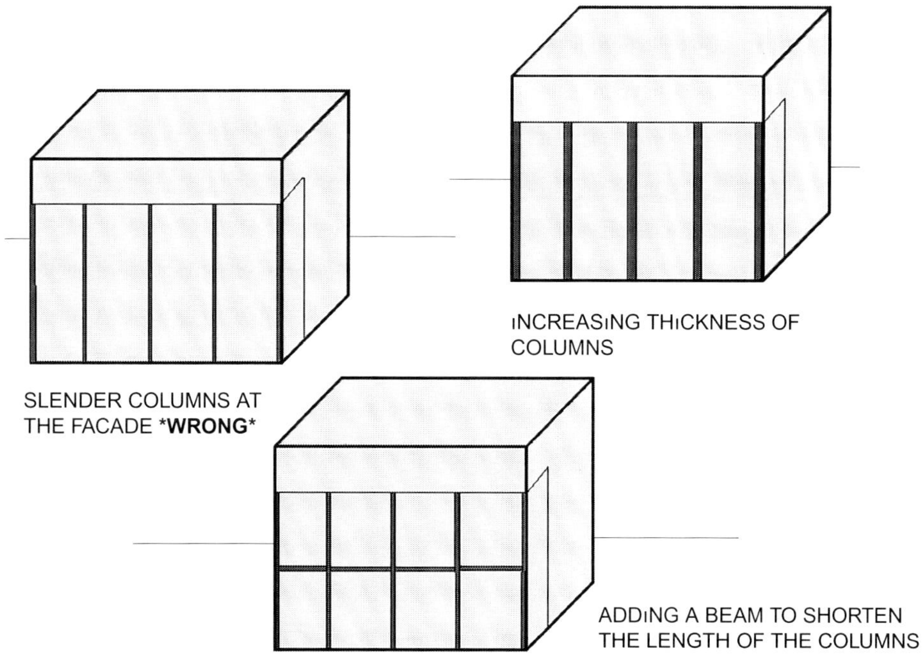

4.20 Reducing column slenderness against buckling

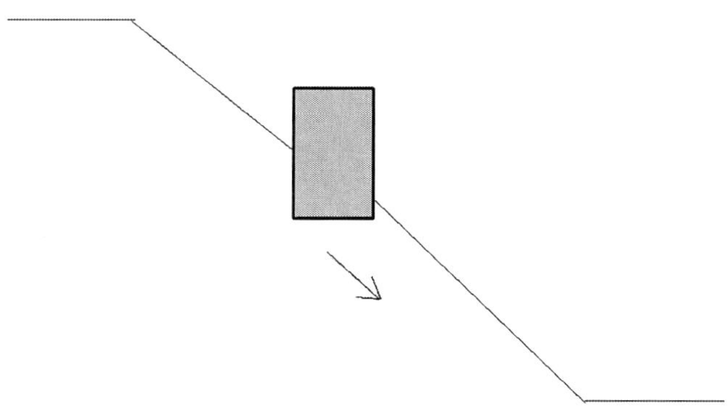

4.21 Sliding

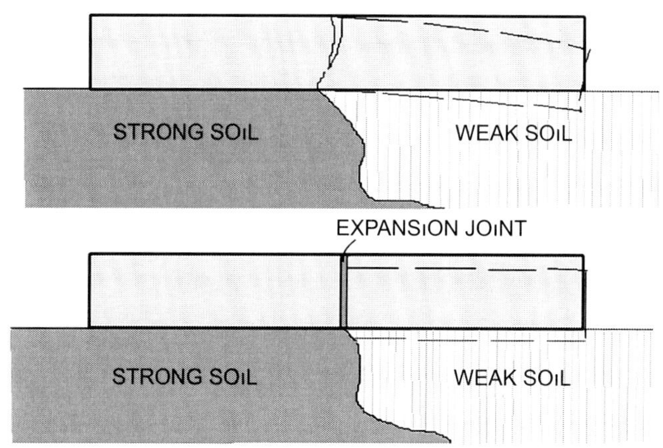

4.22 Uneven settlement

can be recommended either to avoid structures over steep topographies, or to design the foundations of these structures appropriately. The International Building Code suggests that buildings should be 10m away from the top of a cliff and 5m away from the bottom of a cliff (ICC, 2000).

Uneven settlement

All buildings settle after being built. However, uneven settlement occurs if the soil type under the building is not consistent, as shown in Figure 4.22. The left side of the building sits firmly on strong soil, whilst its right side sits on weak soil, which causes settlement of that part. This might cause cracks in the structure at the point where the soil type changes.

To avoid uneven settlement, it is recommended to design an expansion joint along the line where the soil type changes. Then the right side will settle freely and independently from the left side and the occurrence of cracks is prevented. This means that the structure is designed as two independent structures. Expansion joints are explained in chapter 9.

Wind instability

In heavy wind, umbrellas might flap and lose their stability. It then becomes necessary to close the umbrella and re-open it firmly. This happens because umbrellas are not stable against very strong wind. Similarly, all types of structures that work with tension stress have wind instability problems. The cable structure, as shown in Figure 4.23a, can lose its stability easily; but if additional cables are provided as shown in Figure 4.23b, then these cables can take the wind load and the structure remains stable.

The structural requirement concerning wind instability is to avoid it by providing the necessary structural elements that can take the wind load. Wind instability problems of tensile structures will be studied further in chapter 12.

EQUILIBRIUM

Buildings and structures lose their equilibrium when they start to move in any direction. Except for kinetic structures, building structures usually do not move. The supports of the building structure are designed to provide the necessary reactions to stop the movement of the structure. Thus, a structure is a composition of a load, support reactions and the body of the structural system.

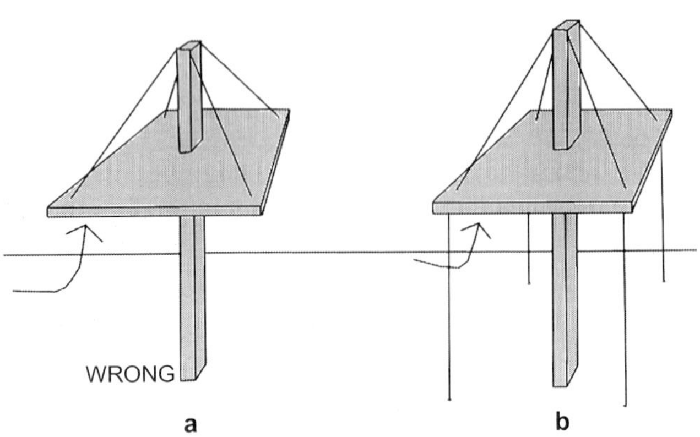

4.23 Wind instability of a tensile structure

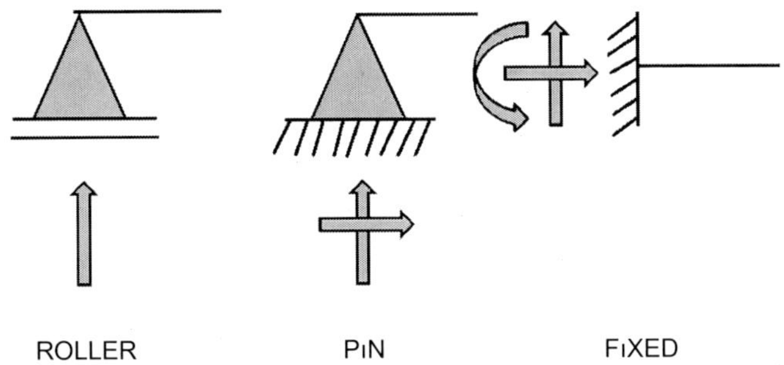

4.24 Support types

There are three types of supports in building structures, as shown in Figure 4.24. These are:

- Roller support
- Pin support
- Fixed support.

Reactions, which can be developed in the supports of two-dimensional systems, are vertical, horizontal and moment reactions. Thus, the movement of two-dimensional systems can be in the vertical or horizontal direction, or the system can turn. The reactions are necessary to stop these three movements.

Roller supports develop only one type of reaction, which is either vertical or horizontal. Any joint that allows two types of movement, such as turning and moving in the horizontal direction, can be categorised as roller support. Pin supports develop two types of reaction by eliminating movements in the vertical and horizontal directions. However, it is possible to turn a structural element around a pin support because it does not develop any moment reaction. Door hinges are an example of pin supports. It is possible to turn a door around the axis of its hinges but it is not possible to move it up or down due to the vertical and horizontal reactions provided in these directions. Fixed supports do not allow any movement to occur because they develop vertical, horizontal and moment reactions simultaneously. The joints of tables, for example, are usually fixed to eliminate movement of the table surface in any direction.

The amount of reactions in simple beams can be found with the help of three equations of equilibrium. A beam is called a simple beam if its supports develop only three reactions. Thus, these unknown reactions can be found with the help of three equations. Avoidance of two-dimensional structures that cannot eliminate the three movements (vertical, horizontal and circular) is a structural requirement. Reactions of another system, which develops more than three reactions, can be found by writing the fourth, fifth, etc., equations or by using another method, such as the Portal Method, which is covered together with frame systems in chapter 9.

Three equations of equilibrium can be presented with the help of the system shown in Figure 4.25. The unknown reactions of VA, VB and HB need to be found in this problem. 1kN is equal to 0.1 tons.

The first equation states that the sum of all horizontal forces in a system is equal to zero. Thus, it guarantees that there is no movement in the horizontal direction.

$$\Sigma F x = 0$$

If the forces to the left are minus and the forces to the right are plus, then;

$$= -HB + 5 = 0$$
$$= +HB = +5kN$$

The second equation states that the sum of all vertical forces in a system is equal to zero. Thus, this equation guarantees equilibrium in the vertical direction.

$$\Sigma F y = 0$$

If the upwards forces are plus and the downwards forces are minus, then;

$$= +VA + VB - 10 = 0$$

Since there are two unknowns in this equation, the third equation has to be written to find VA and VB.

The third equation states that the sum of moments of all forces around a point is equal to zero. Thus, this equation guarantees equilibrium against any possible moment in the system.

$$\Sigma M = 0$$

Writing the moment equation is more complicated than writing the first two equations. The first step to writing the moment equation is to assign a point in order to find the moment of all forces around that point. This can be any point. However, it would be easier to choose one of the support points for this purpose. For example, for this problem, point B can be chosen. The moment

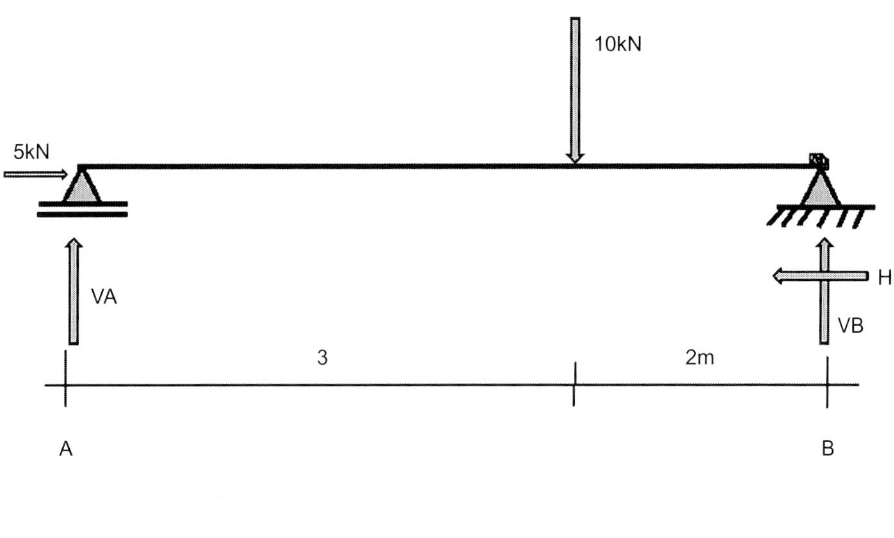

4.25 A problem of finding reactions

Table 4.1 Finding moment around B

M(B)	*F*	*d*
–5VA	VA	5
–	VB	0
–	HB	0
+(10 × 2)	10	2
–	5	0

of each force around B can be found as 'M = F × d' with the help of Table 4.1.

The first step of completing this table is to write all of the forces in the system under the F column. Then the distance of each force to point B should be found and written under the d column. There is a technique which could help in finding the distance of each force to point B. The first step of this technique is to draw a line over the force by extending its line of action. The second step is to draw another line parallel to the first one passing from point B. The distance between these two lines is the distance of the force to point B. For example, if we take the horizontal force of 5kN, one might have difficulties in deciding if its distance to point B is 5m or 0m. However, since the line drawn over the force also passes from point B, the two lines of the technique are overlapping. Thus, the distance of 5kN force to point B is 0m.

The third step to complete the table is to find the moment of each force around point B by multiplying F and d values. This shows that only VA and a 10kN force produces the moment around point B. The fourth and last step is to determine the signs of each moment in order to write them into the moment equation. A second technique can be used to help in finding the signs of each moment. One can imagine a compass and place the needle of the compass to point B and the pencil of the compass on the force (VA) which creates a moment. Then, moving the pencil in the direction of the force one can draw a circle around point B. Since this circle is turning around point B in a clockwise direction, then the sign of this moment is minus. If we apply the same technique to 10kN force, it is seen that the moment of this force is positive because it is turning around point B counter clockwise. It is possible to accept different sign conventions. Thus, the moment equation according to point B becomes:

$$-5VA + (10 \times 2) = 0$$
$$+VA = +4kN$$

By replacing the value of VA into the second equation it can be found that:

$$+VB = +6kN$$

The magnitudes of all unknown reactions have positive values in this problem. This means that the assumed directions of the unknown reactions at the beginning of the problem were correct. If any result is minus, this means that the direction of that reaction should be changed to the opposite direction. For example, the plus sign for HB in this problem does not mean that it should be towards the right.

EXAMPLE 4.1: Find the reactions in the cantilevering beam shown in Figure 4.26.

$$\sum F x = 0 \qquad +HA - 7 = 0 \qquad +HA = 7kN$$
$$\sum F y = 0 \qquad +VA - 10 = 0 \qquad +VA = 10kN$$

All moments are written to the moment column of Table 4.2.

Table 4.2 Finding moment around A for Example 4.1

M(A)	*F*	*d*
–	VA	0
–	HA	0
+MA	–	–
–(10 × 3)	10	3
–	7	0
–5	–	–

$$\sum M(A) = 0 \qquad +MA - (10 \times 3) - 5 = 0 \qquad +MA = +35kNm$$

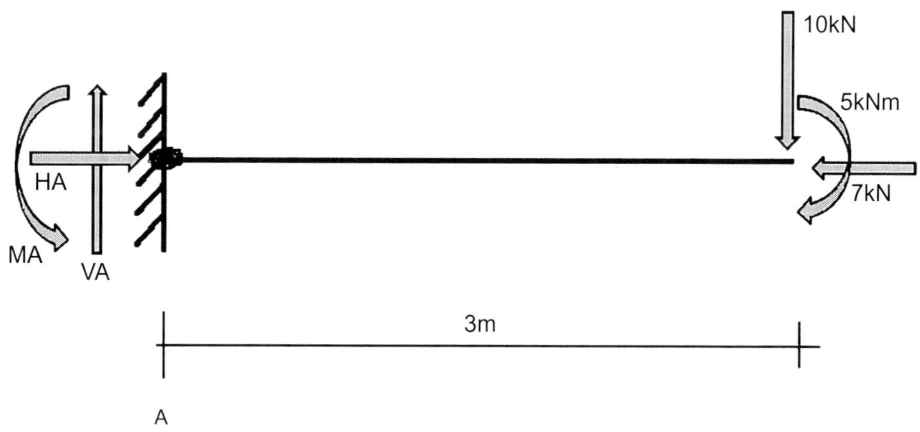

4.26 Example 4.1: finding reactions

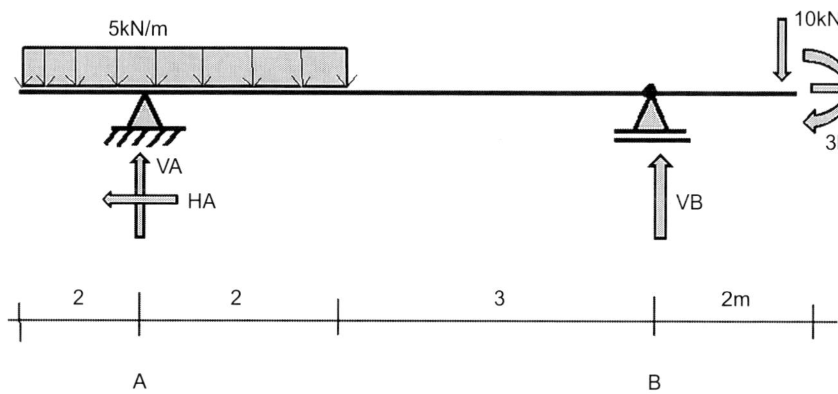

4.27 Example 4.2: finding reactions

EXAMPLE 4.2: Find the reactions in the simple beam shown in Figure 4.27.

$\sum F x = 0$ $-HA + 7 = 0$ $+HA = +7kN$
$\sum F y = 0$ $+VA + VB - 10 - (5 \times 4) = 0$ $+VA + VB - 30 = 0$

All moments are written to the moment column of Table 4.3.

Table 4.3 Finding moment around B for Example 4.2

M(B)	*F*	*d*
–5VA	VA	5
–	VB	0
–	HA	0
–(10 × 2)	10	2
–	7	0
+(5 × 4 × 5)	(5 × 4)	5
–3	–	–

$\sum M(B) = 0$ $-5VA - (10 \times 2) + (5 \times 4 \times 5) - 3 = 0$ $+VA = +15.4kN$
$+VB = +14.6kN$

Another structural recommendation concerning equilibrium of structures is to avoid designing structures that need very high support reactions. By solving problems like the ones given here, one can get an idea about which arrangements and loadings are not good for stable structures.

DEFORMATION LIMIT

Perceivable deformation of structures is disturbing for people because deformation is seen as a sign of danger. Thus, deformation of structures is always controlled by structural engineers.

Deformation due to compression and tension is not as problematic as deflection due to bending. The deflection limit for structural elements in bending can be accepted as L/360, L/240 or L/180 depending on the location of the structural element. Figure 4.28 shows that the deflection limit of a beam for reinforced concrete floors and ceilings is L/360. Attic floors can have a L/240 deflection limit (ACI 318-08 Building Code, 2009). The deflection limit eliminates disturbance due to the perception of the deflection and avoids the occurrence of plaster cracks if the structural element is made out of reinforced concrete. For a beam of 5m, L/360 makes 1.4cm.

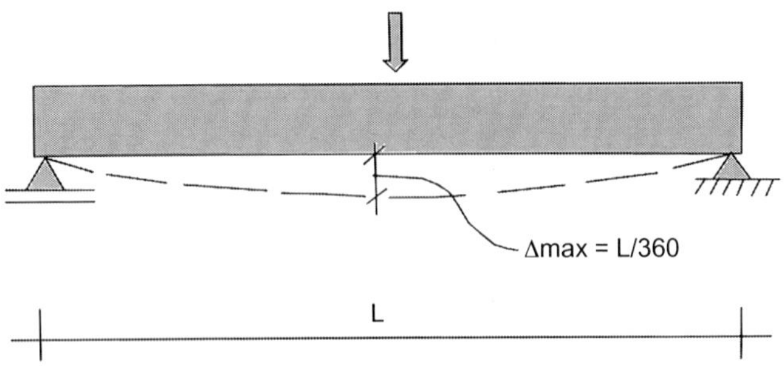

4.28 Deflection limit of a beam

The idea of pre-stressing is based on giving an opposite deflection to the structure. Through pre-stressing reinforcement, resultant deflection is decreased and a thinner structural element is achieved (see Figure 4.29).

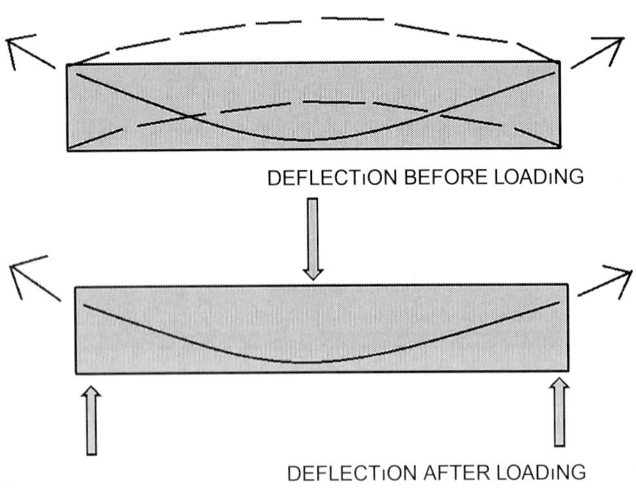

4.29 Pre-stressing of a beam

Deflection of a structural element in bending is proportional to the moment in it. If the moment increases, then the deflection increases. Thus, the structural recommendation is to decrease the moment in the structural system. Drawing the moment diagram of structures is studied in chapter 9, which discusses frame systems.

For large structures, such as suspension bridges and skyscrapers, it is not economic to eliminate the perceivable deformation or deflection. If one walks over a suspension bridge, s/he feels the movement of the bridge. The deflection limit for skyscrapers is usually H/500, as seen in Figure 4.30, where H is the height of the skyscraper (West & Fisher, 2003).

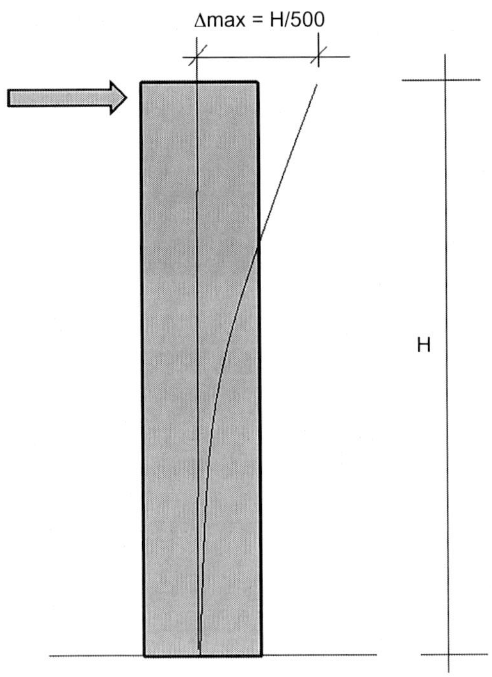

4.30 Deflection limit of a skyscraper

For a 500m high skyscraper, the deflection limit is 1m. This is a perceivable deformation and will affect some interior design decisions. For instance, it is not advisable to use moveable furniture, such as office chairs with wheels, plants with leaves and lights with lampshades because they might move with the movement of the building.

STRUCTURAL GUIDELINES ORIGINATING FROM THESE REQUIREMENTS

The structural requirements and recommendations that have been stated in this chapter are listed in Table 4.4, together with the value system needed to make a judgement about them during design. Some of the recommendations are based on economic reasons, some are based on safety requirements and some are based on the comfort of users.

Table 4.4 shows that most of the structural guidelines are based on economics. It is also seen that structural guidelines concerning safety tend to fulfil the needs of strength, stability and equilibrium. All structural guidelines concerning the deformation limit relate to providing comfort for the users.

Table 4.4 Structural guidelines originating from the requirements of strength, stability, equilibrium and deformation limit and the associated value system

	Structural guidelines	*Value system*
Strength	Decreasing building height in order to decrease wind load.	Economy
	Decreasing the weight of the building in order to decrease earthquake load.	Economy
	Dividing buildings larger than 30m × 30m into pieces in order to avoid temperature load.	Economy
	Elimination of designing high-rise buildings on swampy earth in order to avoid resonance.	Economy
	Having any cross-sectional shape for structural elements in pure tension or compression (without buckling).	Economy
	Having any cross-sectional shape for structural elements in simple shear.	Economy
	Having sufficient cross-sectional area for each type of stress.	Safety
	Increasing the moment of inertia of the structural elements in bending.	Economy
	Avoiding torsion in small structures.	Safety
	Use of hollow circular cross-sectional shapes against torsion.	Economy
Stability	Avoiding all instability types.	Safety
	Avoiding shapes that are weak against overturning.	Economy
	Reduction of slenderness ratio of columns by increasing their thickness or by connecting them with a horizontal element against buckling.	Safety
	Avoiding structures over inclined topographies against sliding.	Economy
	Using expansion joints to avoid uneven settlement.	Economy
	Avoiding wind instability of tensile structures by providing necessary structural elements to take wind load.	Safety
Equilibrium	Providing equilibrium.	Safety
	Avoiding structures which need high-support reactions.	Economy
Deformation limit	Avoiding perceivable deformation for small structures.	Comfort
	Decreasing moment in order to decrease deflection.	Comfort
	Considering presence of perceivable deflection and movement in long-span and high-rise structures during design.	Comfort

PROBLEMS TO SOLVE

4.1: Locate the centre of gravity of the following shapes.

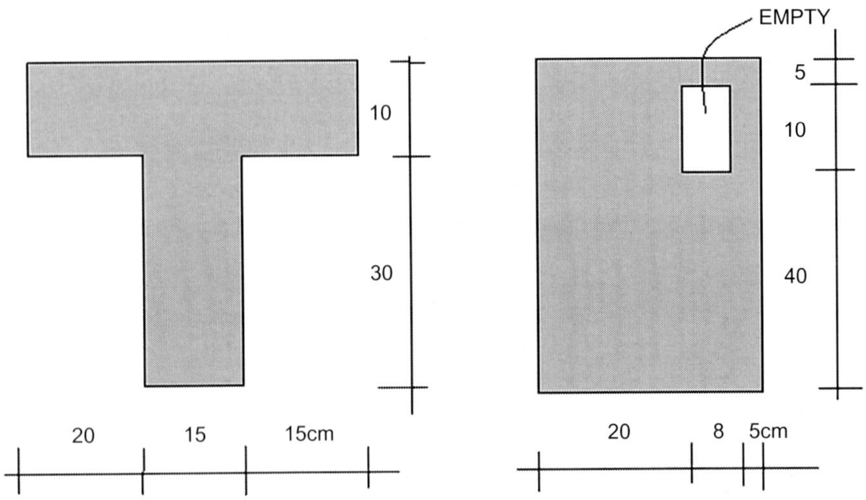

4.2: Find the support reactions in the following systems.

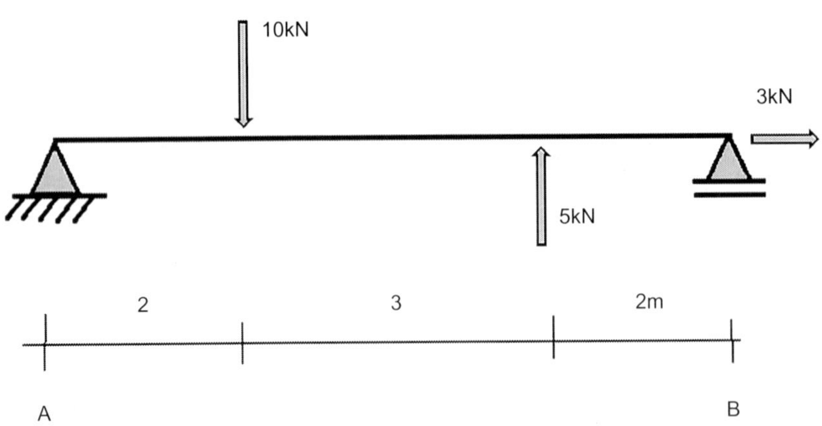

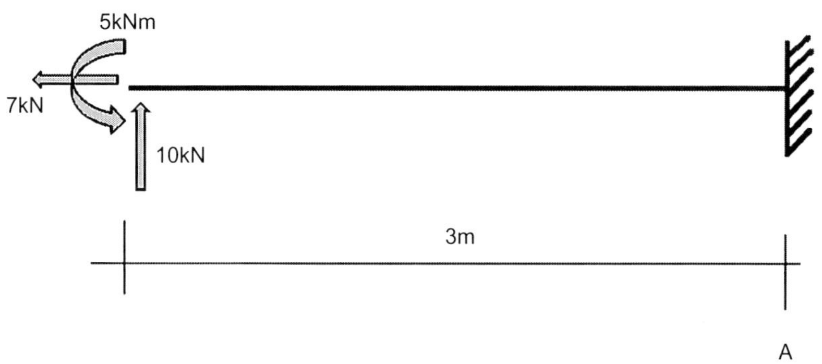

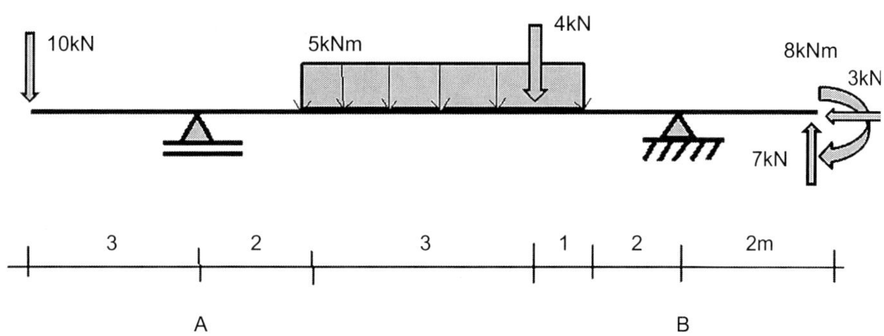

REFERENCES

ACI 318-08 Building Code (2009) *Requirements for Structural Concrete* (viewed 5 September 2013: www.concrete.org/tempComDocs/-28807/r._stehly.pdf)

International Code Council (ICC) (2000) *International Building Code*, 12th edition, International Code Council: Illinois.

West, M., Fisher, J. (2003) *Steel Design Guide, Serviceability Design Considerations*, 2nd edition, American Institute of Steel Construction: Chicago.

URL1 (n.d.) *Back to Bratislava* (viewed 26 September 2014: www.petefox.net/Peter_Fuchs,_Tonmeister/Pictures.html)

5 Structural Guidelines in Relation to Form and Size

In order to discuss structural guidelines in relation to form, it is necessary to classify structural systems from a structural engineering and architectural perspective simultaneously.

CLASSIFICATION OF MODERN STRUCTURAL SYSTEMS

In reviewing the existing literature concerning structural systems, it can be seen that every structure book classifies structures in its own way. There is no common classification system for structural systems. In spite of this, there is a common terminology that helps professionals to understand each other.

Lin and Stotesbury (1981) classify structures as:

- Horizontal subsystems
- Vertical subsystems
- High-rise buildings
- Arch, suspension and shell systems
- Foundation subsystems.

Engel (1997) classifies structural systems as:

- Vector active structures (such as trusses)
- Surface active structures (such as folded plates)
- Section active structures (such as beams)
- Form active structures (such as cables).

Ambrose and Tripeny (2010) classify structural systems according to their materials. They consider timber, steel and reinforced concrete structures.

Place (2007) classifies structural systems as:

- Axial members
- Beams
- Trusses
- Compression structures
- Tensile spanning structures.

In this book, it is preferred to classify modern structural systems according to their stress type and form simultaneously. It can be said that in classifying structural systems according to their stress characteristics, we are also classifying their form. Form is very important for architects, while stress is very important for structural engineers. Thus, this classification creates a common point between architects and structural engineers.

It is accepted in this book that modern structural systems (excluding masonry structures) can be classified into three large families according to their form and the stress type they develop. These are:

- Tensile structures that usually have negatively curved forms.
- Compression structures that usually have positively curved forms.
- Bending structures that are usually compositions of straight lines.

These three large families of modern building structural systems are shown in Figure 5.1.

The structural systems within the family of tensile structures are: cables (or suspension structures), cable trusses, membranes, pneumatic structures and negative curvature shells, as seen in Figure 5.2.

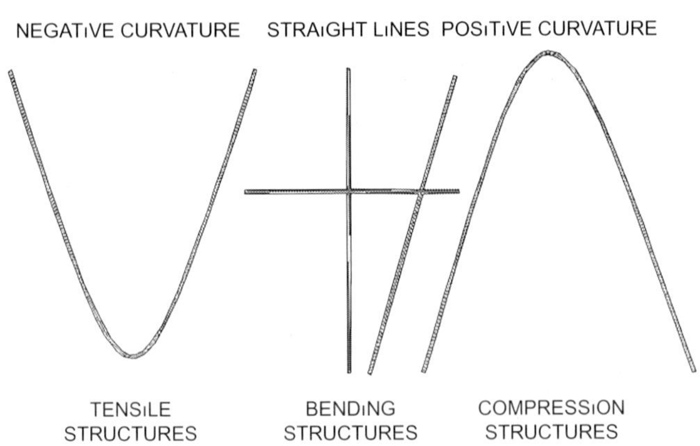

5.1 Three large families of modern building structural systems

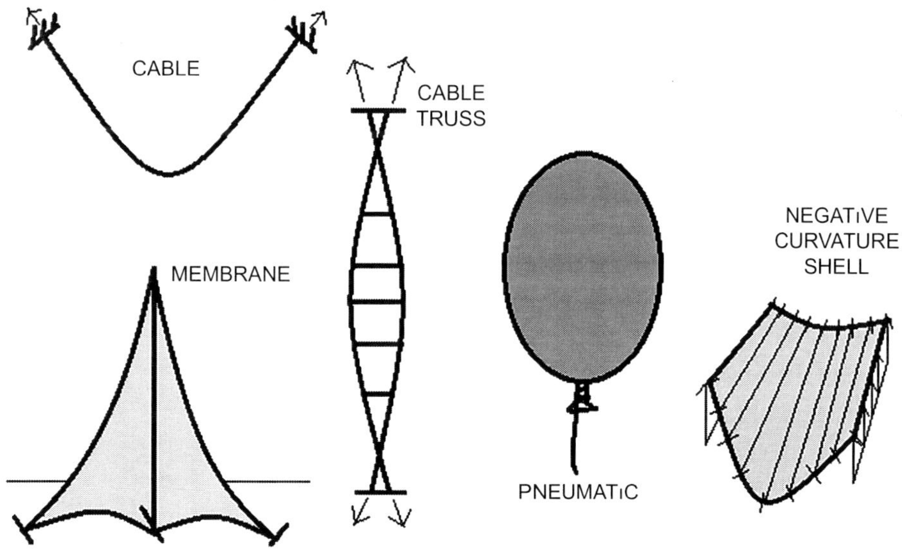

5.2 Tensile structures

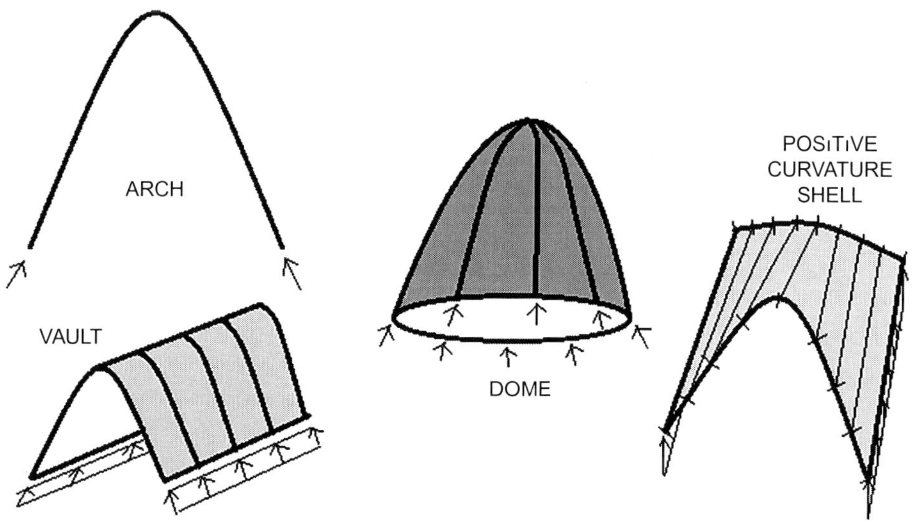

5.3 Compression structures

Cable structures can be seen as the unit structure that repeats in all types of tensile structures. Thus, understanding cables will lead to an understanding of all of the tensile structures. This book, covering tensile structures in chapter 12, discusses cables in detail and explains the differences of other systems as compared to cable systems.

The structural systems within the family of compression structures are: arches, vaults, domes and positive curvature shells, as seen in Figure 5.3.

Arches can be seen as the unit structure that repeats in all types of compression structures. Thus, understanding arches will lead to an understanding of all of the compression structures. This book covers compression structures in chapter 13.

Arches are two-dimensional structures. Vaults and domes are known as compression structures with simple geometric shapes because they are formed by the repetition of the same arch. A vault is achieved by the repetition of the same arch on the same axis. A dome is achieved by turning the same arch around a centre. However, all shells, including negative and positive curvature shells, have complex geometric shapes because the unit structure in them, whether it is a cable or an arch, repeats with variation. As seen in Figures 5.2 and 5.3, shells are achieved by a change in the size of the additive structural unit.

The structural systems that form part of the family of bending structures are divided into two big families: beams and trusses (see Figure 5.4).

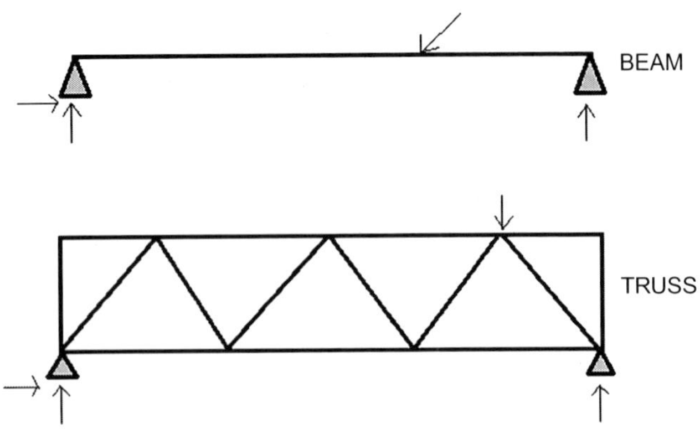

5.4 Bending structures

The beam family comprises: frames, shear walls, tubular structures and folded plates, as seen in Figure 5.5. Frames and shear walls are studied in chapters 9 and 10 of this book and tubular structures are studied in chapter 11 in relation to high-rise building structural systems. Beams are the common unit structure and thus an understanding of these will lead to an understanding of all the other forms in this family.

The truss family contains two- and three-dimensional trusses and space frames, as seen in Figure 5.6. The tectonics of the truss family is studied in chapter 14 of this book. Trusses respond to bending through developing both tension and compression within their members. Thus, they can be categorised together with bending structures and they form resistant structures that work with tension and/or compression. Trusses are the common unit structures in this family. Three-dimensional trusses and space frames are achieved by adding trusses to each other. Thus, understanding trusses will lead to an understanding of all the other elements in this family.

Compression structures, tension structures and trusses are referred to as form-resistant structures in this book because most of their strength comes from their special form. Since the form of folded plates is very important for their strength, they are also

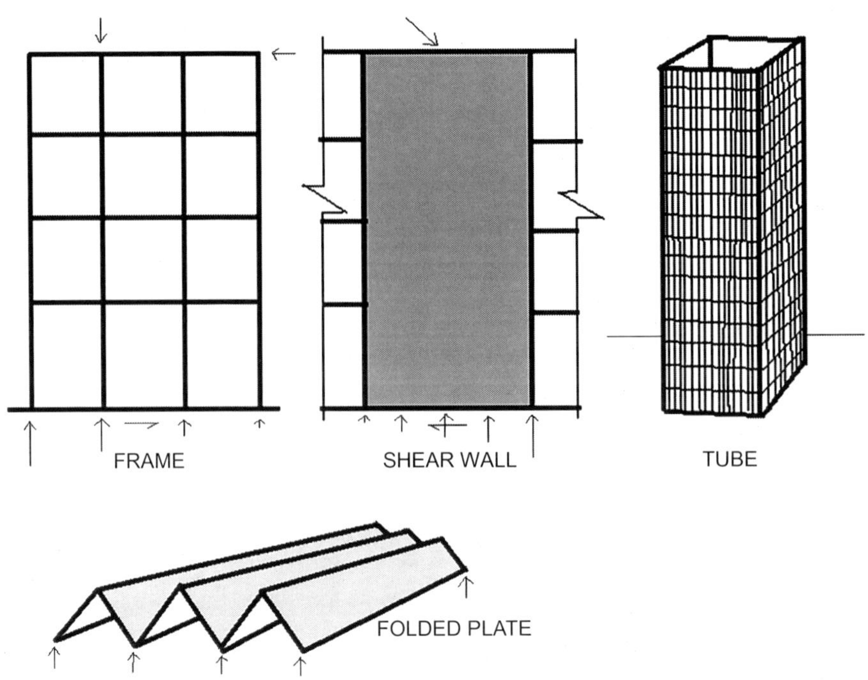

5.5 Structures with beams

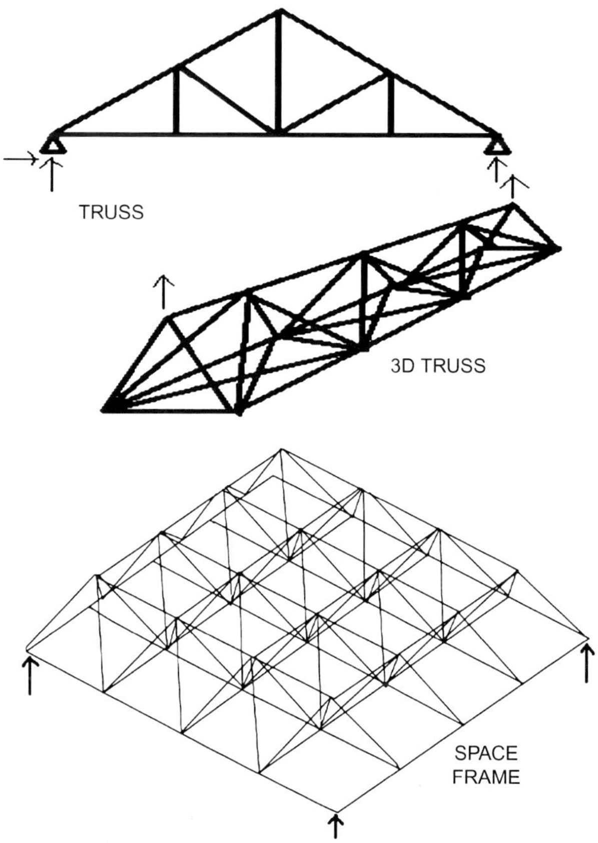

5.6 Structures with trusses

categorised under form-resistant structures. These are all covered in Part 4 of this book (chapters 12–15).

Masonry structures can be seen as a family of structures that is distinct from the families of modern structural systems. The walls that form these structures work mainly in compression. Masonry structures are studied in Part 2 of this book (chapters 6–8).

Each of the above families of structures has inherent structural problems. For example, all compression structures have problems such as buckling, which needs to be considered during the design process. Tensile structures have problems such as wind instability. These common problems for families of structural systems are introduced within the relevant chapters of this book.

EXCEPTIONS NOT FITTING INTO THIS CLASSIFICATION

As already stated, tensile structures usually have forms with negative curvature, bending structures usually have forms with straight lines and compression structures usually have forms with positive curvature. Thus, an architect can recognise the family of most of the structures by looking at its form. Although this rule is true for most structures, there are some exceptions that might create confusion. For example, Richard Rogers' Millennium Dome, UK, has a membrane structure, but it is in the form of a dome (see Figure 5.7). It is therefore a tensile structure with positive curvature. If the Millennium Dome is studied closely, it can be seen that it is a composition of many membrane surfaces and each membrane piece is in tension.

Another exception with respect to the classification of structural systems within this book is the flat arch, as seen in Figure 5.8. This is a compression structure with a straight form. It is shown in chapter 7 that, although flat arches have straight forms, their structural behaviour is still based on positively curved forms.

Thus, the structural guidelines in relation to form are as follows:

- If the form of the building is negatively curved, then it is appropriate to use one of the tension structures with it.

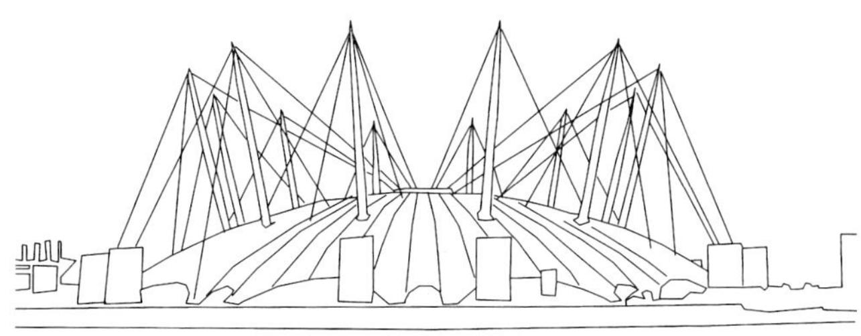

5.7 A sketch of Richard Rogers' Millennium Dome, London, 1999 (drawn with the help of URL1, 2009)

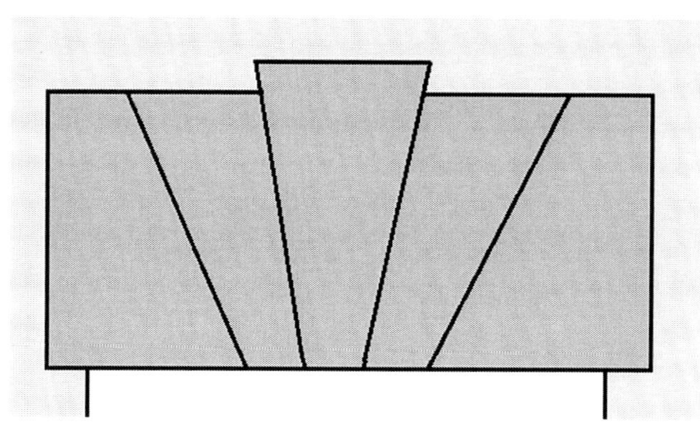

5.8 Flat arch

- If the form of the building is positively curved, then it is appropriate to use one of the compression structures with it.
- If the form of the building elements are straight, then it is appropriate to use one of the bending structures with it.
- When choosing the structural system for a desired form, it should be remembered that there may be exceptions that do not follow the above recommendations.

SIZE AND PROPORTIONS OF STRUCTURAL SYSTEMS

If all dimensions of an 8m long reinforced concrete beam, which has the cross-sectional dimensions of 30cm × 80cm deep, are multiplied by ten, it cannot carry itself any more. Similarly, if all dimensions of a 2mm long mosquito are multiplied by 1,000 so that it is 2m long, as done in some horror films, the mosquito cannot fly or walk anymore, because its wings and legs are not strong enough in their new sizes. Thus, increasing or decreasing all dimensions of structures is not a feasible approach. This is one of the reasons for the presence of many different structural systems.

The appropriateness of structural systems can be discussed according to:

- Their span (the distance between two supports)
- Their height.

If we consider the span of structures first, it can be stated that the structural systems for the shortest spans take place within the family of beams. It is possible to have a high-strength reinforced concrete frame system with a 15m span (Engel, 1997). The size of reinforced concrete slabs can be 4–25m. It is also possible to have a steel frame system that contains 30m long compound beams. Mies van der Rohe's Crown Hall in Chicago, USA (1956), contains steel compound beams of 40m. Two-dimensional steel trusses can easily be used for spans of 15–30m: if designed correctly, they can span up to 80m.

Different structural systems are used for different span ranges. If the span is 25–100m, space frames can be suggested. However, PTW Architects and Ove Arup's Beijing National Aquatics Center in the Republic of China (2008), which is also known as Water Cube, has a special space frame structure spanning 195m. Shells can be used economically for spans of up to 200m. Robert E. Camelot, Jean de Mailly and Bernard Zehrfuss's CNIT Hall, which was built in Paris, France (1958), has a shell structure that spans 240m. There are 500m long 3D trusses. Quebec Bridge, which was built in Canada in 1919, has a 3D trussed structure and it spans 549m. Cable structures are used for the longest structures of the world, such as spans up to 2000m. Akashi Kaikyo Bridge, which was built in Kobe, Japan, in 1998, has a suspension structure and it spans 1991m. This bridge is the longest spanning structure in the world as of 2014. Table 5.1, which is prepared mostly with the help of

Table 5.1 Relationship between structure type and the spans commonly used

Structural system	*Material*	*Span range (m)*
Beam	Timber	4–8
	Laminated timber	10–30
	Reinforced concrete	4–10 (15 for high-strength concrete)
	Steel	7–30
Slab	Reinforced concrete	4–15
Truss	Timber	5–50
	Steel	15–80
3D Truss	Timber	12–25
	Steel	20–80
Space frame	Timber	15–60
	Steel	25–195
Folded plate	Reinforced concrete	10–150
Vault	Timber	20–90
	Reinforced concrete	25–70
	Steel	20–90
Geodesic dome	Timber	40–160
	Steel	50–200
Shell	Reinforced concrete	20–200
Pneumatic	Plastic + metal	10–220
Membrane	Plastic + metal	10–80
Cable	Steel	50–2,000

H. Engel's text (1997), shows the relationship between the structure type and the spans most commonly used.

It can generally be stated that the most effective stress types for long spans are pure tension and pure compression. If a structure is formed to have only tension and/or compression in it, then it can span longer distances economically. It is also suggested that architects should find examples of the long-span structures they are planning to use in order to be sure about the practicality of the span size. This book contains examples of types of structure within each chapter.

The height of the structure also affects selection of the structural system type. Reinforced concrete frames are used up to 20 storeys, whilst steel frames are used up to 30 storeys. If shear walls are added to the structure, the height can reach 40 storeys for reinforced concrete and 50 storeys for steel. Over 50 storeys, tubular structures should be considered (Ali & Moon, 2007). This subject is studied in chapter 11. Table 5.2 shows the relationship between type of structure and economic building height.

The structural guidelines concerning span and height of the structure are dependent upon the requirements of safety, functionality and economy. Certain spans and heights with certain structures can be impossible for safety reasons. For example, having an ordinary reinforced concrete beam longer than 15m is not possible for safety reasons. It might also not be logical to use certain structural systems at certain sizes due to the negative effects of the structural system on architectural space. The structural elements might be so thick that there is no space left for people. On the other hand, some of the suggestions relate purely to economy. It might be very costly to use certain structural systems for certain sizes. For example, the Empire State Building in New York, USA, which was built in 1931, has a steel frame structure and is 102 storeys. However, steel frames with shear walls are usually limited to 50 storeys. This building is famous for being very expensive at the time it was built. It was so uneconomical that people stopped building this type of high-rise building until tubular structures were invented 40 years later. Thus, many structures can be used for heights although they may exceed usual economic limits.

Table 5.2 Relationship between type of structure and economic building height

Type of structure	*Height*
RC frame	Up to 20 storeys
Steel frame	Up to 30 storeys
RC frame + shear wall	Between 20 and 40 storeys
Steel frame + shear wall	Between 30 and 50 storeys
Tubular structures	Over 50 storeys

STRUCTURAL GUIDELINES AND BUILDING FORM

One of the values that determines structural guidelines in relation to form is economy. It is possible to think about finding the best (the most economical) form for a certain loading. The limitations of using certain structural systems might also determine this type of structural recommendation. For example, a structure that is not sufficiently curved cannot be a shell, which is a compression structure: it can only be a slab, which is a bending structure. Thus, it has to be thicker than a shell. However, if somebody wishes to design a flat surface as thin as a shell, this is not safe. Similarly, trusses should be triangulated: if they are not, they cannot be considered as trusses. They can be considered as frames or vierendeel trusses.

The values that determine structural guidelines concerning size of structural systems are safety, functionality and economy. Thus, this book proposes to examine the structural guidelines concerning form and size by studying each structural system in more detail.

REFERENCES

Ali, M.M., Moon, K.S. (2007) 'Structural Developments in Tall Buildings: Current Trends and Future Prospects' *Architectural Science Review*, Vol. 50, No. 3: pp.205–223.

Ambrose, J., Tripeny, P. (2010) *Simplified Engineering for Architects and Builders*, John Wiley and Sons: New York.

Engel, H. (1997) *Structure Systems*, 3rd edition, Gerd Hatje Publishers: Ostfildern, Germany.

Lin, T.Y., Stotesbury, S.D. (1981) *Structural Concepts and Systems for Architects and Engineers*, John Wiley and Sons: New York.

Place, J.W. (2007) *Architectural Structures*, John Wiley and Sons: New York.

URL1 (2009) *The Top Ten Buildings of the Decade* (viewed 26 September 2014: www.theguardian.com/world/gallery/2009/dec/07/best-buildings-noughties)

PART 2
THE TECTONICS OF MASONRY STRUCTURES

Masonry structures are constructed from small units. These units can be stone, brick or adobe and they are connected to each other with various types of mortar; for example, cement mortar or earth-based mortar.

Masonry structures are compositions of structural (load-bearing) walls. The weight of the roof and floors is carried by these walls. Hence, the size and the amount of openings in these structural walls are usually limited.

Many contemporary sources classify masonry structures into two major groups (ACI 530-02/ASCE 5-02/TMS 402-02, 2002; Ambrose, 1991):

- Unreinforced masonry
- Reinforced masonry.

Unreinforced masonry is generally employed in the structure of historical, traditional or vernacular buildings. The components forming the structural walls are joined with mortar. On the other hand, reinforced masonry contains some reinforced concrete or steel elements, which help the structure to behave as one piece. Contemporary masonry applications are usually reinforced masonry. Some building codes accept that unreinforced masonry is weak against earthquakes and hurricanes. (See ACI 530-02/ASCE 5-02/TMS 402-02, 2002.) Some other building codes suggest the use of additional reinforced concrete elements, depending on factors such as the size of spaces. (See the Ministry of Public Works and Settlement Government of the Republic of Turkey, 2007).

Chapters 6 and 7 of this book cover traditional applications of masonry, in other words unreinforced masonry. Chapter 6 deals with wall systems and chapter 7 deals with horizontal systems. Chapter 8 covers contemporary applications of masonry, in other words reinforced masonry. Understanding reinforced masonry requires an understanding of unreinforced masonry because the modern additional elements of reinforced masonry solve some problems of unreinforced masonry. Each of the chapters in Part 2 begins with structural guidelines for the relevant structural system, before discussing associated tectonic qualities. Case studies are used to illustrate successful architectural examples of each system. Finally, Part 2 ends with a discussion concerning the contradictions between building codes and architectural examples.

REFERENCES

ACI 530-02/ASCE 5-02/TMS 402-02 (2002) *Building Code Requirements for Masonry Structures*, Masonry Standards Joint Committee (viewed 5 September 2013: https://engineering.purdue.edu/~ramirez/CE479/FALL05/MasonryBuildingCode1-3-02.pdf)

Ambrose, J. (1991) *Simplified Design of Masonry Structures*, John Wiley and Sons: New York.

Ministry of Public Works and Settlement Government of the Republic of Turkey (2007) *Seismic Performance Evaluation of Dual Reinforced Concrete Systems Design According to Turkish Seismic Code*, trans. E.Y. Karcı (viewed April 2013: www.belgeler.com/blg/22lc/seismic-performance-evaluation-of-dual-reinforced-concrete-systems-design-according-to-turkish-seismic-code)

6 The Tectonics of Traditional Approaches to Masonry Structures

Traditional masonry structures, in other words unreinforced masonry structures, can be classified according to their structural material. This chapter classifies traditional masonry structures into four categories:

- Stone masonry
- Brick masonry
- Adobe masonry
- Timber masonry.

THE TECTONICS OF A TRADITIONAL APPROACH TO STONE MASONRY

Structural characteristics of traditional stone masonry can be derived by analysing examples of historical stone buildings. Analysis of the building codes of the countries in earthquake zones can also be useful in order to make a comparison with the characteristics derived from historical buildings. Codes can be useful in differentiating between the applications recommended in earthquake and in non-earthquake zones. Building codes of the USA, Republic of China, Pakistan, Turkey and Republic of Cyprus are used to support the analysis of historical buildings and to derive structural guidelines for unreinforced masonry.

The USA, Pakistan and Republic of Cyprus' building codes suggest mathematical methods for the design of masonry structures and do not bring any other limitations. Turkish and Chinese building codes use mathematical analysis and bring some physical limitations to the design of masonry structures. Since these limitations help to define the structural guidelines for masonry, this book sometimes refers to Chinese and Turkish building codes to exemplify the various concepts regarding stone masonry. However, it should be clarified from the beginning that all building codes limit the use of unreinforced masonry to those areas with low seismic activity. Using information taken from these codes, this book is able to initiate a concept for stone masonry structures.

Structural guidelines for traditional stone masonry can be presented under two headings:

- Structural guidelines for a stone wall.
- Structural guidelines for a stone building.

Structural guidelines for a stone wall

There are many types of stone wall applications but rubble stone walls and cut stone walls form the two major groups. Rubble stone walls are achieved by using the stone pieces in their natural

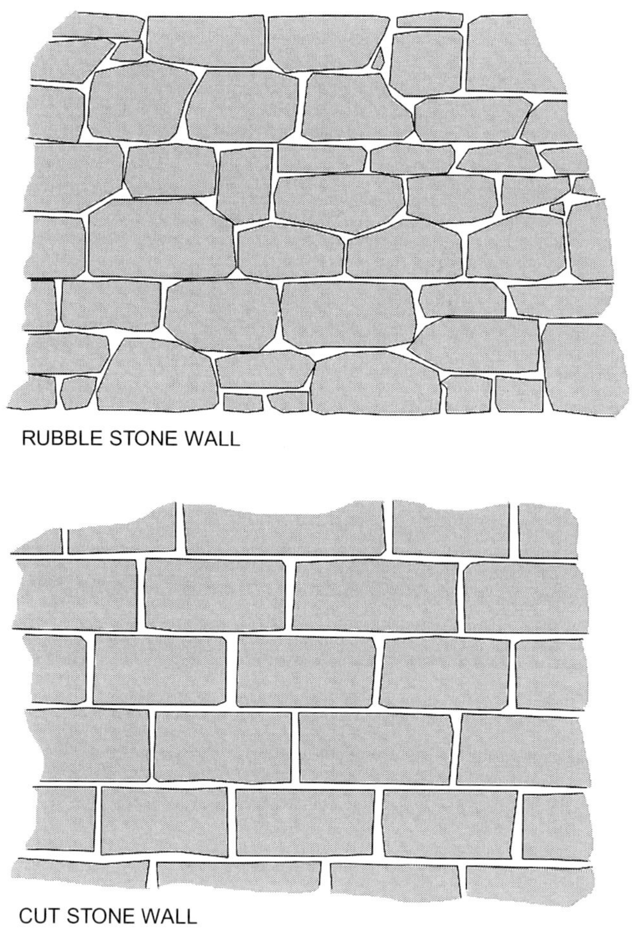

6.1 Rubble stone wall and cut stone wall

forms and cut stone walls are achieved by cutting the stone pieces into rectangular prismatic forms, as seen in Figure 6.1.

Understanding the nature of rubble stone walls leads to understanding the main characteristics of all types of masonry walls. The main characteristics of rubble stone walls can be classified under the following headings:

- Thickness
- Foundation type
- Use of tie-stones
- Organisation of stone pieces
- Types of horizontal layers.

Minimum thickness of stone masonry walls is more than the thickness of walls of other types of masonry. Some building codes only stipulate for calculations of wall thickness depending on the loads applied to the walls, but other codes define minimum thickness as well as defining calculation methods. Turkish building code defines the minimum thickness for stone walls as 50cm (Ministry of Public Works and Settlement Government of the Republic of Turkey, 2007), whilst Republic of Cyprus building code defines it as minimum 35cm (Eurocodes Committee – Scientific and Technical Chamber of Cyprus under a Ministry of Interior's Program, 2004).

Stone walls can have stone or reinforced concrete continuous foundations, as seen in Figure 6.2. Some building codes, such as the Turkish building code, no longer allow the use of stone foundations (Ministry of Public Works and Settlement Government of the Republic of Turkey, 2007). However, stone foundations are still being used in the rural areas of many countries. Continuous foundations should reach the level of firm soil and they should be under frost level.

It is easier to put middle-sized stone pieces to two sides of the wall and fill in the gap between them with smaller pieces of stone during the construction of a wall. However, a wall like this tends to separate into two from the middle. In order to eliminate this problem, tie-stones are used to connect the two sides of the wall, as seen in Figure 6.3. A tie-stone is larger than other stones. Tie-stones are put into every stone row with a certain distance, such

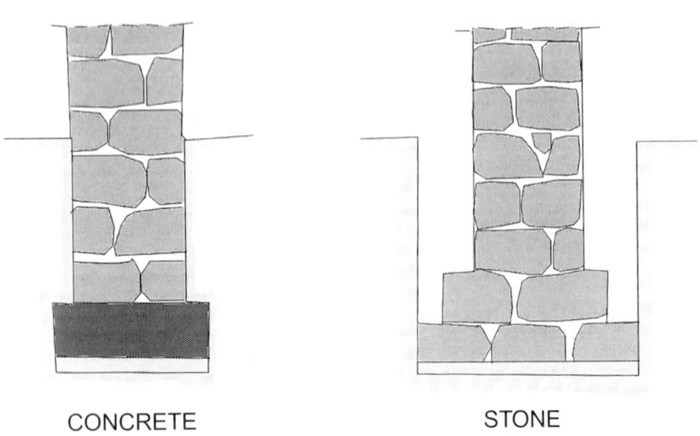

6.2 Reinforced concrete and stone continuous foundations

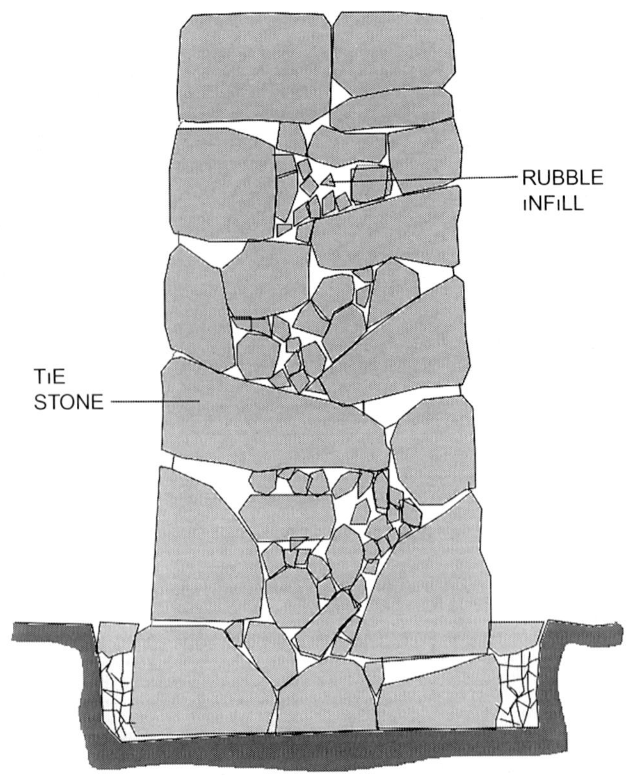

6.3 Organisation of stone pieces in a section of stone wall

as 150cm, between them. Tie-stones of adjacent rows should not overlap. In other words, tie-stones are well distributed in stone walls.

Stone pieces should be organised within a wall in such a way that their joints do not form continuous vertical lines, as seen in Figure 6.2. Continuous vertical joints can easily cause separation of the wall from those joints.

Rows of stone should form straight layers at certain intervals, such as 150cm. These horizontal layers can be arranged in many different ways, as seen in Figure 6.4. These are:

- Organisation of rubble stone pieces to form horizontal layers.
- Use of cut stone rows.
- Use of brick rows.
- Use of timber elements.
- Use of reinforced concrete elements.

The purpose of these layers is to provide an even distribution of load throughout the length of the wall. If these layers do not exist, it may result in high-stress concentrations at certain parts of the wall.

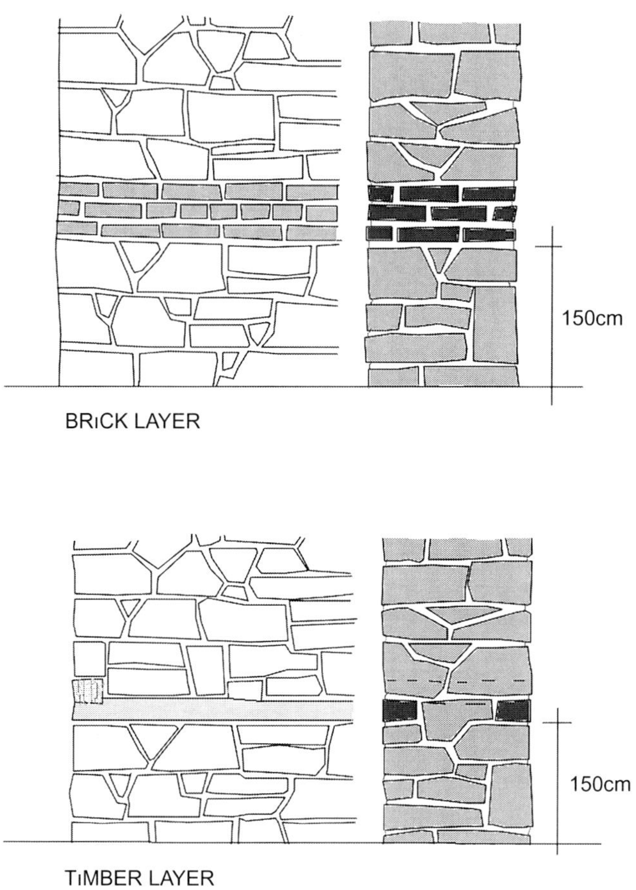

6.4 Horizontal layers in a rubble stone wall

Structural guidelines for a stone building

A stone building consists of some rooms, which are defined by stone walls. Determination of the dimensions of these rooms depends on the structure of the ceiling of the room. If we imagine a timber structure with an earth roof over it, the shorter dimension of the room can be determined according to the size of the timber elements. It is better to use the timber elements of the structure across the shorter span of the room. If the size of these elements, which can be found in the nearby environment, is 4m, then the shorter dimension of the room should be slightly less than 4m in order to provide an end bearing to the elements. Figure 6.5 shows the arrangement of timber beams on the stone walls of the room.

The other dimension of the room is also limited because it is not advisable to have long masonry walls. According to the Turkish building code, if wall length exceeds 5.5m in regions with seismic danger, that wall has to be supported with reinforced concrete vertical tie-beams, which transform the unreinforced masonry structure into reinforced masonry. If there is no seismic danger, maximum wall length can be 7.5m (Ministry of Public Works and Settlement Government of the Republic of Turkey, 2007).

The distance between the timber elements that support the roof depends on the dimensions of the construction element which covers the top of these timber elements. If we imagine a flat earth roof on top of the timber elements, according to the Cypriot tradition of earth roofs, there can be layers of earth, straw, bamboo sticks and mashes (crushed dry plants) on timber elements

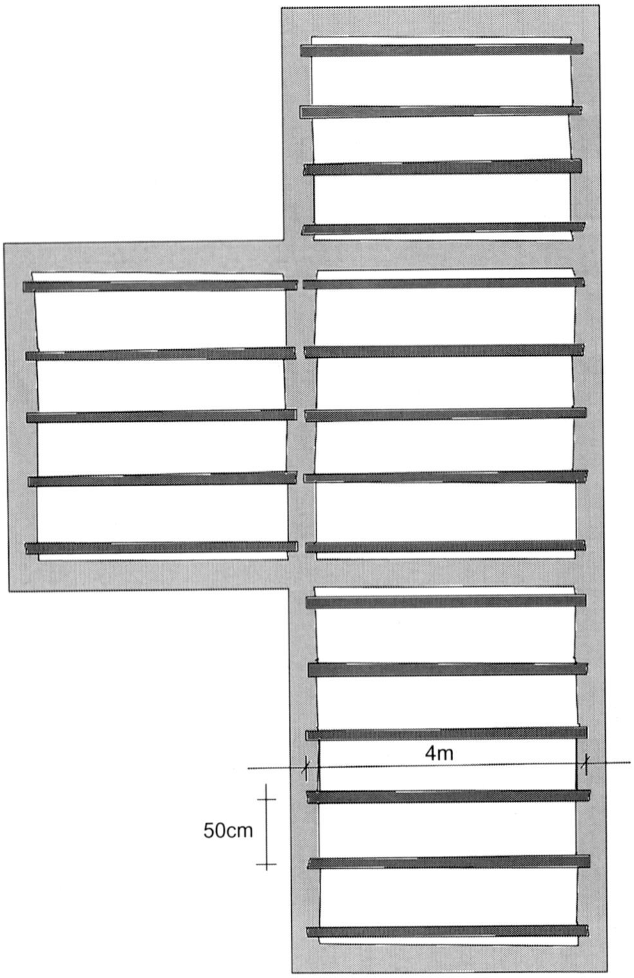

6.5 Determination of the dimensions of a room according to slab structure

from top to bottom (Dincyurek et al., 2003). Here, the size of the bamboo sticks determines the distance between the two timber elements (joists). It is possible to imagine a probable distance, such as 50cm, between the two timber elements.

If the shorter dimension of the room needs to be larger than 4m, it becomes more appropriate to use trusses, as seen in Figure 6.6. These trusses support the rafters, which support the other layers of the roof. In this case, the distance between the two trusses should be less than 4m.

The plan of stone masonry buildings should be either symmetrical or close to symmetrical in order to avoid twisting due to earthquake loads. This recommendation was made by the Roman architect Vitruvius (15 BC) in his book *Ten Books on Architecture* (1914). The building codes of China and Turkey also make this structural recommendation (GB-50011, 2001; Ministry of Public Works and Settlement Government of the Republic of Turkey, 2007). The subject of twisting instability due to earthquakes is studied in chapter 9.

The height of a stone building can be determined according to the degree of seismic activity risk. The building height should

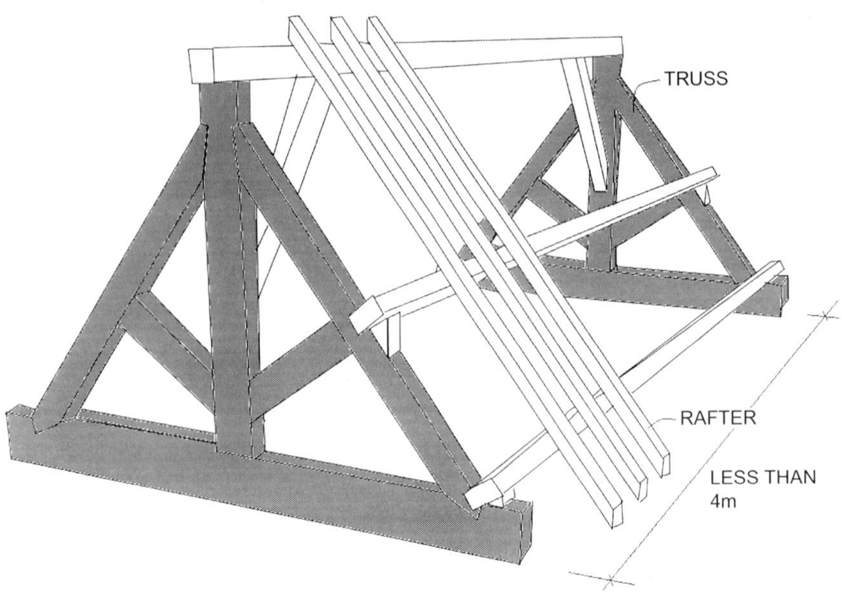

6.6 Use of trusses

be kept low if there is earthquake risk. According to the Turkish building code, if there is earthquake risk, the height of cut stone buildings can be maximum two storeys (including the entrance level). This can be increased to four storeys if there is no risk. However, rubble stone walls are weaker than cut stone walls, so the use of rubble stone walls is allowed only at basement and entrance levels. This means that the upper levels can be cut stone or brick, etc. The thickness of stone walls also increases depending on the number of storeys (Ministry of Public Works and Settlement Government of the Republic of Turkey, 2007). The building code of China limits building height to one, two or three levels, depending on the severity of earthquake danger (GB-50011, 2001).

The height of stone masonry walls is also limited. Turkish building code limits them to 3m (Ministry of Public Works and Settlement Government of the Republic of Turkey, 2007). It is possible to have higher walls, but this might not be economical.

If the building has more than one level, the upper load-bearing walls should be supported by the lower load-bearing walls. Continuous foundations can be used under these walls, as shown in Figure 6.7. The walls are interlocked with each other at the wall intersections. The building corners are very frequently interlocked with each other using cut stone pieces (see Figure 6.8).

Since the load-bearing walls carry the weight of the building and other loads that affect the building, the dimensions and arrangement of openings in the walls are limited. Turkish building code defines the limitations concerning openings as follows (Ministry of Public Works and Settlement Government of the Republic of Turkey, 2007):

- Distance of the opening from the building corner.
- Length of one opening.
- Distance between two openings.
- Ratio of the total length of openings on a wall to the wall's total length.

Corners of buildings take the largest share from lateral loads. The distance of openings from the building corner should therefore be limited to 1.5m if the earthquake risk is high. If earthquake risk is low, the limit can decrease to 1m. Openings are also kept away from other wall intersections. Openings can be 0.5m away from wall intersections.

The maximum opening size should be limited to 3m. The distance between openings can be 1m if there is high earthquake risk, or 0.8m if the risk is lower. The ratio of the total length of

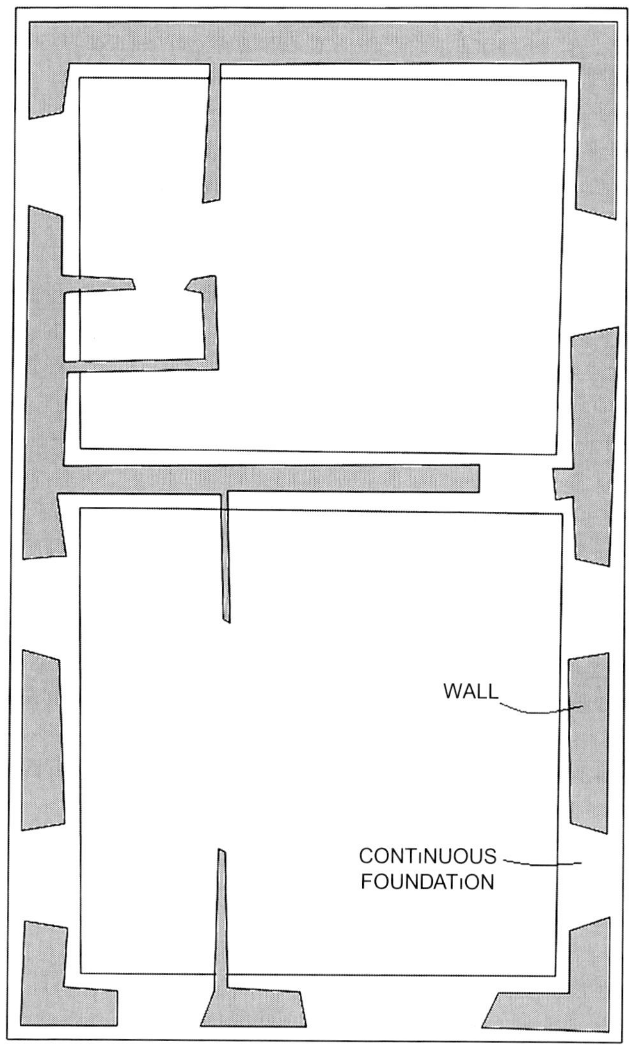

6.7 Continuous foundations under a building

6.8 Use of cut stone corners

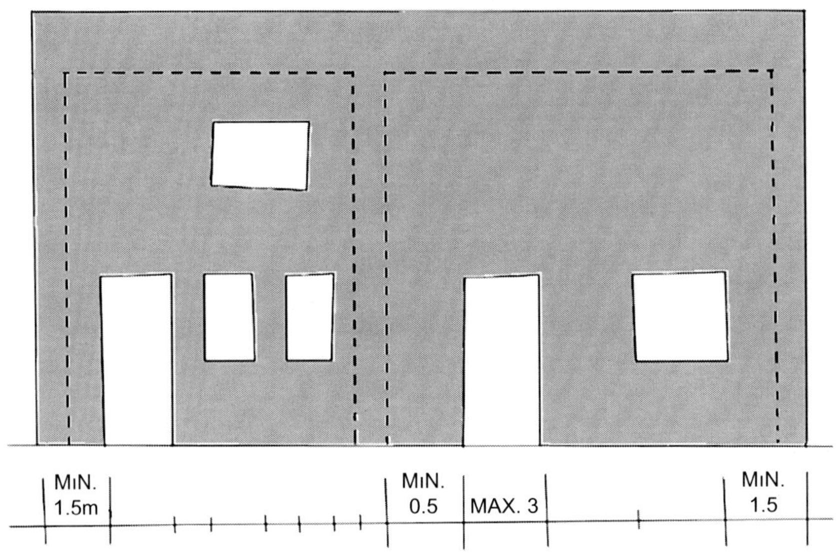

6.9 Arrangement of openings on a masonry wall

openings on a wall to the wall's total length should be limited to 40% for a masonry structure. Most of these recommendations about openings can be considered together, as shown in Figure 6.9.

The openings can have either lintels or arches over them (see Figure 6.10). These lintels or arches support the wall over the opening. There can be stone, reinforced concrete, steel or timber lintels. The size of opening is dependent upon how the wall over the opening is supported. Lintels and flat arches are usually used for small openings, whilst arches are used for larger openings. Arches are studied in chapter 7 of this book.

All the walls in a stone building are connected to each other with the help of horizontal tie-beams (or bond beams), as seen in Figure 6.11. Horizontal tie-beams should be minimum 20cm deep according to Turkish building code. According to the building codes of China and Turkey, horizontal tie-beams should exist at every floor level (GB-50011, 2001; Ministry of Public Works and Settlement Government of the Republic of Turkey, 2007). Plastering also increases the strength of a stone masonry wall by connecting the stone pieces to each other.

Earthquake resistance of traditional stone masonry structures can be increased by using cross-walls and buttresses (counterforts), as seen in Figure 6.12. The building code of Pakistan (Pakistan Engineering Council, 2007) suggests the use of these elements for

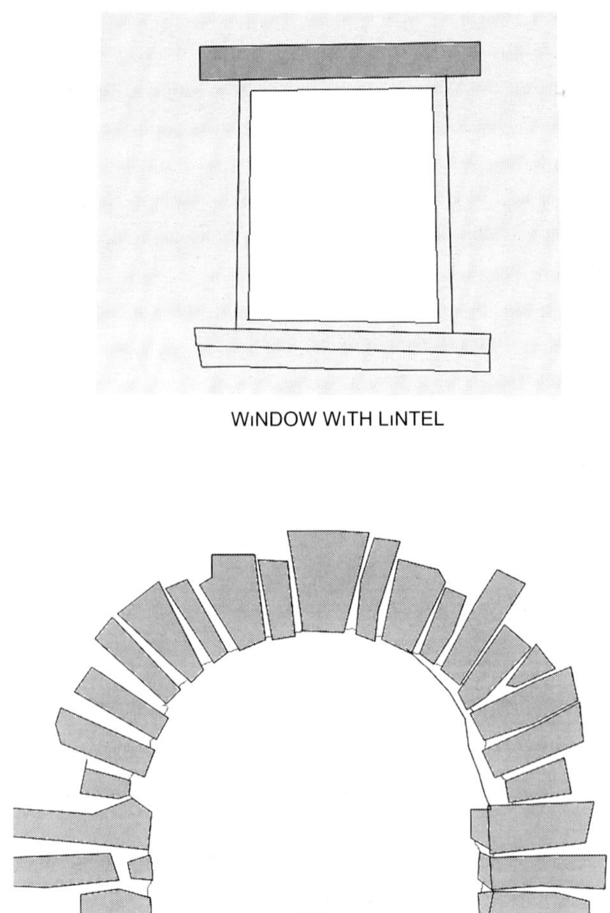

6.10 Lintel/arch over an opening

lateral support. Cross-walls are usually used at interior wall junctions and buttresses are used at exterior wall junctions, especially at the corners of buildings.

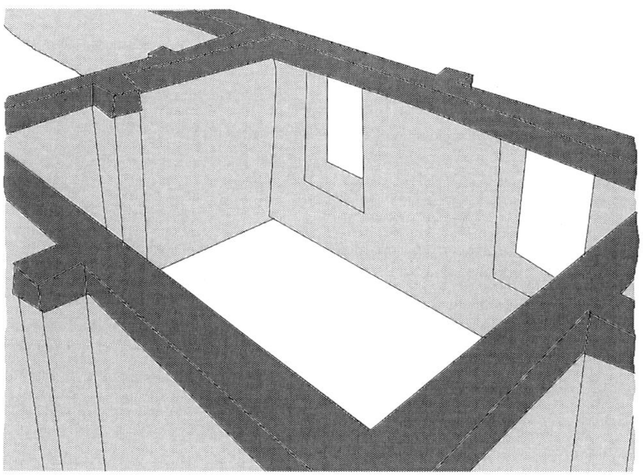

6.11 Use of horizontal tie-beams

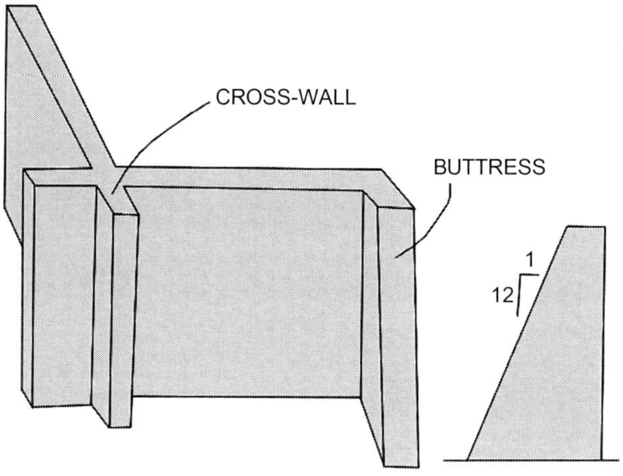

6.12 Cross-walls and buttresses

Structural guidelines for stone masonry

Structural guidelines for stone masonry structures and the value systems determining them are listed in Table 6.1. Most of the structural guidelines are determined according to the requirement for safety. However, these guidelines may not be followed if the problems can be alleviated by other means.

Table 6.1 Structural guidelines for stone masonry structures and the associated value system

Structural guidelines	*Value system*
Minimum thickness of a stone wall can be around 35–50cm.	Safety
Continuous foundations should be used under stone walls.	Safety
Tie-stones should be used regularly to connect two sides of rubble stone walls.	Safety
Stone pieces should be organized in order not to form continuous vertical joints.	Safety
Stone pieces can be organized in order to form horizontal layers in every 150cm height of the wall.	Safety
Length of stone walls between two wall intersections can be around 5.5m long in earthquake regions and 7.5m long if there is no earthquake risk.	Safety
Plan of the building should be symmetrical or close to symmetrical in order to avoid twisting due to earthquakes.	Economy
Unsupported height of a stone wall can be around 3m.	Economy
Stone masonry buildings can be around two storeys high in earthquake regions and four storeys high if there is no earthquake risk.	Safety
All stone walls must be supported by other stone walls or foundations.	Safety
Cut stone pieces should be interlocked at the corners of buildings.	Safety
Total length of openings on a stone wall can be around 40% of the length of the wall.	Safety

Table 6.1 *continued*

Structural guidelines	*Value system*
Maximum opening size should be around 3m.	Safety
Distance of openings from the corner of the building can be around 1.5m in earthquake regions and 1m if there is no earthquake risk.	Safety
Openings can be 0.5m away from wall intersections.	Safety
Distance between two openings can be around 1m in earthquake regions and 0.8m if there is no earthquake risk.	Safety
Openings can have lintels or arches over them.	Safety
Stone walls should be connected to each other with horizontal tie-beams at every floor level.	Safety
Thickness of reinforced concrete horizontal tie-beams can be around 20cm.	Safety
Cross-walls or buttresses can be used against horizontal forces.	Economy

6.13 A sketch of case study 1: Cologne Cathedral, Cologne, Germany, 1248–1880 (drawn with the help of URL1, 2014)

Case study 1: Cologne Cathedral, Germany

The aim of case studies covering different types of structural systems in buildings is to discover the relationship between structural guidelines and the tectonics of architecturally successful buildings. Cologne Cathedral was chosen for a stone masonry case study because it represents the tectonic qualities of Gothic cathedrals.

Figure 6.13 is a sketch of Cologne Cathedral and Figure 6.14 shows its plan and section. This building was designed to praise religion. It has a strong impact on its environment with its dimensions, ornamentation and architectural elements. The interior space was also designed to give the required mystical character to the space. The height of the space and verticality of the linear elements contribute to the characteristics of the interior space. It is clear that structure is the dominant physical entity in the development of the tectonic characteristics of Cologne Cathedral.

By studying the plan of Cologne Cathedral, it is seen that most of the structure is on the exterior of the building, which mirrors a high-rise tubular structure (studied in chapter 11). The stone columns inside the cathedral are very slender in comparison to the dimensions of the outer walls: with its extraordinary height, this large building needs support against lateral loads. Although structural guidelines advise against high stone buildings, a very high space was needed in order to give a religious mystical meaning to the cathedral. The walls surrounding the building therefore contain many vertical ribs in order to support these high walls.

The structural elements in this stone structure are more similar to columns than to walls. The required areas of outer stone elements have been achieved by increasing the dimensions of them perpendicular to the axis of the facade. This was necessary because there are many openings on the facades, which is against structural guidelines for stone masonry structures. Thus, the areas of stone elements have been enlarged in order to make many openings.

Gothic cathedrals are light structures in comparison to other stone structures. This tectonic characteristic of them is due to the strategic placement of structural materials on the plan and the presence of many openings. The plan of the building is in the shape of a Latin cross and it is almost symmetrical in order to eliminate torsion due to horizontal loads. Thus, it can be stated

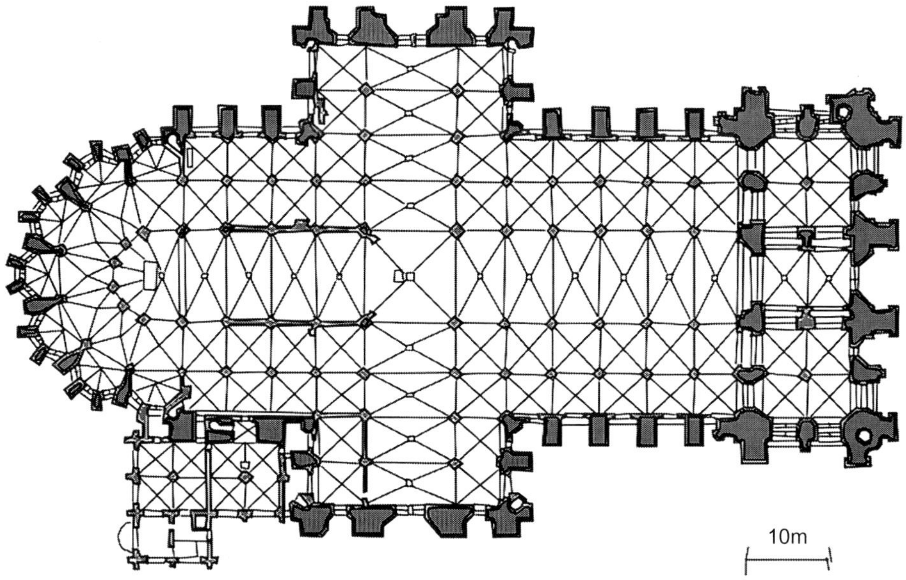

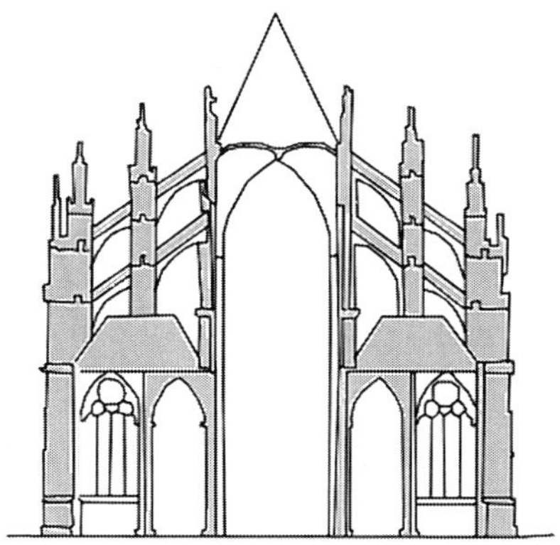

6.14 Plan and section of Cologne Cathedral (drawn with the help of URL2, n.d. and URL3, n.d.)

that the meaningful shape of the plan is following structural guidelines. The vault system forming the roof of the structure is studied in chapter 7. The flying buttresses support the horizontal loads generated by these vaults and contribute to the tectonics of the light stone structure. The openings in the walls are made with the help of many different types of arches. The surfaces are tectonically articulated with these arches, ribs and ornamentation.

In summary, it can be stated that the design of Cologne Cathedral was against the structural guidelines concerning building height and ratio of openings on the facades. However, the mystic quality from the height of the building and the lightness of the building from the many openings are amongst the most important architectural achievements of Gothic cathedrals. Although structural recommendations concerning the use of a symmetrical plan shape and the use of buttresses against horizontal forces were followed, Cologne Cathedral can still be identified with its contravening relationship to structural guidelines.

Case study 2: The stone tower at Norman Castle, UK

The stone tower at Norman Castle, which is seen in Figures 6.15 and 6.16, was chosen because its thick walls form a contrast to the lightness of Gothic buildings. The architectural concept of the stone tower is to form a comparably high tower that takes place within the context of a castle and provides continuity with it. The indoor spaces are large and the thick walls surrounding these spaces are articulated with openings and smaller spaces. Structure is the dominant physical entity in the development of the tectonic characteristics of the stone tower at Norman Castle.

Although it is a tower, this building only has three storeys. Yet the unsupported height of the stone walls is more than 3m. This is against structural guidelines for stone structures because the upper levels are timber and cannot support these heavy walls against horizontal loads.

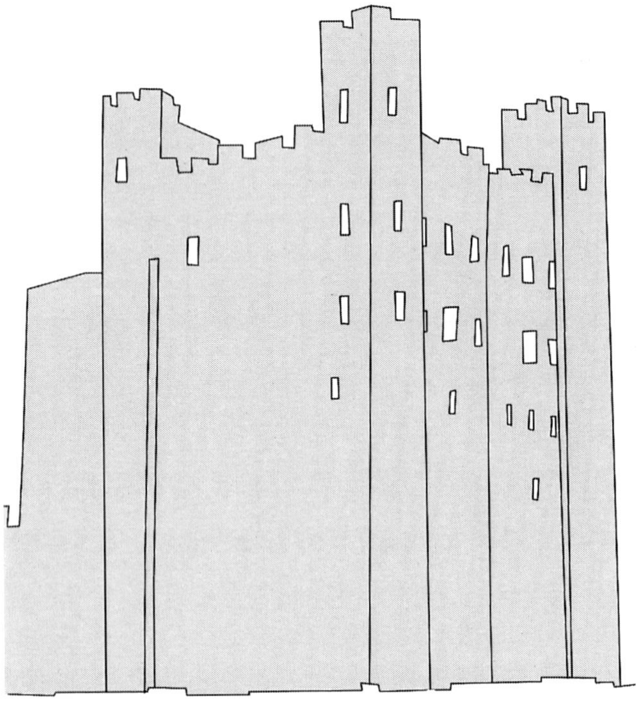

6.15 A sketch of case study 2: the stone tower at Norman Castle, Rochester, UK, twelfth century (drawn with the help of URL4, n.d.)

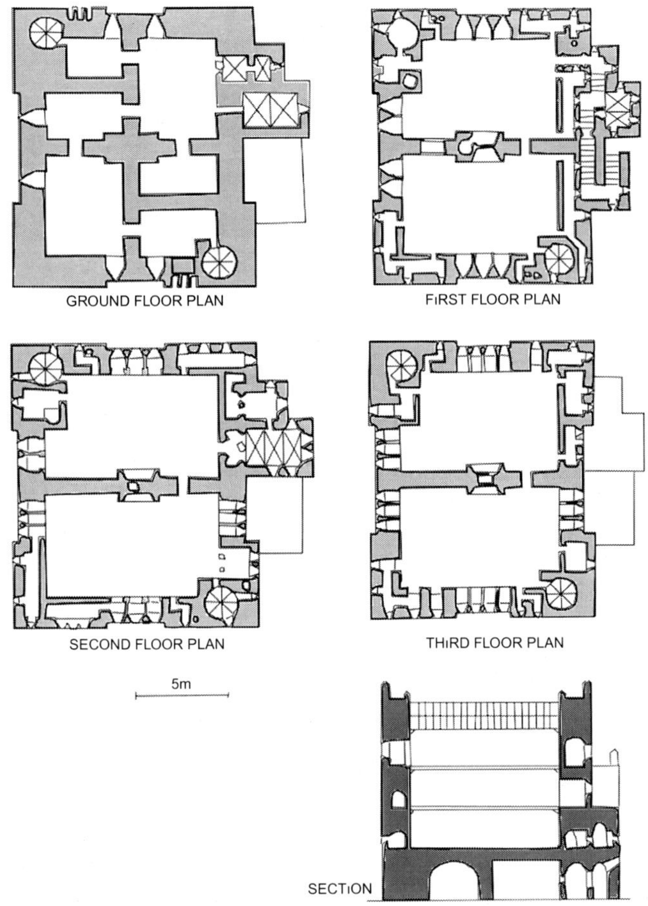

6.16 Plans and section of the stone tower at Norman Castle (drawn with the help of URL4, n.d.)

There are some small spaces, which take place in the stone walls, and there are some larger spaces within the tower. The dimensions of these larger spaces are more than 7.5m, meaning these dimensions do not follow structural guidelines for stone structures.

The symmetry of the structure is disturbed only by the additional spaces for stairs. However, if one considers the amount of structural materials, it can be seen that the structural materials are balanced on both sides. Thus, the structural recommendation concerning symmetry was followed during the design of the tower.

Having high walls and large spaces takes place within the initial concept of the building, and so the problems created by these are compensated by thick walls of more than 2m. The spaces within these thick walls give a tectonic quality to the interior space. This is the strongest tectonic quality achieved by the building. The ground floor walls have fewer openings for both security reasons and the necessary stone spanning elements. The openings in the walls of upper levels are grouped together and these groups' length is around 3–4m. Therefore, the total length of openings is not more than 40% of the total length of the walls.

In summary, it can be stated that the design of this tower was against structural guidelines concerning the height of stone walls and room dimensions. However, this is compensated by the use of thick walls determined by the tectonic quality of indoor spaces.

Although structural recommendations concerning plan symmetry, building height and organisation of openings were followed, the stone tower at Norman Castle still has a contravening relationship with structural guidelines.

THE TECTONICS OF A TRADITIONAL APPROACH TO BRICK MASONRY

Most of the structural guidelines for unreinforced stone masonry are also valid for unreinforced brick masonry. Therefore, only the different characteristics and recommendations will be discussed. These differences concern the types of bricks, and the thickness of load-bearing and non-load-bearing brick walls.

Types of bricks

There are many types of bricks. Bricks can be classified according to:

- Their material
- Their load-bearing capacity
- The places in which they are used.

There are various types of fired red clay bricks. There are also concrete bricks, autoclaved aerated concrete blocks (AAC), cellular lightweight concrete blocks (CLC), and fly ash blocks, etc. Some bricks are load-bearing and some cannot be used for structural purposes. Non-load-bearing bricks have more hollow parts in them in comparison to load-bearing bricks. Load-bearing bricks' solid parts should be kept vertical. On the other hand, hollow parts of non-load-bearing bricks are usually kept horizontal when they take place in a wall. Materials are also different: load-bearing bricks have a higher resistance to forces than non-load-bearing bricks. Thus, non-load-bearing bricks should not be used for structural purposes. There are also fire bricks and brick veneer, which are used for totally different purposes than the above brick types.

Thickness of brick walls

Some building codes, such as ACI 530-02/ASCE 5-02/TMS 402-02 (2002), specify the strength of different masonry types, require mathematical analysis of the masonry structure and do not limit any characteristic of the building. However, other building codes, such as the Turkish building code, contain structural recommendations relating to the thickness of brick walls. The Ministry of Public Works and Settlement Government of the Republic of Turkey (2007) states that thickness of a load-bearing brick wall depends on the buildings' number of storeys. If there are more than two storeys, the thickness of the load-bearing walls of the upper two levels is suggested to be minimum one brick (1 brick is 19cm) and the thickness of the load-bearing walls of the remaining lower levels is suggested to be minimum 1.5 bricks. Thickness of non-load-bearing walls is limited to 10cm.

However, the Monadnock Building (built in Chicago, USA, in 1891) is brick masonry with iron vertical tie-beams. This building has 17 storeys and the thickness of the walls at ground floor is nearly 2m. It is possible to have high masonry buildings with thick walls, but this is not efficient because most of the space is occupied by the thick walls.

Structural guidelines for brick masonry

Structural guidelines for brick masonry structures and the value systems determining them are listed in Table 6.2.

Table 6.2 Structural guidelines for brick masonry structures and the associated value system

Structural guidelines	*Value system*
Only load-bearing bricks should be used for building load-bearing brick walls. Non-load-bearing bricks can be used only in non-load-bearing partition walls.	Safety
Minimum thickness of a load-bearing brick wall can be 20cm, and should be increased if the building has more than two levels. Minimum thickness of a non-load-bearing brick wall can be 10cm.	Safety
Continuous foundations should be used under load-bearing brick walls.	Safety
Length of brick walls between two wall intersections can be around 5.5m long in earthquake regions and 7.5m long if there is no earthquake risk.	Safety
Plan of the building should be symmetrical or close to symmetrical in order to avoid twisting due to earthquakes.	Economy
Unsupported height of a brick wall can be around 3m.	Economy
Brick masonry buildings can be around two storeys high in earthquake regions and four storeys high if there is no earthquake risk.	Safety
All load-bearing brick walls must be supported by other load-bearing walls or foundations.	Safety
Total length of openings on a brick wall can be around 40% of the length of the wall.	Safety
Maximum opening size should be around 3m.	Safety

Table 6.2 *continued*

Structural guidelines	*Value system*
Distance of openings from the corner of the building can be around 1.5m in earthquake regions and 1m if there is no earthquake risk.	Safety
Openings can be 0.5m away from wall intersections.	Safety
Distance between two openings can be around 1m in earthquake regions and 0.8m if there is no earthquake risk.	Safety
Openings can have lintels or arches over them.	Safety
Brick walls should be connected to each other with horizontal tie-beams at every floor level.	Safety
Thickness of reinforced concrete horizontal tie-beams can be around 20cm.	Safety
Cross-walls or buttresses can be used against horizontal forces.	Economy

Case study 3: Summer House, Finland

Alvar Aalto's Summer House (Experimental House), as seen in Figures 6.17 and 6.18, is the only case study in this book concerning brick masonry. It was chosen because many types of bricks were used together for tectonic purposes. Most of the surfaces were built with red bricks, including walls and floor covers. Use of very different types of bricks affected the tectonic quality of the building. Thus, the dominant physical entity determining the tectonic characteristics of Summer House is the structural material. Summer House was designed to take place within nature. Its continuity with the surrounding natural environment is its main architectural characteristic.

Apart from the separate room at the back of the building, the plan of the building is square in shape. However, the placement of load-bearing walls within this square is not symmetrical. Load-bearing and non-load-bearing walls in the building can easily be differentiated when looking at the plan of the building. The lintels on the windows and horizontal tie-beams over the walls are hidden behind brick surfaces.

The length of the walls between two wall intersections is usually more than 7.5m. Length of openings is more than 40% of the total length of the wall at many places. Maximum opening size is also more than 3m at three places. These are the openings around the courtyard, which is the main space in the building with forest and mountain views.

The concept of the building was to have a small building in nature. This led to the use of red bricks, which is a more natural material than many modern materials. There occurs a conflict between the building material and the decision to open the building to nature, and this is resolved with large openings. Thus, it can be said that Alvar Aalto did not follow structural guidelines concerning symmetry, length of walls between intersections of walls, and amount and size of openings in order to realise the concept of Summer House. Summer House therefore has a contravening relationship with structural guidelines.

6.17 A sketch of case study 3: Summer House, Muuratsalo, Finland, 1953 (drawn with the help of URL5, n.d.)

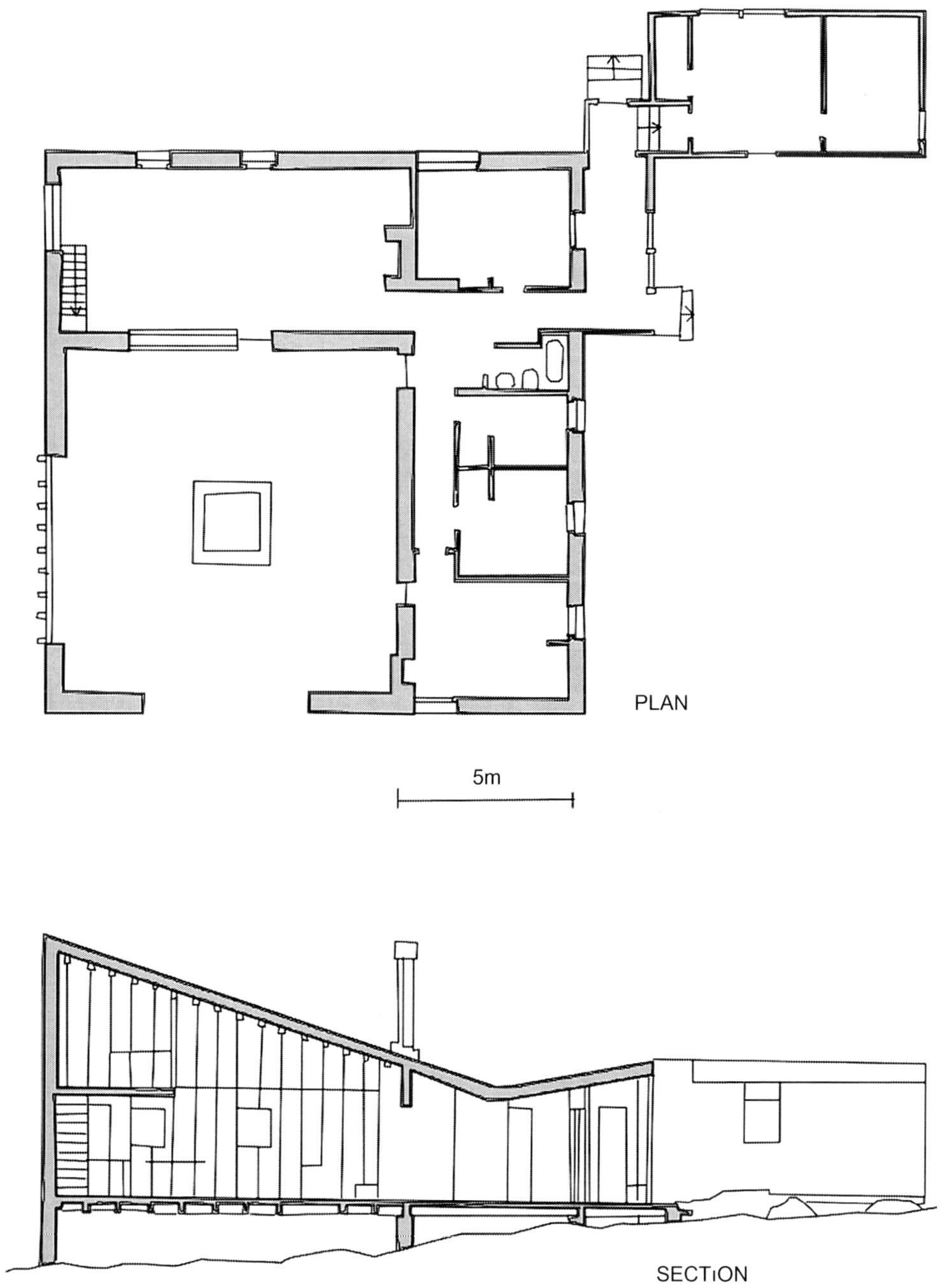

6.18 Plan and section of Summer House (drawn with the help of Diaz, n.d.)

THE TECTONICS OF A TRADITIONAL APPROACH TO ADOBE MASONRY

Structural guidelines for adobe masonry structures can be analysed by classifying adobe masonry wall types and by stating the differences between adobe masonry structures and stone masonry structures.

Types of adobe masonry walls

The most common applications of adobe are:

- Adobe bricks
- Compressed earth blocks
- Rammed earth
- Stabilised adobe.

Adobe bricks are handmade bricks that do not have any holes in them. They are usually connected with earth-based mortar.

Compressed earth blocks are produced with the help of simple machines. They have holes in them and they are also connected with earth-based mortar. Rammed earth is achieved by pouring adobe into formwork and pressurising it for every 70cm thickness. Stabilised adobe (alker) is achieved by adding lime and gypsum to earth. This triples the compressive strength of adobe. The durability and workability of adobe also increase through stabilisation. Stabilisation solves the moisture problem of adobe by increasing its water-resistance properties (Isık & Tulbentci, 2008).

This book discusses the traditional approach to adobe masonry by explaining structural guidelines regarding the use of traditional adobe bricks.

Differences between adobe and stone masonry structures

Traditional approaches to adobe masonry structures are very different to traditional approaches to stone masonry structures. These differences are in:

- Number of storeys
- Height of each storey
- Wall thickness
- Length of wall between two intersecting walls
- Arrangement of openings
- Use of horizontal tie-beams
- Roof structure
- Lateral support.

Building codes in some countries include adobe masonry within all other structures. The Turkish building code is one of these, but other countries such as Mexico and New Zealand have special building codes for adobe masonry structures. This book refers to these specialised building codes in order to illuminate the differences between adobe and stone masonry structures.

Structural guidelines limiting the number of storeys for adobe buildings differ from those of stone buildings. Turkish building codes limit the number of storeys for adobe structures to only one level in all earthquake regions (Ministry of Public Works and Settlement Government of the Republic of Turkey, 2007). The building code of Mexico, however, allows a maximum of two levels (Construction Industries Division of the Regulation and Licensing Department, 2009), whilst the building code of New Zealand limits total height of the building to 6.5m (New Zealand Edict of Government, 1998b).

Turkish building codes limit the height of each storey for adobe buildings to 2.7m (Ministry of Public Works and Settlement Government of the Republic of Turkey, 2007), whilst the New Zealand building code advises special consideration for storey heights of more than 3.3m (New Zealand Edict of Government, 1998b). The New Zealand building code also relates wall thickness to storey height by limiting the slenderness ratio of adobe walls and columns. The thickness of walls can be 1/6 of the wall height if there is a high seismic risk, and 1/10 of the wall height in regions with less risk. Similarly, adobe column slenderness is limited to 1/3 of column height in regions with seismic risk, and 1/4 of column height if there is less risk.

Wall thickness of adobe masonry buildings differs from wall thickness of stone and brick masonry buildings. Turkish building code advises minimum 29cm for exterior load-bearing walls and minimum 19cm for interior load-bearing walls (Ministry of Public Works and Settlement Government of the Republic of Turkey, 2007). The Mexican building code recommends 35cm thickness for the ground floor walls of adobe buildings with two storeys and 25cm for the first floor walls (Construction Industries Division of the Regulation and Licensing Department, 2009). For high-risk earthquake regions, it also recommends that the wall thickness should not be less than both the length of the wall divided by ten and the height of the wall divided by six, whichever is the higher (Blondet et al., 2011).

The length of the wall between intersections for adobe masonry buildings is limited to 4.5m by the Turkish building code. If the wall length is more, the use of reinforced concrete vertical tie-beams is suggested at every 4m (Ministry of Public Works and Settlement Government of the Republic of Turkey, 2007).

Distance of openings from the corners of abode buildings is specified as 1m by both Turkish and Mexican building codes

(Ministry of Public Works and Settlement Government of the Republic of Turkey, 2007; Construction Industries Division of the Regulation and Licensing Department, 2009). Both the distance between two openings and the length of openings are also limited. Turkish building code limits both the distance between two openings and the length of openings to 1m (Ministry of Public Works and Settlement Government of the Republic of Turkey, 2007). However, the building code of New Zealand allows 1.8m wide arches (New Zealand Edict of Government, 1998a). The Turkish building code also limits the heights of openings: doors must be maximum 1.9m and windows 1.2m (Ministry of Public Works and Settlement Government of the Republic of Turkey, 2007).

Horizontal tie-beams of adobe buildings can be either reinforced concrete or timber. The Turkish building code advises two 10cm × 10cm timber pieces or a 20cm deep reinforced concrete tie-beam (Ministry of Public Works and Settlement Government of the Republic of Turkey, 2007). The Mexican building code advises 15cm deep reinforced concrete or timber horizontal tie-beams (Construction Industries Division of the Regulation and Licensing Department, 2009).

It is safer to have lightweight roof systems in adobe masonry buildings. The Turkish building code recommends the use of a lightweight roof system, rather than an earth roof, in regions with high seismic risk. If the seismic risk is lower, it becomes possible to have earth roofs no thicker than 15cm (Ministry of Public Works and Settlement Government of the Republic of Turkey, 2007).

Different building codes guarantee lateral support in different ways. The Turkish building code limits the amount of openings (Ministry of Public Works and Settlement Government of the Republic of Turkey, 2007). The New Zealand building code advises making calculations for the length of bracing walls (New Zealand Edict of Government, 1998a). The Mexican building code recommends having cross-walls or buttresses at every 8m (Construction Industries Division of the Regulation and Licensing Department, 2009). There are also some sources, such as Blondet et al. (2011), which recommend reinforcing adobe walls with the use of canes or barbed wire, etc.

Structural guidelines for adobe masonry

Structural guidelines for adobe masonry structures and the value systems determining them are listed in Table 6.3.

Table 6.3 Structural guidelines for adobe masonry structures and the associated value system

Structural guidelines	*Value system*
Minimum thickness of a load-bearing adobe exterior wall can be 29cm, and increased to 35cm if the building has more than one level. Minimum thickness of a load-bearing adobe interior wall can be 19cm. Slenderness ratio of adobe walls should also be considered when determining thickness. Wall thickness should not be under either the length of the wall divided by ten or the height of the wall divided by six, whichever is the higher.	Safety
Continuous foundations should be used under load-bearing adobe walls.	Safety
Length of adobe walls between two wall intersections can be around 4.5m long in earthquake regions and 7.5m long if there is no earthquake risk (where vertical tie-beams can be used for support).	Safety
Plan of the building should be symmetrical or close to symmetrical in order to avoid twisting due to earthquakes.	Economy
Unsupported height of an adobe wall can be around 2.7–3.3m, depending on thickness.	Economy
Adobe masonry buildings can be one storey high in earthquake regions and two storeys high if there is no earthquake risk.	Safety
Total length of openings on an adobe wall can be around 40% of the length of the wall.	Safety
Maximum opening size should be around 1m if the opening has a lintel. If there is an arch, maximum opening size can increase to 1.8m.	Safety
Distance of openings from the corner of the building can be around 1m.	Safety
Openings can be 0.5m away from wall intersections.	Safety

Table 6.3 *continued*

Structural guidelines	*Value system*
Distance between two openings can be around 1m.	Safety
Openings can have lintels or arches over them.	Safety
Adobe walls should be connected to each other with horizontal tie-beams at every floor level.	Safety
Thickness of reinforced concrete horizontal tie-beams can be around 20–25cm. Timber tie-beams can be formed with two 10cm × 10cm pieces.	Safety
Cross-walls, buttresses and other reinforcement, such as canes, can be used against horizontal forces.	Economy
A lightweight roof structure can be selected for adobe masonry buildings.	Safety

Case study 4: Harran Houses, Turkey

Harran Houses, as seen in Figures 6.19 and 6.20, were chosen as a case study because the tectonics of these buildings are directly related to the life difficulties faced by the inhabitants. The origin of these houses goes back to 2500–600 BC. Taxes in the past were based on the number of houses in a village. Villagers therefore built houses that could be easily dismantled before the visit of the tax collector, and easily reconstructed again after the visit. Thus, villagers would collect adobe masonry from the ruined buildings around them and build the new corbelled domes within half a day. Since there was no wood available in such a dry climate, villagers were forced to use adobe masonry and stone (Ozdeniz et al., 1998). The architectural concept behind these buildings is therefore based on functionality and practicality. These buildings also provide strong continuity with the context surrounding them. The dominant physical entity in these buildings is structure because of their ability to be dismantled.

When Harran Houses are analysed according to their relationship with structural guidelines for adobe masonry structures, it

6.19 A sketch of case study 4: Harran Houses, Urfa, Turkey, 2500–600 BC (drawn with the help of URL6, 2013 and painted by Nicholas Wilkinson)

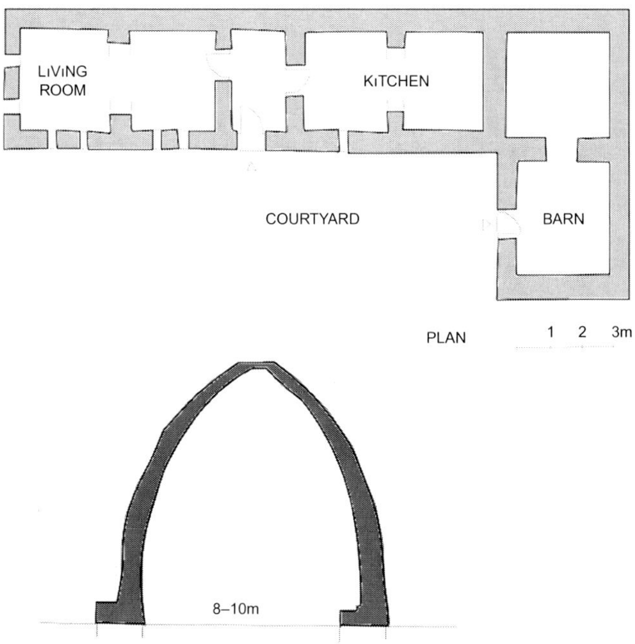

6.20 Plan and section of Harran Houses (drawn with the help of Ozdeniz et al., 1998)

is seen that most of the spaces are a maximum of 5m in plan, are either symmetrical or almost symmetrical, and have openings (holes at the top and sides of the dome) that are no more than 30–40cm diameter. However, it was not possible to use lightweight roof systems because of the lack of wood in the environment. Foundations and the intersections between the domes and the walls were built from stone. The domes could rise 3–5m (Ozdeniz et al., 1998). It is very clear that the difficulties in achieving larger spaces, lighter roof systems and lower domes are due to social problems and the lack of materials. Thus, Harran Houses have a contravening relationship to structural guidelines.

Case study 5: Great Mosque of Djenne, Mali

Great Mosque of Djenne, as seen in Figures 6.21 and 6.22, was chosen as a case study due to the extraordinary tectonic use of lateral stability elements, including buttresses, cross-walls and wooden reinforcement. The main architectural concept of Great Mosque of Djenne is to provide a large and high space for worshippers to gather. The major architectural qualities of the building are the continuity the building provides with the context and the rhythmic order of its walls. The dominant physical entities determining the tectonic characteristics of the building are the structure and the use of adobe as the structural material.

Great Mosque of Djenne is raised on a platform to protect it from floods. Its roof is carried by masonry columns and pointed arches. Its towers are around 15m high. Thus, the building has long walls and its unsupported height is considerably greater than structural guidelines for adobe structures. The location of the towers also disturbs the symmetry of the structure plan.

The presence of high and long walls is part of the initial concept of this building. In order to compensate for not following structural guidelines in relation to height and length, the adobe

6.21 A sketch of case study 5: Great Mosque of Djenne, Mali, 1907 (drawn with the help of Kamiya, 2009)

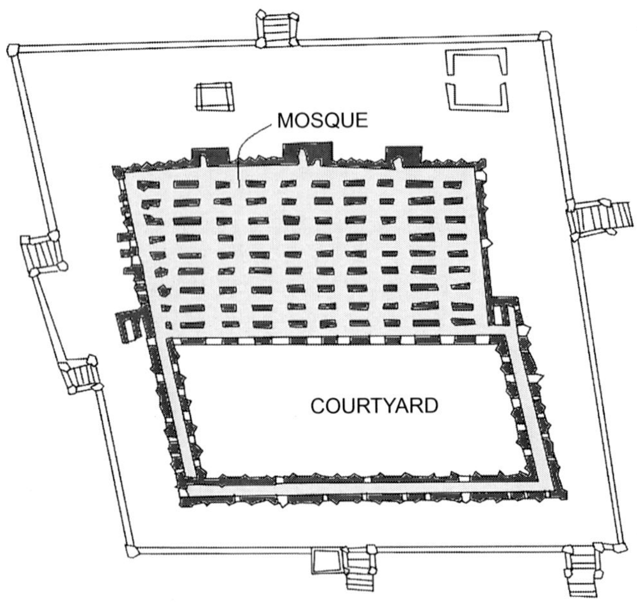

6.22 Plan of Great Mosque of Djenne (drawn with the help of Kamiya, 2009)

walls of the mosque are made very thick, reaching 1m at certain parts of the building. The walls are also supported by many cross-walls and buttresses, and they contain wooden sticks as reinforcement against lateral forces. These cross-walls, buttresses and reinforcement give the aforementioned rhythm to the facade of the building.

The openings are very limited and small. Earth-based plaster, which was applied to the surface, increases lateral stability as well as improving the tectonic qualities of the building by creating curved forms and eliminating sharp corners. Yet the relationship of Great Mosque of Djenne to structural guidelines is still contravening.

THE TECTONICS OF A TRADITIONAL APPROACH TO TIMBER MASONRY

Since this book categorises masonry structures that contain steel reinforcement as contemporary masonry structures, many types of timber masonry applications can be considered here, including:

- Log cabin structures, which are built by interlocking timber logs at the corners of the building.
- Traditional timber masonry that contains horizontal, vertical and diagonal elements.
- The timber balloon frame, which is the contemporary interpretation of traditional timber masonry going back to the early nineteenth century.

Figure 6.23 shows these three types of timber masonry structures. This book will concentrate on the last two types.

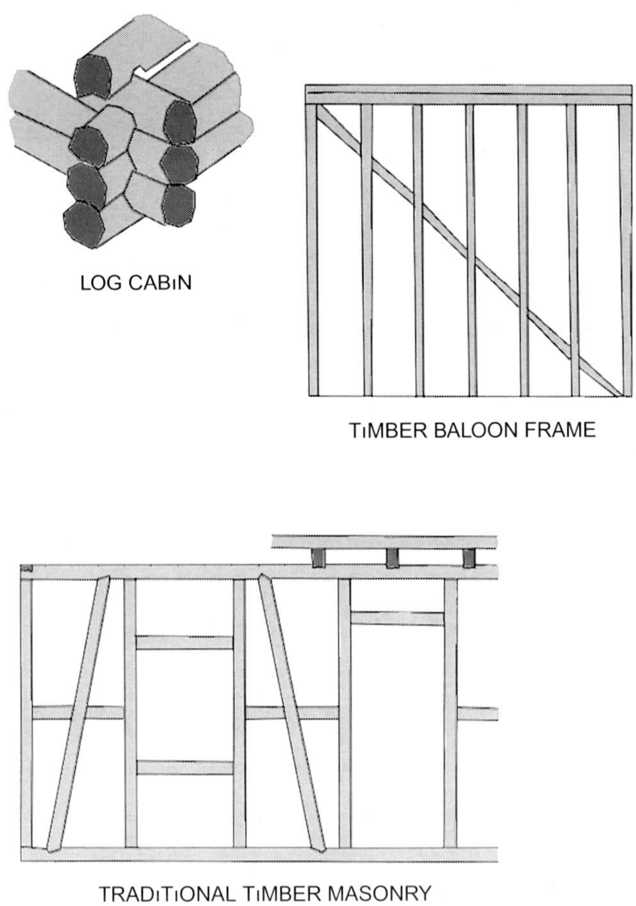

6.23 Types of timber masonry

Some countries' building codes do not include timber structures, whereas other countries have specific building codes concerning them. For example, Eurocode 5 relates to the design of timber structures (EN 1995-1-1, 2004). However, this building code is based on a mathematical analysis and does not contain any structural recommendations that can be directly used by architects. On the other hand, the American Wood Council (2003) makes some recommendations about the dimensions of heavy timber columns and beams[1] only. According to them, the plan dimensions of timber columns should be a minimum of 20cm × 20cm and the depth of beams should be a minimum of 25cm. This information can be more meaningful for the design of timber frame structures, which have larger spans in comparison to timber masonry structures. Thus, this book concentrates on other literature concerning timber masonry, and traditional examples.

In order to discuss the differences between structural guidelines for timber and stone masonry, a comparison is made between traditional timber masonry and the timber balloon frame. These are both timber masonry structures because they both have frequently placed vertical elements that form the masonry wall. After making this comparison, the structural role of the elements in traditional timber masonry is presented.

Comparing traditional timber masonry with timber balloon frames

A timber balloon frame is a contemporary interpretation of traditional timber masonry. Figure 6.24 illustrates types of elements seen in traditional timber masonry. Although they both have vertical structural elements, the distance between the vertical structural elements is 40cm in a balloon frame and 1–1.5m in traditional timber masonry. These verticals are placed at the corners of the building, at wall intersections, and at both sides of openings. The distance between the vertical elements is kept fairly constant, and the empty sections are filled. Dimensions for the vertical elements are usually 10cm × 10cm for traditional timber masonry and 5cm × 10cm for balloon frames. The height of vertical elements is one storey for traditional timber masonry, but timber balloon frames can be two storeys (Eser, 1977). According to Gulkan and Langenbach (2004), the thickness of the walls of

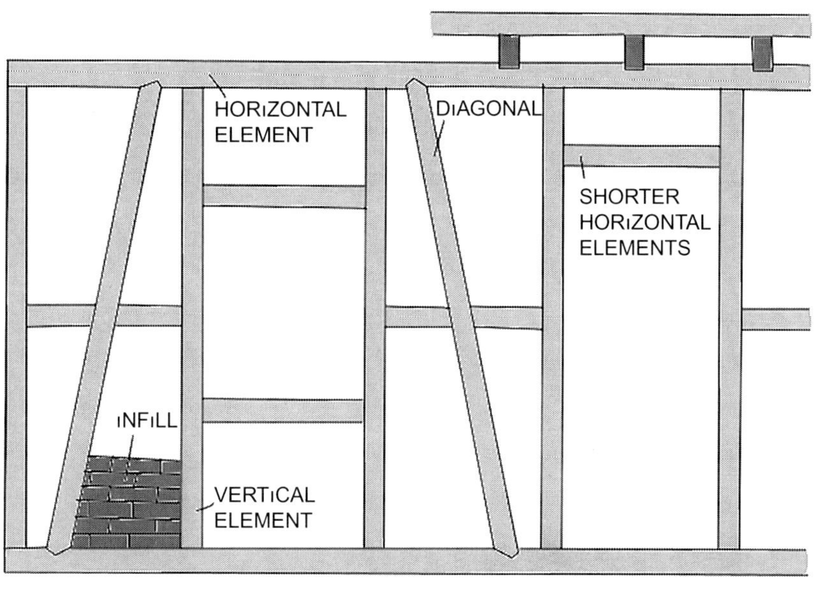

6.24 Structural elements of traditional timber masonry and timber balloon frames

traditional timber masonry buildings is 10–12cm, which means that timber masonry structures have very thin walls in comparison to other masonry structures.

The diagonal elements take place between two vertical elements in traditional timber masonry. This affects the angle of the diagonal element and decreases its performance. However, the diagonals' place is not restricted in balloon frames and thus they can have angles such as 45° or 60°. Traditional timber masonry has some short horizontal elements, which define the upper and lower borders of openings and connect the two adjacent verticals to each other. However, the timber balloon frame does not contain these short horizontal elements, and is usually covered with timber surfaces from both sides. The distances between the timber structural elements of traditional timber masonry can be filled in with stone pieces, red bricks and adobe bricks, or the timber structure can be covered with timber surfaces from both sides (Eser, 1977).

The floor systems of traditional timber masonry and timber balloon frames are usually made out of timber elements. Traditional timber masonry structures usually have continuous stone foundations. The ground floor structure of many traditional timber masonry structures was built as stone masonry. However, the contemporary examples of timber balloon frames are usually supported with steel frames and reinforced concrete individual footings.

Traditional timber frame structures are usually handmade, and so the distances between the elements of a traditional timber masonry structure can differ. The small variations in sizes of elements, and the distances between them, give a particular aesthetic quality to traditional timber masonry structures. In contrast, timber balloon frames can be produced in factories: thus, these frames are more precise structures.

The role of structural elements in traditional timber masonry

The structural elements in traditional timber masonry are as follows:

- Vertical structural elements.
- Horizontal structural elements that exist at every storey level.
- Diagonal structural elements.
- Shorter horizontal structural elements between the vertical structural elements.
- Infill material between the timber elements.
- Plaster.

The vertical structural elements form the timber wall, carrying the loads in a vertical direction. They define openings by taking place on both sides. The infill materials are packed together by the vertical structures and the short horizontal elements.

Horizontal structural elements, which connect the vertical elements from top and bottom, equally distribute the load of slabs or roof to the vertical structural elements. The horizontal elements, taking place between the foundation walls and the vertical elements, provide an easy connection between the foundations and the timber structure.

Diagonal structural elements transfer the horizontal loads directly to the foundations or vertical structural elements. They are also used to adjust any deficiencies made during construction, which results in slight deformations in building form (Eser, 1977).

The shorter horizontal elements define the top and bottom of openings. Together with the vertical structural elements, they also pack the infill material together. Infill materials take place between the timber elements and contribute to the formation of the timber masonry wall. Plaster increases the strength of traditional timber masonry structures.

Traditional timber masonry structures have shown good performances in earthquakes. Gulkan and Langenbach (2004) state that this success of traditional timber masonry in earthquakes is due to its flexibility and energy dissipation properties, rather than strength and stiffness.

Structural guidelines for timber masonry

Structural guidelines for timber masonry structures and the value systems determining them are listed in Table 6.4. (Table 6.1, showing the structural guidelines for stone masonry structures, is used as a basis.)

Table 6.4 Structural guidelines for timber masonry structures and the associated value system

Structural guidelines	*Value system*
Minimum thickness of a timber masonry wall can be around 10–12cm.	Safety
Distance between vertical elements should be 1.5m for traditional timber masonry and 40cm for balloon frame.	Safety
Continuous foundations or individual footings can be used under timber masonry walls.	Safety
Length of timber masonry walls between two wall intersections can be around 5.5m long in earthquake regions and 7.5m long if there is no earthquake risk.	Safety
Plan of the building should be symmetrical or close to symmetrical in order to avoid twisting due to earthquakes.	Economy
Unsupported height of a timber masonry wall can be around 3m.	Economy
Timber masonry buildings are usually two storeys high in earthquake regions but there are examples of three or four storeys if there is no earthquake risk.	Safety
Total length of openings on a timber masonry wall can be around 40% of the length of the wall.	Safety
Length of openings is usually around 1m.	Safety
Distance of openings from the corner of the building can be around 1–1.5m.	Safety
Openings can be 0.5m away from wall intersections.	Safety
Distance between two openings can be around 1m.	Safety
Vertical structural elements should be connected to each other with horizontal structural elements at every floor level.	Safety
Diagonal structural elements should be used against horizontal forces.	Safety

Case study 6: İsmail Hacı Çakır House, Turkey

İsmail Hacı Çakır House, as seen in Figure 6.25, was chosen as a case study because it is one of the typical examples of traditional timber masonry. The architectural concept behind İsmail Hacı Çakır House is based on functionality. It also provides strong continuity with the context surrounding it. The dominant physical entities in this building are structural system and structural material because the use of stone and timber masonry provide a contrast to each other. The stone masonry ground floor is heavy and closed, whilst the timber masonry upper levels are light and open.

It is very common in Turkey to have the ground floor of houses built with stone because privacy at the ground level is critical. Thus, people use stone walls with no openings at the ground level. Timber masonry is used only at the first and second floors. Stone and timber masonry is mixed at the first and second floors. Timber masonry is preferred around the main room and sofa, which can be opened in summers and closed in winters. Building one half of the upper levels with stone, which is heavy, and the other half with timber masonry, which is light, means that the distribution of structural material on the plan is uneven. This situation might cause twisting problems in earthquakes: Kütahya is located within a secondary earthquake region of Turkey.

The number of openings in the main room is not against structural guidelines. However, there are too many openings around the sofa and the structure around this area is more similar to timber frames than to timber masonry structures. This is explained by the need for having a large open living area in summer. (These living areas are closed with shutters in winter.) The contrast between the heavy walls of the ground floor and the lightness of the top floor reflects the contrast between the need for privacy and the need for being open to nature. İsmail Hacı Çakır House has a contravening relationship with structural guidelines because of its large openings at the upper levels.

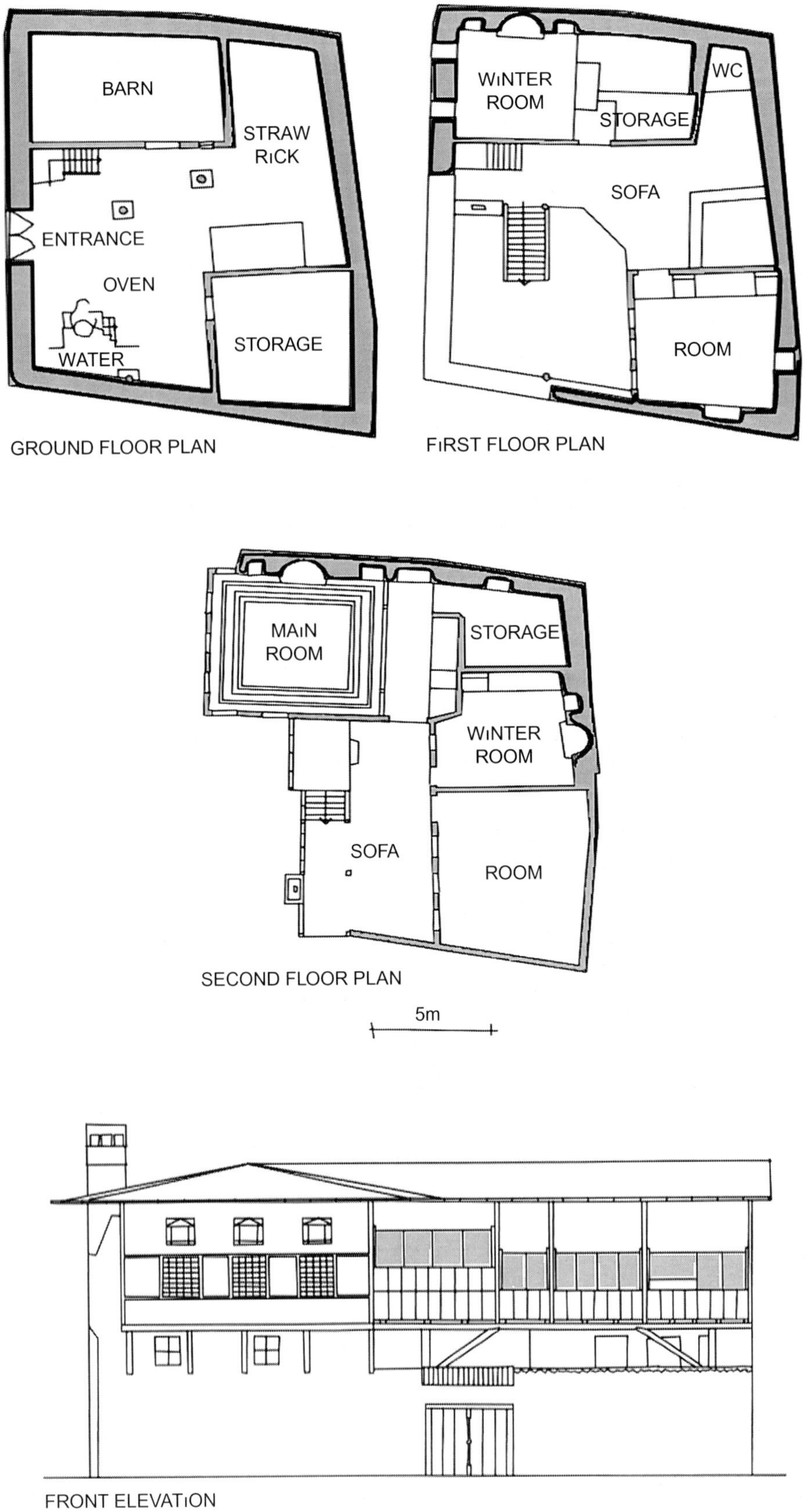

6.25 Plans and elevation of case study 6: İsmail Hacı Çakır House, Kütahya, Turkey, 1781 (drawn with the help of Eser, 1954)

CONCLUSIONS

The analysis of structural guidelines for different types of traditional masonry shows that the heaviest masonry structure type is stone masonry and the lightest type is timber masonry. When the degree of restriction is studied, especially around openings, it can be stated that the most restricted masonry structure is adobe masonry. It should also be stated that in comparison to other structural systems, such as frame structures, masonry structures are very restricted.

Analysis of the case studies shows that structural guidelines are not strictly followed. None of the case studies followed all of the structural guidelines. The buildings therefore all have a contravening relationship with structural guidelines. Most of the case studies are old examples; modern building codes did not exist at the time they were built. However, there are always architectural reasons for not following the guidelines, including:

- Achieving symbolic meanings
- Achieving functional requirements
- Lack of materials
- Social pressure
- Values.

Cologne Cathedral is a good example for not following structural guidelines in order to achieve symbolic meanings. The height and the lightness of the cathedral defy some structural guidelines but this is how the building gains its symbolic meaning: the architectural concept of Gothic cathedrals is high and light.

The stone tower at Norman Castle and Great Mosque of Djenne are good examples for not following structural guidelines in order to achieve functional requirements. The stone tower at Norman Castle needed height to become a tower. On the other hand, Great Mosque of Djenne needed long and high walls in order to collect the community under one ceiling.

Harran Houses are a good example for not following structural guidelines because of the lack of materials. The hot and dry climate means that there was no wood: thus, they had to use adobe for the roof structure. Harran Houses also reject structural guidelines as a result of social pressure. In order to avoid paying high tax, the houses needed to be easily dismantled and rebuilt: thus, the preference was for corbelled domes.

Summer House and İsmail Hacı Çakır House are good examples for not following structural guidelines in order to achieve specific values. Both buildings are designed to be integrated with nature. The Summer House is built with natural materials and it contains large openings to catch the mountain and forest views. İsmail Hacı Çakır House reconciles the needs for privacy and for being open to nature. Thus, the ground floor is designed to have stone walls with no openings, whilst the upper floors have timber masonry walls and very large openings.

The structural guidelines for traditional masonry structures (not always followed during the design of these case studies) are mostly concerned with the length and height of walls and the number of openings.

NOTE

1 Heavy timber elements have a high resistance to fire.

REFERENCES

ACI 530-02/ASCE 5-02/TMS 402-02 (2002) *Building Code Requirements for Masonry Structures*, Masonry Standards Joint Committee (viewed 5 September 2013: https://engineering.purdue.edu/~ramirez/CE479/FALL05/MasonryBuildingCode1-3-02.pdf)

American Wood Council (2003) *Heavy Timber Construction*, 3rd edition, American Wood Council: Washington.

Blondet, M., Villa Garcia, G., Brzev, S., Rubinos, A. (2011) *Earthquake Resistant Construction of Adobe Buildings: A Tutorial*, 2nd edition (viewed 15 November 2013: www.world-housing.net/wp-content/uploads/2011/06/Adobe_Tutorial.pdf)

Construction Industries Division of the Regulation and Licensing Department (2009) *New Mexico Earthen Building Materials*

Code (viewed 29 October 2013: www.nmcpr.state.nm.us/nmac/parts/title14/14.007.0004.htm)

Diaz, A. (n.d.) *Analysis of Aalto's House* (viewed 26 September 2014: www.alexdiaz.us/Analysis-of-Aalto-s-House)

Dincyurek, O., Mallick, F., Numan, I. (2003) 'Cultural and Environmental Values in the Arcaded Mesaorian Houses of Cyprus' *Building and Environment*, Vol 38: pp.1463–1473.

EN 1995-1-1 (2004) *Eurocode 5: Design of Timber Structures*, Part 1-1 (viewed 15 November 2013: ftp://law.resource.org/eur/ibr/en.1995.1.1.2004.pdf)

Eser, L. (1954) *Kutahya Evleri* (Houses of Kutahya), ITU Docentlik Calismasi: Istanbul.

Eser, L. (1977) *Geleneksel ve Gelismis Geleneksel Yapi 1* (Traditional and Modern Structures), Istanbul Teknik Universitesi Baski Atolyesi: Istanbul.

Eurocodes Committee: Scientific and Technical Chamber of Cyprus under a Ministry of Interior's Program (2004) CYS Annex to CYS EN 1998-1. *Eurocode 8: Design of Structures for Earthquake Resistance* (viewed 20 October 2013: www.cys.org.cy/images/public/eurocodes)

GB-50011 (2001) *Code for Seismic Design of Buildings*, Construction Ministry of P.R. China and National Bureau for Quality Supervision, Inspection and Quarantine: Beijing.

Gulkan, P., Langenbach, R. (2004) 'The Earthquake Resistance of Traditional Timber and Masonry Dwellings in Turkey' *13th World Conference on Earthquake Engineering*, 1–6 August, Paper No. 2297: Vancouver, Canada.

Isık, B., Tulbentci, T. (2008) 'Sustainable Housing in Island Conditions Using Alker – Gypsum Stabilized Earth: A Case-study from Northern Cyprus' *Building and Environment*, Vol. 43: pp.1426–1432.

Kamiya, T. (2009) *Islamic Architecture in Mali* (viewed 26 September 2014: www.kamit.jp/27_mali/mal_eng.htm)

Ministry of Public Works and Settlement Government of the Republic of Turkey (2007) *Seismic Performance Evaluation of Dual Reinforced Concrete Systems Design According to Turkish Seismic Code*, trans. E.Y. Karcı (viewed 26 October 2013: www.belgeler.com/blg/22lc/seismic-performance-evaluation-of-dual-reinforced-concrete-systems-design-according-to-turkish-seismic-code)

New Zealand Edict of Government (1998a) NZS 4299. *Earth Buildings Not Requiring Specific Design* (viewed 29 October 2013: https://law.resource.org/pub/nz/ibr/nzs.4299.1998.pdf)

New Zealand Edict of Government (1998b) NZS 4297. *Engineering Design of Earth Buildings* (viewed 29 October 2013: https://law.resource.org/pub/nz/ibr/nzs.4297.1998.pdf)

Ozdeniz, M.B., Bekleyen, A., Gonul, I.A., Gonul, H., Sarigul, H., Ilter, T., Dalkilic, N., Yildirim, M. (1998) 'Vernacular Domed Houses of Harran, Turkey' *Habitat International*, Vol. 22, No. 4: pp.477–485.

Pakistan Engineering Council (2007) *Building Code of Pakistan* (viewed 2 November 2013: www.pec.org.pk/building code.aspx)

Vitruvius (15 BC, 1914) *Ten Books On Architecture*, trans. M.H. Morgan, Harvard University Press: Cambridge, Massachusetts. (viewed on 3 November 2013: www.gutenberg.org/files/20239/20239-h/29239-h.htm)

URL1 (2014) *Amazing Cologne Cathedral in Germany* (viewed 26 September 2013: http://raredelights.com/amazing-cologne-cathedral-germany/)

URL2 (n.d.) *Churches & Cathedrals* (viewed 26 September 2013: www.pinterest.com/martijnfa/churches-cathedrals-maps-and-drawings/)

URL3 (n.d.) *Cologne Cathedral* (viewed 26 September 2013: www.f1online.pro/en/image-details/5411029.html)

URL4 (n.d.) *Norman Stone Keeps* (viewed 26 September 2013: www.castrabritannica.co.uk/texts/text06.html)

URL5 (n.d.) *Muuratsalo Experimental House, Alvar Aalto* (viewed 26 September 2013: www.archdaily.com/214209/ad-classics-muuratsalo-experimental-house-alvar-aalto/always-credit-a%C2%A9-nico-saieh-as-author-of-these-photographs-40/)

URL6 (2013) *Harran, Şanlıurfa ve Mardin panoramio fotoğraf galerisi* (Photo gallery of Harran, Şanlıurfa and Mardin) (viewed 26 September 2013: www.illerarasimesafe.com/mardin_sanliurfa-harran/fotograflar)

7 The Tectonics of Masonry Roof Structures

Masonry roof structures can be studied under the following classifications:

- Arch
- Vault
- Dome.

An arch can be seen as the unit structure, which is repeated to achieve vaults and domes. Thus, understanding arches is the key to understanding vaults and domes.

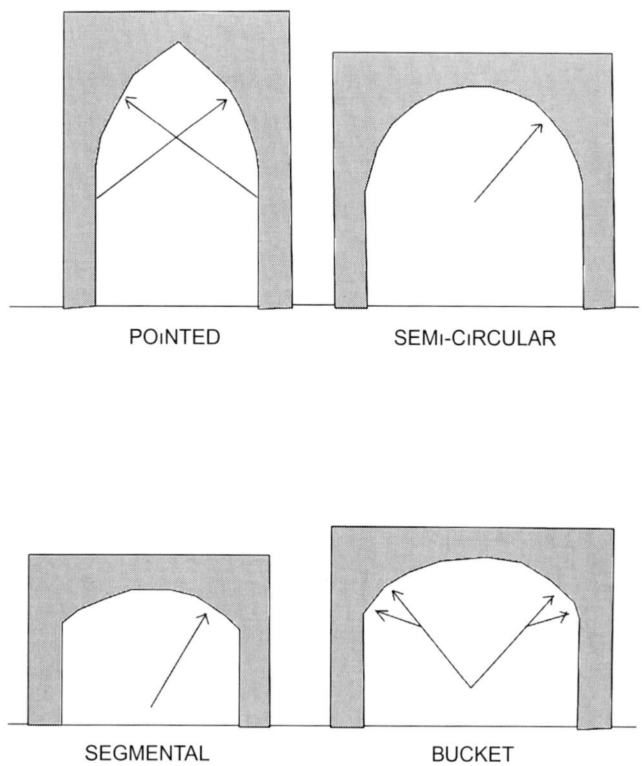

7.1 Forms of arches

MASONRY ARCH

Masonry arches can be studied under the following headings:

- Form
- Structural behaviour
- Construction process
- Span of stone, brick and adobe arches
- Structural guidelines for masonry arches.

This section will then conclude with an analysis of a case study involving the use of masonry arches.

Form of masonry arches

Arches are two-dimensional structural elements and they usually have a positive curvature. There can be many forms of arches. As seen in Figure 7.1, there can be semi-circular and segmental arches with one centre, and pointed and bucket arches with multiple centres.

Structural behaviour of masonry arches

The structural behaviour of arches can be understood by explaining why the key-stone of an arch cannot fall down. This is due to both the shape of the key-stone and the shape of the gap it is in. The top of the key-stone is wider than the bottom: thus, it is not possible for the key-stone to fall through the gap, as seen in Figure 7.2.

Since gravity forces are pulling each stone downwards, friction occurs between the surfaces of the adjacent stone pieces. However, these friction forces are not in the same direction as the gravity forces. Instead, they are parallel to the touching surfaces of the stone pieces. The weight of each stone piece is transferred to the adjacent stone with a force that is perpendicular to the surface between the stone pieces. These forces also consist of vertical and horizontal components. The sum of the vertical components of these forces (see Figure 7.2) should be equal to the gravity force. Additionally, the horizontal forces should balance each other in magnitude.

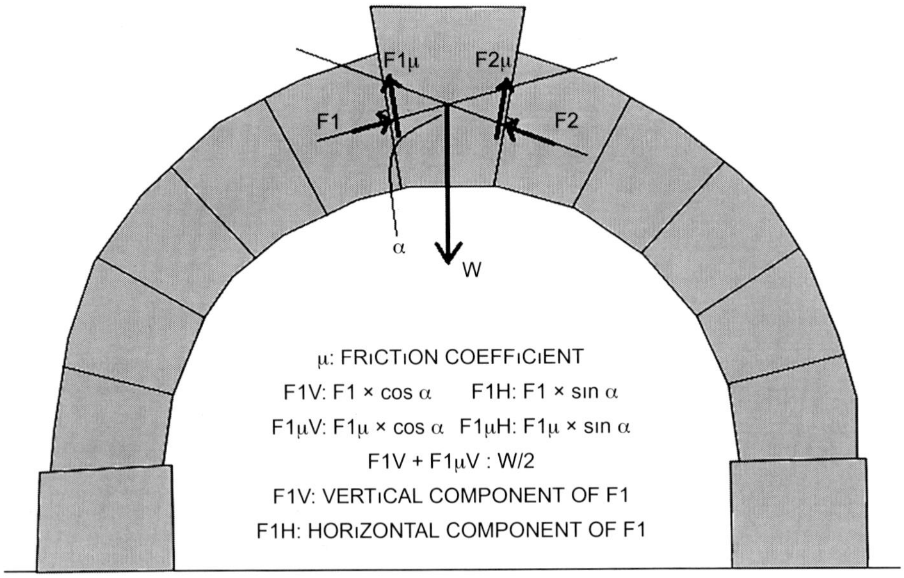

7.2 Forces affecting the key-stone of stone arches

It is possible to have a flat stone arch if the stone pieces in it have the proper form, as seen in Figure 7.3. Thus, the behaviour of a stone arch is more related to the form of its pieces rather than the overall form of the arch. In earthquake regions, stone arches can be designed as seen in Figure 7.4. Changing the form of the stone pieces in this way eliminates the fall of stone pieces with the movement of the earth.

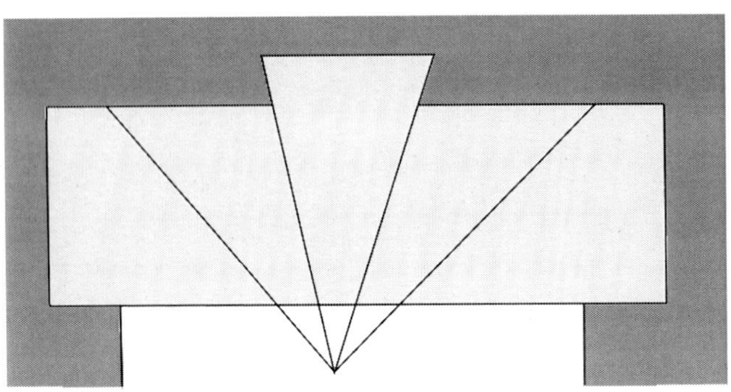

7.3 Flat stone arch

Stone arches work mainly with compression, and the direction of this compressive internal force is always parallel to the curve of the arch. Thus, the weight of an arch is transferred to its supports with two forces parallel to the legs of the arch, as seen in Figure 7.5.

These forces transferring the weight of the arch to its supports consist of vertical and horizontal components. The sum of the vertical components is equal to the total weight of the arch and the total weight carried by the arch. These vertical components are balanced by the reaction coming from the support. However, the horizontal components need to be balanced by some other

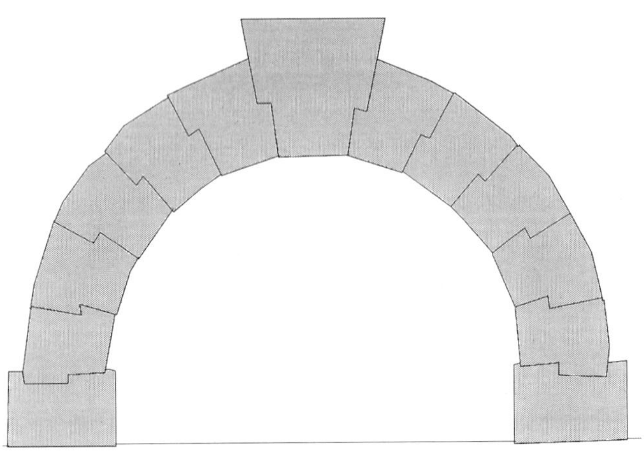

7.4 Example of a stone arch in earthquake regions

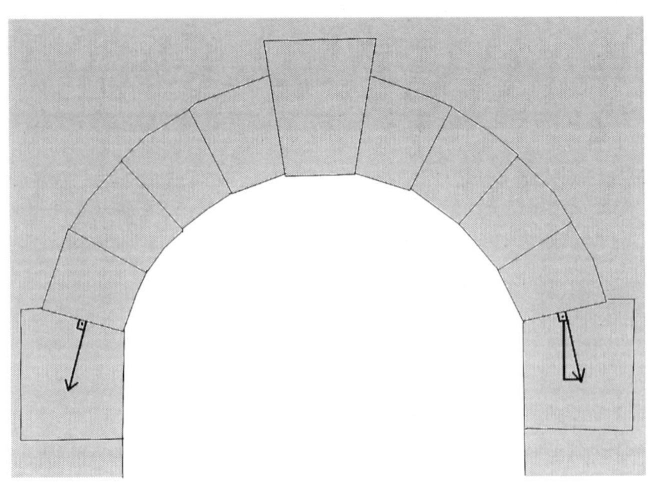

7.5 Transfer of arch weight

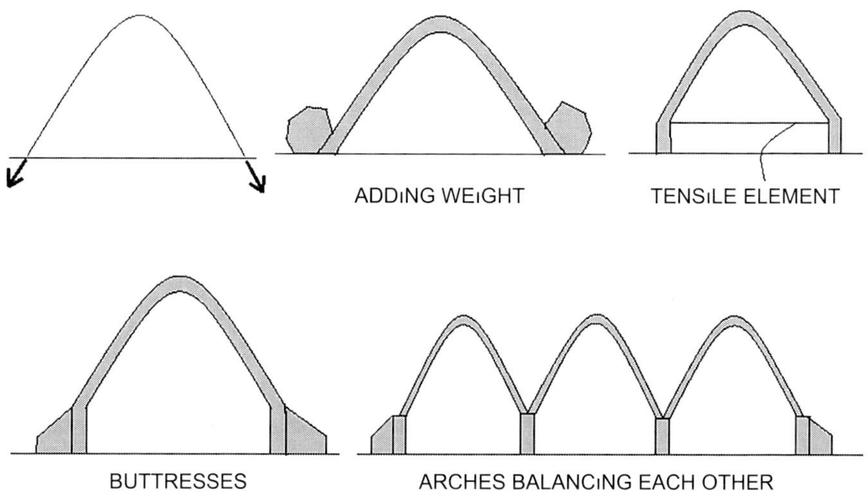

7.6 Horizontal force problems in arches and associated solutions

means. This issue can be termed the horizontal force problem in compression structures. An architect should solve this problem in the design of a compression structure. These horizontal forces will be higher in shallow arches and lower in high arches. Thus, it can be recommended to design shallow arches only for short distances, such as 2m.

Horizontal force problem in arches can be solved by adding weight on both legs of the arch, by providing buttresses, by tying the legs of the arch with a tensile element, and by using arches side by side, as seen in Figure 7.6.

Construction process of masonry arches

Arches can be constructed in two different ways. These are:

- Construction of arches without centring.
- Construction of arches with centring.

Stone arches can be constructed without centring by shifting pieces of stone inwards in each row, as seen in Figure 7.7. These types of arches are known as a corbelled arch or false arch.

7.7 Construction of arches without centring

Corbelled arches are usually high and they do not behave as stone arches, as described above. Arches can also be built with the help of centring, as seen in Figure 7.8. First the centring is placed; then the stone pieces and the key-stone are placed; finally the centring is removed. The top temporary support in Figure 7.8 is mostly used for professional applications, whereas the bottom temporary support is more common in villages and rural places.

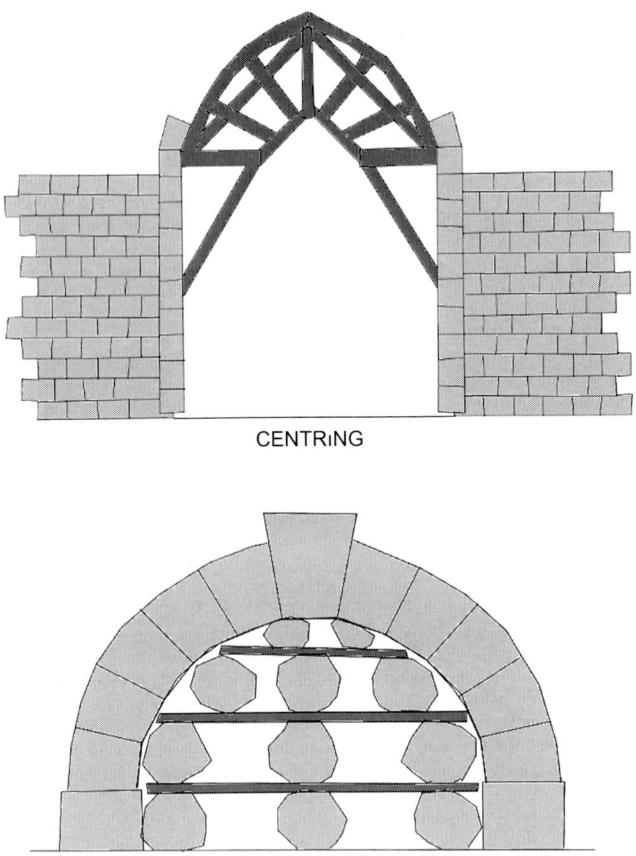

7.8 Construction of arches with centring

Span of masonry arches

The capacity of masonry arches to span distances can be discussed with the help of existing examples. The unreinforced concrete arch of Syratalviadukt in Germany, which was built in 1905, spans 90m (URL1, 2010). The stone arch of Over Bridge in the UK, which was designed by Thomas Telford in 1825, spans 100m (URL2, 2004). The brick railway bridge of Marc Brunel at Maidenhead in the UK, which was built in 1838, spans 39m: this is a shallow arch which rises only 7m (URL3, n.d.).

Structural guidelines for masonry arches

Structural guidelines for masonry arches are listed in Table 7.1.

Table 7.1 Structural guidelines for masonry arches and the associated value system

Structural guidelines	*Value system*
Positively curved forms can be preferred for masonry arches.	Economy
Shallow arches can be used for shorter spans (in comparison to high arches) due to the effect of horizontal forces.	Safety
Stone arches can span up to 100m and brick arches can span up to 39m.	Economy
Problems of horizontal force can be solved by using buttresses, tension ties, etc.	Economy

Case study 7: The arches at Cordoba Mosque, Spain

Cordoba Mosque, as seen in Figures 7.9 and 7.10, is the only case study concerning traditional masonry arches in this book. This building was chosen because of the extraordinary tectonic qualities of its arches. The interior space of Cordoba Mosque is rhythmic, an effect produced by the presence of columns and the white and red colour contrast in the stones of the arches. Perspective, which can be seen between two columns, makes people appreciate the dimensions of the space. The dominant physical entity determining the tectonic characteristics of the building is the structure.

There are horseshoe arches as well as multi-lobed arches and these are used on top of each other in many places. The colour contrast between red and white is achieved through using alternating stone and bricks for the construction of the arches. The arches stand over the capitals of slender stone columns. The height of columns and span of the arches are around 4m. This span is much less than the maximum achievable span of masonry arches.

Since the arches are used side by side, they balance the horizontal forces at the base of each other. Only the arches at the end of

7.9 A sketch of case study 7: the arches at Cordoba Mosque, Cordoba, Spain, eighth century (drawn with the help of URL7, 2014)

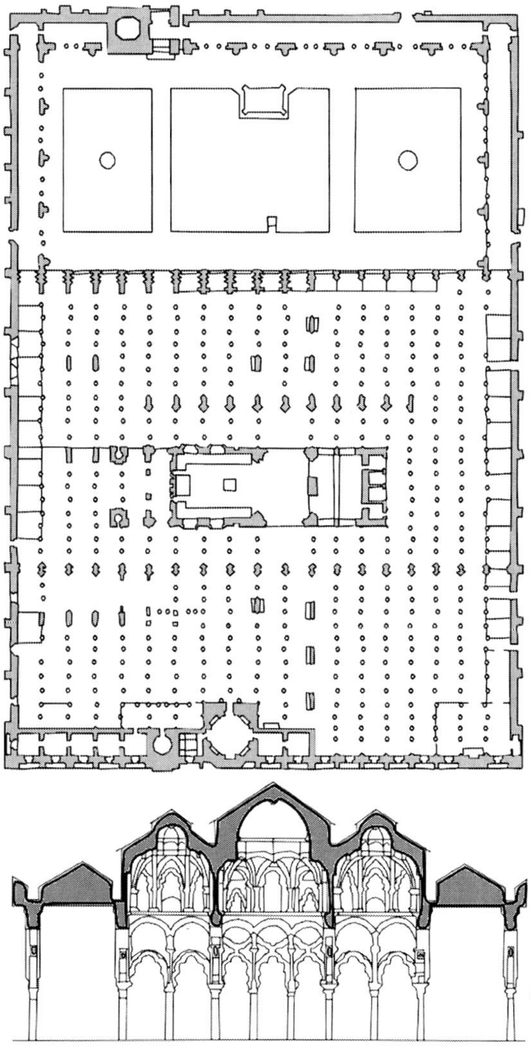

7.10 Plan and section of the arches at Cordoba Mosque (drawn with the help of URL8, 2009 and URL9, 2010)

each row need support for balancing the horizontal forces at their bases. For this reason, the outer walls of the mosque have some buttresses, which also affects the tectonics of the building facade.

All structural guidelines for arches are followed in Cordoba Mosque and the tectonic qualities of the arches are achieved without contravening structural guidelines. Thus, the relationship of the arches at Cordoba Mosque to structural guidelines is affirmative.

MASONRY VAULT

Masonry vaults can be studied under the same headings as masonry arches: form; structural behaviour; construction process; span of stone, brick and adobe vaults; structural guidelines for masonry vaults; and the examination of a case study. However, since there are many similarities with masonry arches, only the main differences for masonry vaults are explained.

Form of masonry vaults

There are various forms of vaults: the variation is especially wide because these three-dimensional structures can be added to each other to form more composite forms. The terminology concerning the forms of vaults can be found in Figure 7.11. A simple vault can be imagined as many arches added to each other along a linear axis. However, the masonry pieces of two adjacent arches should be interlocked (see Figure 7.12).

Structural behaviour of masonry vaults

The structural behaviour of masonry vaults is very similar to the structural behaviour of masonry arches. However, if they are raised above ground level, the two sides of the vault should be supported by beams, as seen in Figure 7.12.

The problem of horizontal forces in masonry vaults can be solved with the help of buttresses placed every few metres

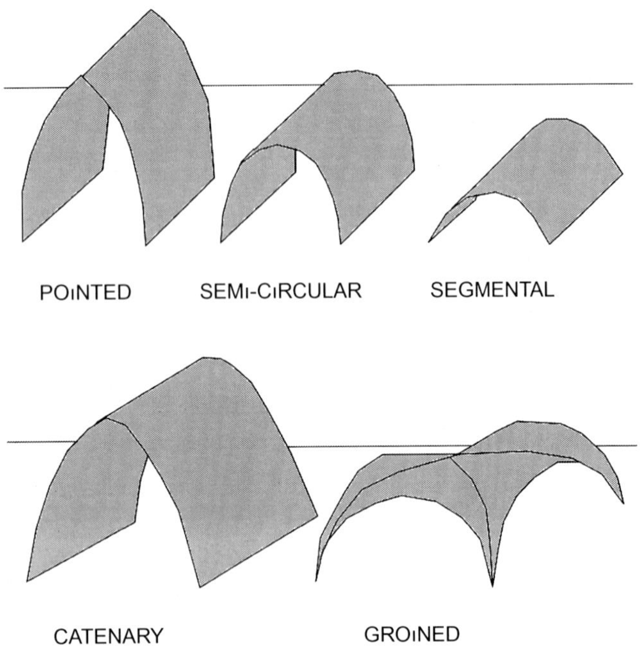

7.11 Forms of vaults

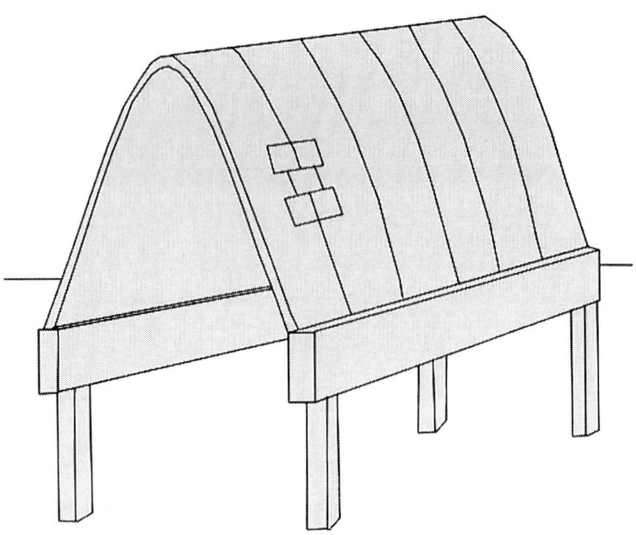

7.12 Raising the vault above ground level with the help of beams

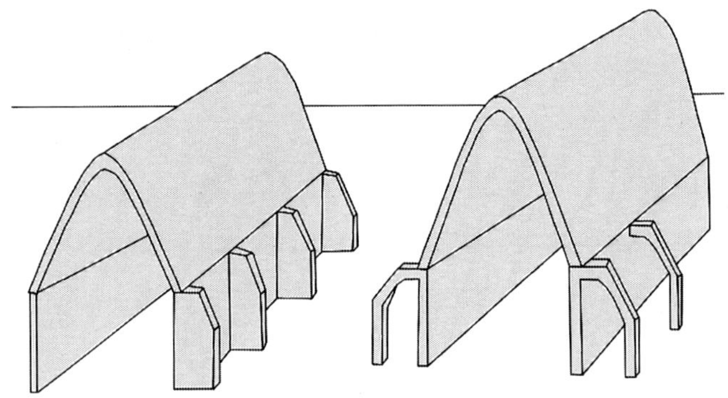

7.13 Solutions to horizontal force problems in masonry vaults

(such as every 4m) and flying buttresses, as seen in Figure 7.13. However, the use of tension rods to resist horizontal forces may not be recommended because they will compromise the quality of internal space.

Construction process of masonry vaults

The construction process of simple masonry vaults has similarities to the construction process of masonry arches. They can be built with or without centring. For example, corbelled vaults can be built without centring. Other types of simple vaults can be built with the help of sliding centring, as seen in Figure 7.14, instead of building a large temporary formwork for the whole vault. More complex forms of vaults, such as groined vaults, can be built by building ribs at the intersection of two vaults, as seen in Figure 7.15.

Span of masonry vaults

When existing masonry vaults are studied, it can be seen that they can span around 20–25m. The parabolic barrel vault of the Ctesiphon Palace, which was built in Iraq in AD 531 with fired bricks, spans 26m with a thickness of 1.5m and its height is 37m (URL4, 2013). The stone Mallorka (Palma) Cathedral, which was built in

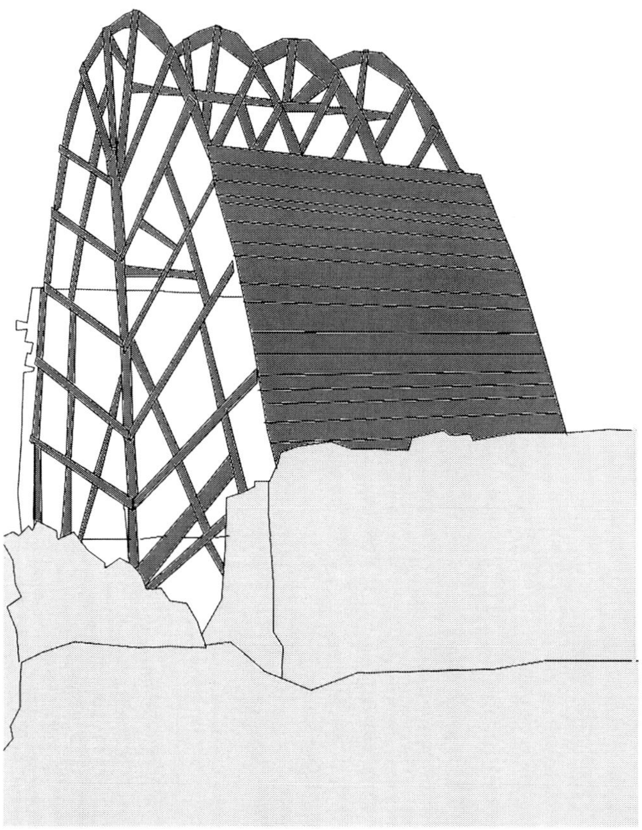

7.14 Construction of vaults with centering

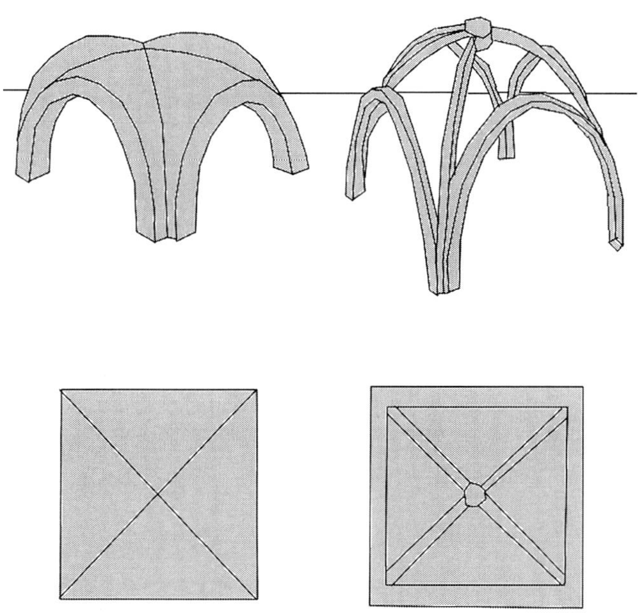

7.15 Vaults with and without ribs

1350 in Majorca, Spain, spans 17.8m with a thickness of 20cm (Roca, 2001). The stone vault of Girona Cathedral, which was built in the eleventh century in Catalonia, Spain, spans 22m (Roca, 2001).

Structural guidelines for masonry vaults

Structural guidelines for masonry vaults are listed in Table 7.2.

Table 7.2 Structural guidelines for masonry vaults and the associated value system

Structural guidelines	*Value system*
Positively curved forms can be preferred for masonry vaults.	Economy
Shallow vaults can be used for shorter spans (in comparison to high vaults) due to the effect of horizontal forces.	Safety
Masonry vaults can span up to 20–25m.	Economy
Problems of horizontal force can be solved by using buttresses, flying buttresses, etc.	Economy

Case study 8: The vaults at Cologne Cathedral, Germany

The stone vaults at Cologne Cathedral are illustrated in Figure 7.16. The main conceptual characteristic behind the design of the vaults is the emphasis made on verticality, height and the continuity of lines in the vertical direction. This contributes to the mystical atmosphere inside the building. The dominant physical entity determining the tectonic characteristics of the building is therefore structure.

The vaults at Cologne Cathedral are ribbed groined vaults. The span of the longer vaults is approximately 20m (URL5, 2013). The horizontal forces created by the vaults are taken by the flying buttresses outside the building. Thus, it can be stated that all guidelines for stone vaults have been followed at Cologne Cathedral: the vaults have an affirmative relationship to structural guidelines. (Note that Cologne Cathedral is also a case study related to stone masonry structures in chapter 6.)

7.16 A sketch of case study 8: the vaults at Cologne Cathedral, Cologne, Germany, 1248–1880 (drawn with the help of URL10, 2014)

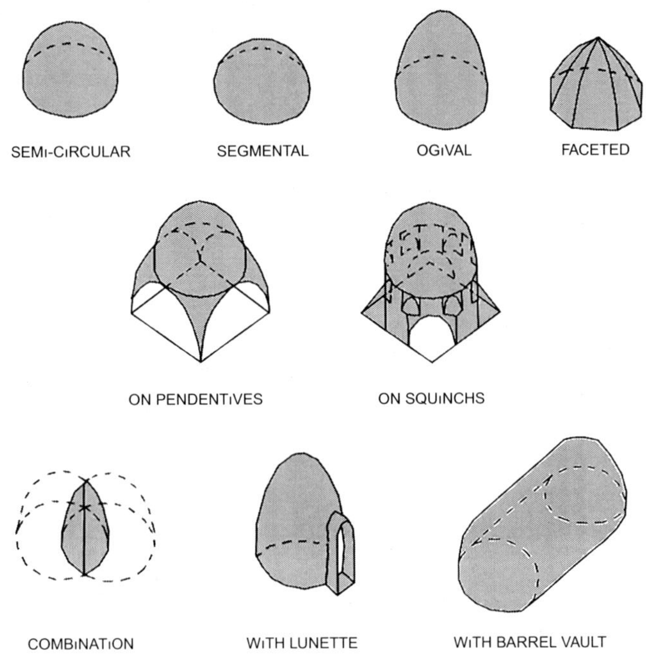

7.17 Forms of domes

MASONRY DOME

Similar to masonry arches and vaults, masonry domes can be studied under the headings of: form; structural behaviour; construction process; span of stone, brick and adobe domes; structural guidelines for masonry domes; and the case study examination. However, since there are many similarities with masonry arches, only the main differences from arches are explained here.

Form of masonry domes

A dome can be described as arches added to each other around a centre. However, in order to form a dome, the masonry pieces of two adjacent arches should be interlocked. There are various forms of domes: except for the categories of shallow and high domes, they can have similar forms to arches. The terminology concerning the forms of domes can be found in Figure 7.17.

Structural behaviour of masonry domes

There are two types of domes according to their structural behaviour. These are:

- Shallow domes
- High domes.

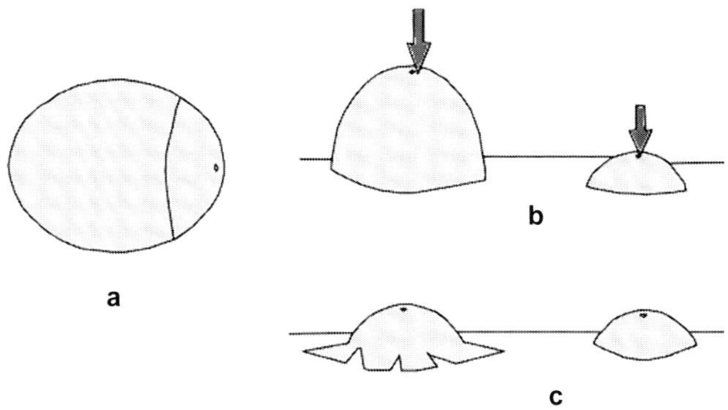

7.18 Understanding shallow and high domes with the help of an orange

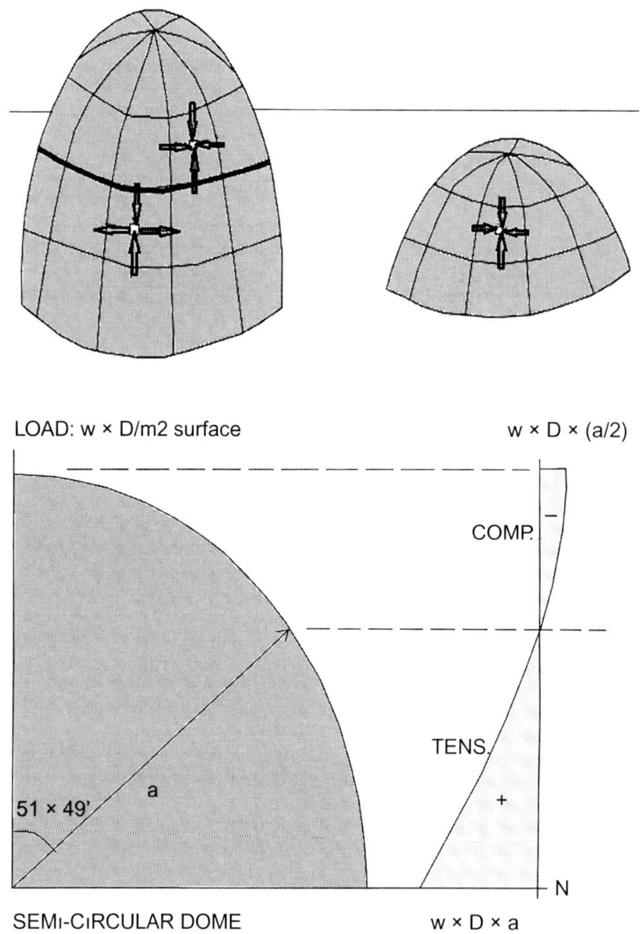

7.19 Stress types in high and shallow domes

In order to understand the difference between the structural behaviour of these two types of dome, an easy experiment can be carried out with a big orange. Cut the orange in order to achieve a shallow and a high dome as seen in Figure 7.18a. Squeeze the juice out of both orange domes. Then put them on the table and press them from the top as seen in Figure 7.18b. You will see that the two types of domes deform in a different manner. The shallow dome stands firm but the high dome splits around its base, as seen in Figure 7.18c.

A high dome consists of two parts: its upper part is a shallow dome, but it also has a bottom part. On the other hand, a shallow dome has only one part. Drawing meridians and parallels on a shallow and a high dome can help to understand what happens within these parts of the shallow and high domes. The internal forces in the parallels are known as hoop forces. The shallow parts in both domes are under compression in both the meridians and parallels, as seen in Figure 7.19. However, the bottom part of the high dome is under compression only in the meridians. It is under tension in the parallels. These tensile hoop forces in the parallels are the reason for the splitting at the base of the high dome. Figure 7.19 also shows that the line separating the shallow and the high parts of a semi-circular dome passes from 51.49° angle from the origin (Engineering Rome, 2014).

The weight of the dome is transferred to the building structure with the help of forces that are parallel to the bottom part of the dome. The magnitude of the horizontal component within these forces depends on the angle of the bottom part of the dome. The horizontal component should be balanced by an equal and opposite horizontal force, as seen in Figure 7.20. This can be realised by weight towers, buttresses, and half domes, etc. An example of the application of weight towers is the Hagia Sophia in Istanbul, Turkey. An example of the application of half domes is the Selimiye Mosque in Edirne, Turkey.

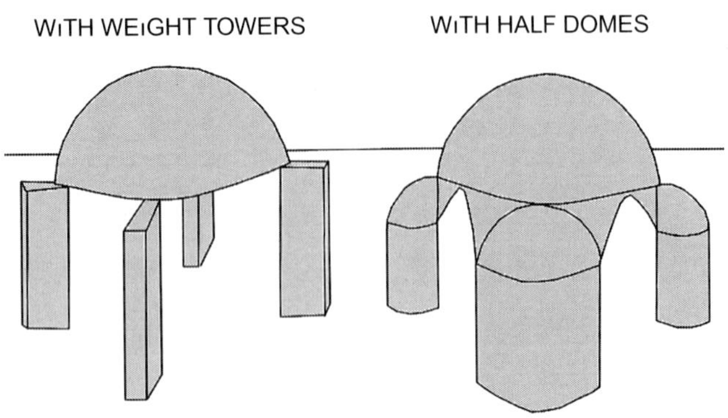

7.20 Horizontal force problems of domes and associated solutions

7.21 Construction of domes without centering

Construction process of masonry domes

The construction process of masonry domes has some similarities to the construction process of masonry arches and vaults. They can be built with or without centring. For example, corbelled domes can be built without centring. Shallow masonry domes can also be built without centring due to the presence of compressive hoop forces (see Figure 7.21). Using stone pieces with a special shape might help in fitting the pieces in the incomplete ring of the dome into the lower complete ring (Cipriani & Lau, n.d.). However, since the hoop forces at the bottom part of high domes are tensile, these parts can only be built with the help of centring (Lau, 2006). Other types of domes can be built with the help of full or partial centring (Lancaster, 2005). Domes can also be built by building ribs inside them, as seen in Figure 7.22.

7.22 A dome with ribs: the dome of San Lorenzo Church in Turin, Italy (drawn with the help of URL11, 2014)

Span of masonry domes

The old brick and adobe masonry domes can span around 35–45m. The largest dome is within the Baths of Caracalla, built in Rome, Italy, in AD 216, with clay hollow-ware, and which spans 35m (URL6, 2013). The brick dome of Hagia Sophia, which was built in Istanbul, Turkey, in AD 360, spans 30.3m (Parker, 2010). The brick dome of Santa Maria del Fiore, which was built in Florence,

Italy, between 1436 and 1881, spans 42m (Farfan, 2001). The adobe dome of Dhyanalinga Meditation Shrine, which was built in Tamil Nadu, India, in AD 1999, spans 22.2m with maximum thickness 53cm at the bottom and minimum thickness 21cm at the top (Auroville Earth Institute, n.d.). The stone dome of the Global Vipassana Pagoda, which was built in Mumbai, India, in AD 2006, spans 85.15m (Pachoriya et al., 2013).

Structural guidelines for masonry domes

Structural guidelines for masonry domes are listed in Table 7.3.

Table 7.3 Structural guidelines for masonry domes and the associated value system

Structural guidelines	*Value system*
Positively curved forms can be preferred for masonry domes.	Economy
Shallow domes can be used for shorter spans (in comparison to high domes) due to the effect of horizontal forces.	Safety
Precautions should be taken at the bottom parts of high domes due to the presence of tensile hoop loads.	Safety
Adobe and brick domes can span around 35–40m. Contemporary stone domes can span up to 85m.	Economy
Problems of horizontal force can be solved by using weight towers, buttresses, flying buttresses, etc.	Economy

Case study 9: The dome at Pantheon, Italy

The dome at Pantheon, as seen in Figures 7.23 and 7.24, was chosen as a case study due to its contribution to the tectonics of the building.

The Emperor Hadrianus wanted the Pantheon to house all the pagan gods and goddesses in order to achieve political power over the whole world. Therefore the main architectural concept behind the building concerned its dimensions. Every niche in the rotunda was dedicated to a pagan god or goddess. The building had a strong geometric order, which represented the power of the emperor (Sennett, 1996). This strong geometric order also continues in the dome. The dominant physical entity determining the tectonic qualities of the dome at Pantheon is structure.

The dome at Pantheon, which spans an extraordinary distance of 43.2m (Parker, 2009), is made up of unreinforced concrete

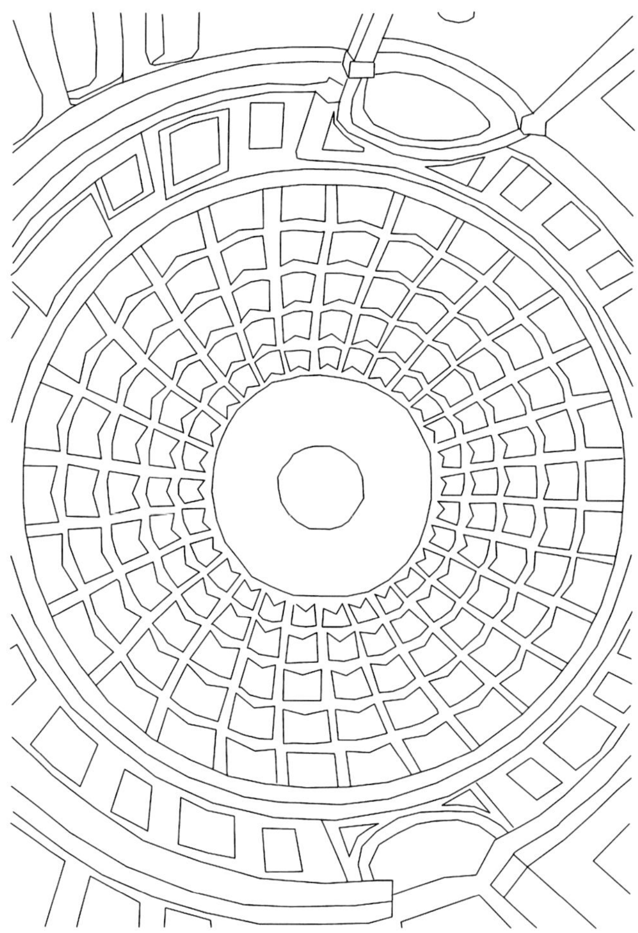

7.23 A sketch of case study 9: the dome at Pantheon, Rome, Italy, AD 126 (drawn with the help of URL12, 2010)

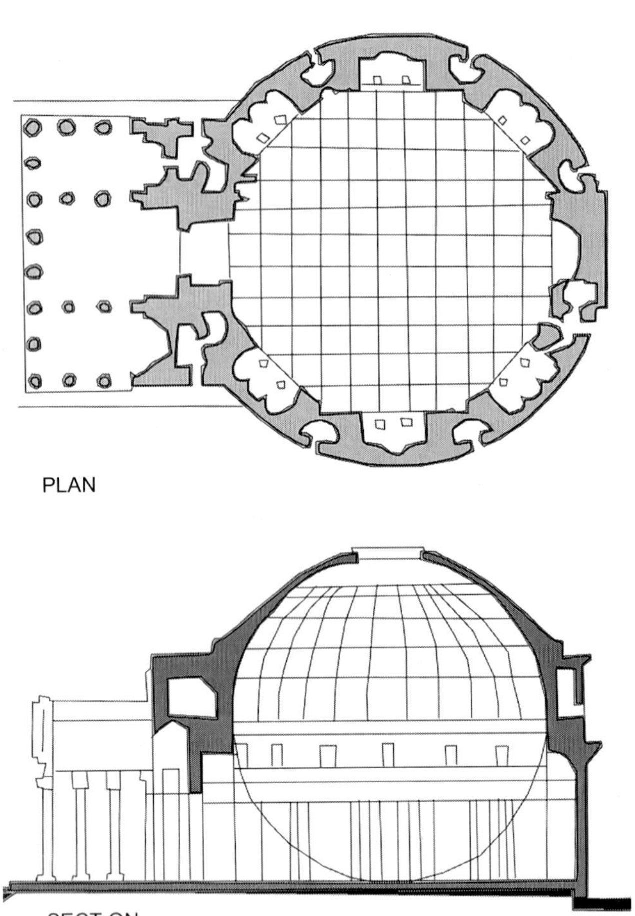

7.24 Plan and section of the dome at Pantheon (drawn with the help of URL13, 2014)

mixed with heavier and stronger materials at the bottom and weaker and lighter materials at the top. Travertine was mixed with concrete at the bottom, followed by concrete mixed with travertine and tufa, then mixed with tufa and brick, and, finally, brick was mixed with concrete at the top part of the dome (Parker, 2009; Moore, 1995). The thickness of the dome is approximately 7m at the base and 0.7m around the oculus (Building Big, 2001). Although this value seems unbelievable, studies on these sections of the building show it to be true. The lower half of the dome is cracked as a result of hoop tension because concrete is weak against tension and there is no reinforcing at the bottom part of the dome (Moore, 1995).

The dome contains five rows of 28 coffers, which makes it lighter and tectonically articulates the dome. It terminates with an oculus, which is the only source of light to the rotunda. This oculus was possible due to the presence of compressive hoop forces. These hoop forces mean that domes do not need elements such as the key-stones in arches, and can therefore be empty.

The dome rests on eight vaults at the base. The walls under these vaults act as buttresses that are perpendicular to the dome's circumference and thus react against the horizontal forces. These walls are necessary in order to support the high walls of the temple. The arrangement of the walls under the dome also adds to the tectonic qualities of the building. These walls provide articulated spaces and niches, besides taking the horizontal forces of the dome.

It can be stated that structural guidelines concerning the span of masonry domes and the height of masonry walls were not followed during the design of Pantheon. A very large dome was built in order to provide shelter for pagan gods. The materials used for the dome were therefore selected according to their combined strength and weight, although many coffers and a large oculus were provided to decrease the weight of the dome. Thus, it can be stated that the dome at Pantheon has a contravening relationship to structural guidelines.

CONCLUSIONS

The design of most of the masonry roof structures in this chapter follow structural guidelines concerning span, except for the dome at Pantheon. For most of the roof structures, tectonic qualities were achieved through following structural guidelines; however, technical innovation was at the forefront of the design of Pantheon. Contemporary developments in the area of reinforced masonry are reflected in the design of most contemporary masonry arches, vaults and domes.

REFERENCES

Auroville Earth Institute (n.d.) *Dome of the Dhyanalinga Meditation Shrine* (viewed 13 December 2013: www.earth-auroville.com/dhyanalinga_dome_en.php)

Building Big (2001) *Pantheon* (viewed 13 December 2013: www.pbs.org/wgbh/buildingbig/wonder/structure/pantheon.html)

Cipriani, B., Lau, W.W. (n.d.) *Construction Techniques in Medieval Cairo: The Domes of Mamluk Mausolea (AD 1250–1517)* (viewed 14 December 2013: www.arct.cam.ac.uk/Downloads/ichs/vol-1-695-716-cipriani.pdf)

Engineering Rome (2014) *Evolution of the Roman Dome* (viewed 31 March 2014: https://engineeringrome.wikispaces.com/Evolution+of+the+Roman+Dome)

Farfan, M.P. (2001) *Dome Structures: Santa Maria del Fiore (Florence)* (viewed 13 December 2013: www.arch.mcgill.ca/prof/sijpkes/arch374/winter2001/sfarfa/ensayo1.htm)

Lancaster, L.C. (2005) *Concrete Vaulted Construction in Imperial Rome – Innovations in Context*, Cambridge University Press: New York.

Lau, W.W. (2006) *Equilibrium Analysis of Masonry Domes*, unpublished Master thesis, Massachusetts Institute of Technology (viewed 13 December 2013: https://dspace.mit.edu/bitstream/handle/1721.1/34984/71791712.pdf?sequence=1)

Moore, D. (1995) *The Pantheon* (viewed 13 December 2013: www.romanconcrete.com/docs/chapt01/chapt01.htm)

Pachoriya, M., Namdeo, N., Bajaj, R. (2013) *Global Pagoda – Mumbai* (viewed 13 December 2013: www.slideshare.net/RahulBajaj9/global-pagoda-mumbai)

Parker, F. (2009) *The Pantheon – Rome* (viewed 13 December 2013: www.monolithic.com/stories/the-pantheon-rome-126-ad)

Parker, F. (2010) *Hagia Sophia in Istanbul, Turkey* (viewed 13 December 2013: www.monolithic.com/stories/hagia-sophia-in-istanbul-turkey)

Roca, P. (2001) 'Studies on the structure of gothic cathedrals' in eds. P.B. Lourenco & P. Roca, *Historical Constructions*, University of Guimarães: Portugal.

Sennett, R. (1996) *Flesh and Stone: The Body and the City in Western Civilization*, W.W. Norton & Company: New York.

URL1 (2010) *List of Longest Masonry Arch Bridge Spans* (viewed 4 March 2014: www.infosources.org/what_is/List_of_longest_masonry_arch_bridge_spans.html)

URL2 (2004) *Over Bridge* (viewed 4 March 2014: www.english-heritage.org.uk/daysout/properties/over-bridge/)

URL3 (n.d.) *Isambard Kingdom Brunel Portal* (viewed 7 December 2013: www.ikbrunel.org.uk/index.php?id=1)

URL4 (2013) *Ctesiphon – Fallen City of the Sassanid Kings* (viewed 4 March 2014: http://slingsandarrowsblog.blogspot.com.tr/2012/12/ctesiphon-fallen-city-of-sassanid-kings.html)

URL5 (2013) *Cologne Cathedral* (viewed 4 March 2014: www.princeton.edu/~achaney/tmve/wiki100k/docs/Cologne_Cathedral.html)

URL6 (2013) *Thermae of Caracalla* (viewed 13 December 2013: www.greatbuildings.com/buildings/Thermae_of_Caracalla.html)

URL7 (2014) *Caliphate of Cordoba* (viewed 26 September 2014: http://en.wikipedia.org/wiki/Caliphate_of_C%C3%B3rdoba)

URL8 (2009) *Cordoba la Mezquita* (viewed 26 September 2014: http://67daniel.blogspot.com.tr/2009/01/al-andalus-cordoba-la-mezquita.html)

URL9 (2010) *Cordoba Mosque* (viewed 26 September 2014: www.infocordoba.com/spain/andalusia/cordoba/info/mosque/mosque_visitor_guide.htm)

URL10 (2014) *Cologne Cathedral – Vault* (viewed 26 September 2014: http://faeriedivine.deviantart.com/art/Cologne-Cathedral-vault-45457545)

URL11 (2014) *Dome of the Baroque Church of San Lorenzo in Turin* (viewed 26 September 2014: www.shutterstock.com/pic-30721420/stock-photo-dome-of-the-baroque-church-of-san-lorenzo-in-turin-torino.html)

URL12 (2010) *Rome* (viewed 26 September 2014: www.pontuali.com/marco/en/tours/rome/473-roma-barocca-segreti-storie-e-legende-in-mezza-giornata-en-gb-1.html)

URL13 (2014) *Roman Art Test* (viewed 26 September 2014: www.studyblue.com/notes/note/n/roman-art-test-2/deck/748882)

8
The Tectonics of Contemporary Approaches to Masonry Structures

What makes contemporary masonry different to traditional masonry is the application of steel reinforcement between the masonry units. This technology is known as reinforced masonry. It can be argued that the presence of steel and reinforced concrete within this technology means that it should not be termed masonry. This chapter contains stone, brick and adobe reinforced masonry and hybrids of reinforced masonry and frames.

TYPES OF REINFORCED MASONRY

If masonry is reinforced with steel and cement grout or concrete, the outcome is called reinforced masonry. J. Ambrose (1991) classifies reinforced masonry into two main groups:

- Reinforced grouted masonry
- Reinforced hollow unit masonry.

Reinforced grouted masonry is formed by leaving a cavity between the masonry units and filling this with steel reinforcement and grout. Reinforced hollow unit masonry, however, is formed by using special masonry units that provide continuous vertical and horizontal cavities within the wall. These cavities are filled with steel reinforcement and concrete (Ambrose, 1991: pp.30–31).

Fodi and Bodi (2011) make another classification for reinforced masonry. According to them, the three types of reinforced masonry are as follows:

- Putting horizontal mesh or zig-zag reinforcement into the horizontal layer of mortar between masonry units in every three rows of units.
- Putting continuous vertical reinforcement within the mortar between the units and arranging the units in a manner that provides continuity to this reinforcement.
- Leaving vertical cavities between masonry units and filling these with reinforcement and concrete to form vertical tie-beams.

This book summarises the classification of reinforced masonry into three divisions, as seen in Figure 8.1:

- Filling the cavity wall with reinforced concrete.
- Placing reinforcement between the masonry pieces.
- Using vertical tie-beams (confined masonry).

It is clear that there are many different types of applications of reinforced masonry.

The strength of reinforced masonry with vertical tie-beams depends on how frequently the vertical tie-beams are placed

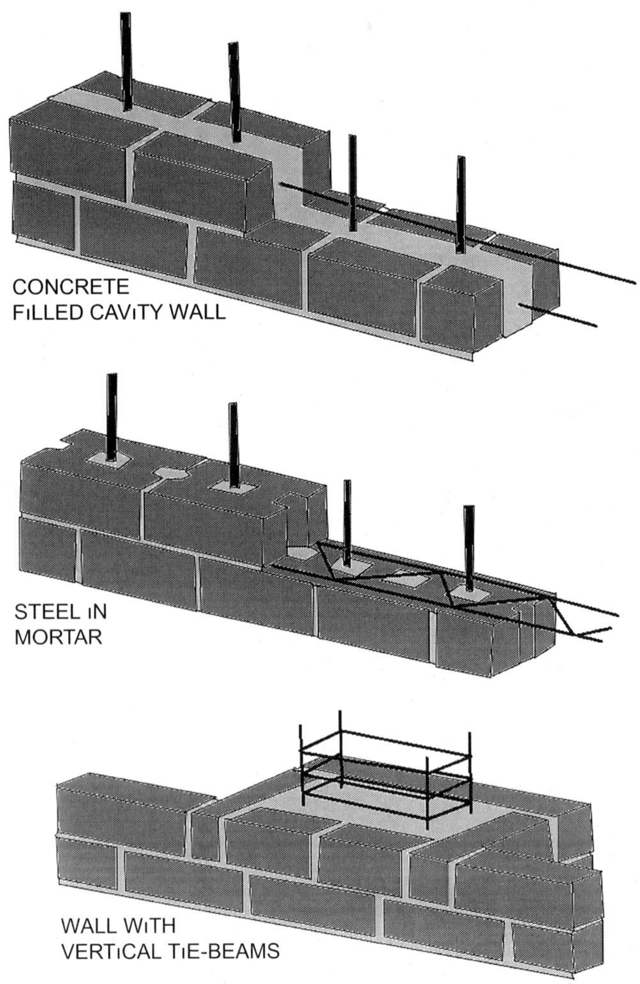

8.1 Types of reinforced masonry

within the masonry walls. They can be at the main corners of the building only, at the intersections of all the walls, or at all wall intersections plus at the two sides of all openings. The latter of the three is stronger in comparison to the first two strategies. However, the type of reinforced masonry in which reinforced concrete is used for filling the cavity between masonry units is the strongest.

The construction of the first and third types of reinforced masonry is realised by building the masonry walls first and then introducing the reinforced concrete. Except for the difference in construction process, the third type of reinforced masonry is very similar to reinforced concrete frame systems.

FLEXIBILITY PROVIDED BY REINFORCED MASONRY

Unlike unreinforced masonry, the flexibility that can be achieved by using reinforced masonry depends upon the following factors:

- The type of reinforced masonry.
- The building code requirements of the relevant country.

The flexibility of the architectural design is dependent upon the amount of reinforced concrete elements within the reinforced masonry. Building codes can be classified into two groups relating to flexibility of design in reinforced masonry structures:

- The building codes that do not bring any physical restrictions but require mathematical analysis of the structure as evidence of its strength.
- The building codes that bring physical restrictions in addition to requiring mathematical analysis of the structure.

The first group of building codes, such as those from the USA and the European Union, provide the maximum flexibility that can be achieved through architectural design supported by engineering analysis (ACI 530-02/ASCE 5-02/TMS 402-02, 2002; Eurocodes Committee, 2004). However, if there are other restrictions brought by the building codes, then the flexibility of the structure becomes dependent on that building code. Codes can restrict the dimensions and number of openings as well as the type of reinforced masonry that can be used. A building code that imposes many restrictions can easily reduce the advantages of reinforced masonry in comparison to traditional masonry.

Although the author of this book does not defend restrictive building codes, she thinks that the contradiction between the restrictive and flexible building codes represents the nature of freedom in design. It is better to know the restrictions in order to imagine the possible flexibilities. Thus, structural guidelines for reinforced masonry are mainly based on the restrictions that exist in various building codes, and present a conservative approach to the architectural design of reinforced masonry.

The restrictions brought by the building codes to the design of reinforced masonry buildings can be categorised into four groups:

- Eurocodes do not suggest the use of unreinforced masonry in areas with high seismic activity (Eurocodes Committee, 2004).
- Chinese building codes suggest the height of a reinforced masonry building can be up to four storeys if frequent vertical tie-beams are used in areas with high seismic activity. The height can rise to eight storeys if the seismic risk is low (Tsionis et al., 2010). Eurocodes, however, limit height to one storey for confined masonry and two storeys for cavity-type reinforced masonry in areas with high seismic activity (Eurocodes Committee, 2004).
- Turkish building codes suggest the wall length in plan can be up to 16m with the use of vertical tie-beams at all wall intersections and at the two sides of all openings (Ministry of Public Works and Settlement Government of the Republic of Turkey, 2007). This brings a considerable increase to the dimensions of spaces within a masonry building.
- Turkish building codes suggest the ratio of openings can be increased to 20% and the distance between the openings can be decreased to 0.5m for reinforced brick and stone masonry structures. For reinforced adobe masonry however, the distance between openings can be 0.8m (Ministry of Public Works and Settlement Government of the Republic of Turkey, 2007).

This means that the use of more reinforced concrete elements with more reinforcement increases flexibility. The use of reinforced masonry brings some flexibility to the dimensions of spaces and the number of openings compared to unreinforced traditional masonry.

STRUCTURAL GUIDELINES FOR REINFORCED MASONRY STRUCTURES

On the basis of the information above, it is possible to define a conservative set of structural guidelines concerning the use of reinforced stone, brick and adobe structures. The three groups of guidelines, which are presented in Tables 8.1, 8.2 and 8.3, are based on the tables of traditional masonry types outlined in chapter 6 of this book. It is also possible to increase the flexibility of timber masonry structures by incorporating some steel elements to make reinforced timber masonry.

HYBRIDS OF REINFORCED MASONRY AND REINFORCED CONCRETE FRAME SYSTEMS

In order to increase the design flexibility of reinforced masonry, this system can be integrated with reinforced concrete frame systems. Figure 8.2 shows how a reinforced masonry wall can be connected to reinforced concrete columns and beams with the help of reinforced concrete vertical and horizontal tie-beams.

As seen in Figure 8.2, the reinforced stone masonry wall contains reinforced concrete vertical tie-beams. On the top of the wall there is a reinforced concrete horizontal tie-beam. The plan dimensions of the vertical tie-beams are minimum 20cm × 20cm, because this is the minimum required dimension for a worker to put his/her hands into the cavity in order to place reinforcement. The depth of reinforced concrete horizontal tie-beams can also be 20cm. However, the dimensions of the columns and beams of the frame system will be different to the dimensions of the tie-beams. Columns and beams will be designed according to structural guidelines for reinforced concrete frame systems.

Table 8.1 Conservative structural guidelines for reinforced stone masonry structures and the associated value system

Structural guidelines	*Value system*
Minimum thickness of a stone wall can be around 35–50 cm.	Safety
Continuous reinforced concrete foundations should be used under reinforced stone walls.	Safety
Tie-stones should be used at regular intervals to connect two sides of the rubble stone walls.	Safety
Stone pieces should be organized in such a way that they avoid forming continuous vertical joints.	Safety
Stone pieces should be organized so that they form horizontal layers every 150cm.	Safety
Length of reinforced stone walls between two wall intersections can be a maximum of around 16m, if vertical reinforced concrete tie-beams are used at all wall intersections and at the two sides of all openings.	Safety
Plan of the building should be symmetrical or close to symmetrical in order to avoid twisting due to earthquakes.	Economy
Unsupported height of a reinforced stone wall can be a maximum of around 3m.	Economy
Reinforced stone masonry buildings can be around two storeys high in earthquake regions and four storeys high if there is no earthquake risk.	Safety
All stone walls must be supported by other stone walls or foundations.	Safety
Total length of openings on a stone wall can be a maximum of around 60% of the length of the wall.	Safety
Maximum opening size should be around 3m.	Safety
Distance of openings from the corner of the building can be a maximum of around 1.5m in earthquake regions and 1m if there is no earthquake risk.	Safety
Openings can be a minimum of 0.5m away from wall intersections.	Safety
Distance between two openings can be a minimum of around 0.5m.	Safety
Openings can have lintels or arches over them.	Safety
Reinforced stone walls should be connected to each other with reinforced concrete horizontal tie-beams at every floor level.	Safety
Thickness of reinforced concrete horizontal tie-beams can be around 20cm.	Safety

Table 8.2 Conservative structural guidelines for reinforced brick masonry structures and the associated value system

Structural guidelines	*Value system*
Only load-bearing bricks should be used for building load-bearing brick walls. Non-load-bearing bricks can be used only in non-load-bearing partition walls.	Safety
Minimum thickness of a load-bearing brick wall can be 20cm, and should be increased if the building has more than two levels. Minimum thickness of a non-load-bearing brick wall can be 10cm.	Safety
Continuous reinforced concrete foundations should be used under reinforced brick walls.	Safety
Length of reinforced brick walls between two wall intersections can be a maximum of around 16m, if vertical reinforced concrete tie-beams are used at all wall intersections and at the two sides of all openings.	Safety
Plan of the building should be symmetrical or close to symmetrical in order to avoid twisting due to earthquakes.	Economy
Unsupported height of a reinforced brick wall can be around 3m.	Economy
Reinforced brick masonry buildings can be a maximum of two storeys high in earthquake regions and four storeys high if there is no earthquake risk.	Safety
All load-bearing brick walls must be supported by other load-bearing walls or foundations.	Safety
Total length of openings on a brick wall can be a maximum of around 60% of the length of the wall.	Safety
Maximum opening size should be around 3m.	Safety
Distance of openings from the corner of the building can be a minimum of around 1.5m in earthquake regions and 1m if there is no earthquake risk.	Safety
Openings can be 0.5m away from wall intersections.	Safety
Distance between two openings can be a minimum of around 0.5m.	Safety
Openings can have lintels or arches over them.	Safety
Reinforced brick walls should be connected to each other with reinforced concrete horizontal tie-beams at every floor level.	Safety
Thickness of reinforced concrete horizontal tie-beams can be a minimum of around 20cm.	Safety

Table 8.3 Conservative structural guidelines for reinforced adobe masonry structures and the associated value system

Structural guidelines	*Value system*
Minimum thickness of a load-bearing adobe exterior wall can be around 29cm, and should be increased up to 35cm if the building has more than one level. Minimum thickness of interior load-bearing adobe walls can be around 19cm. Slenderness ratio of adobe walls should also be considered when determining thickness.	Safety
Continuous reinforced concrete foundations should be used under reinforced adobe walls.	Safety
Length of reinforced adobe walls between two wall intersections can be a maximum of around 16m, if vertical reinforced concrete tie-beams are used at all wall intersections and at the two sides of all openings.	Safety
Plan of the building should be symmetrical or close to symmetrical in order to avoid twisting due to earthquakes.	Economy
Unsupported height of a reinforced adobe wall can be a maximum of around 2.7–3.3m.	Economy
Reinforced adobe masonry buildings can be one storey high in earthquake regions and two storeys high if there is no earthquake risk.	Safety
Total length of openings on an adobe wall can be a maximum of around 40% of the length of the wall.	Safety
Maximum opening size should be around 1m if the opening has a lintel. If there is an arch, maximum opening size can increase to 1.8m.	Safety
Distance of openings from the corner of the building can be a minimum of around 1m.	Safety
Openings can be 0.5m away from wall intersections.	Safety
Distance between two openings can be a minimum of around 0.8m.	Safety
Openings can have lintels or arches over them.	Safety
Reinforced adobe walls should be connected to each other with reinforced concrete horizontal tie-beams at every floor level.	Safety
Thickness of reinforced concrete horizontal tie-beams can be around 20–25cm.	Safety
A lightweight roof structure can be selected for adobe masonry buildings.	Safety

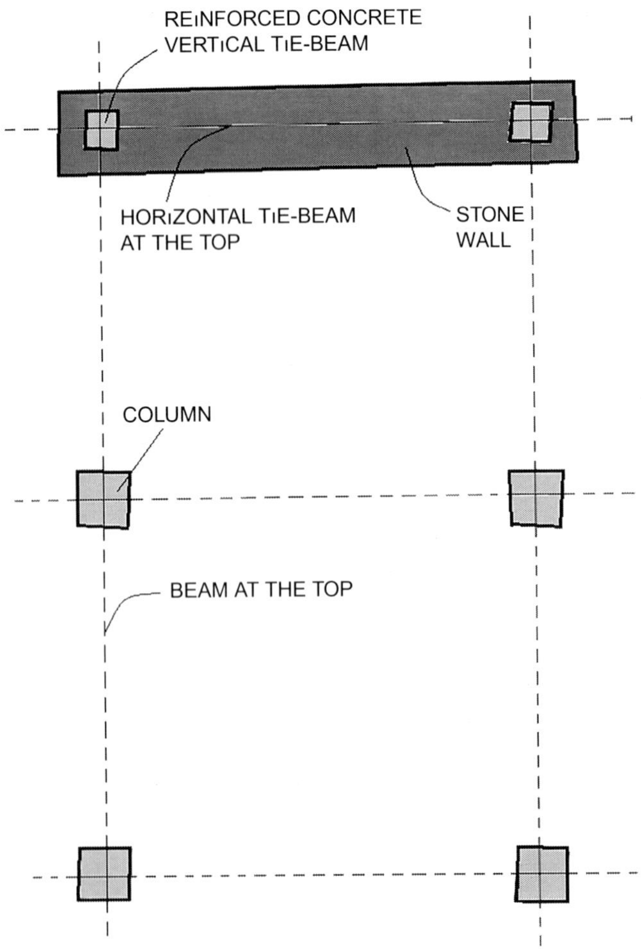

8.2 Integration of frame elements with a reinforced masonry structure

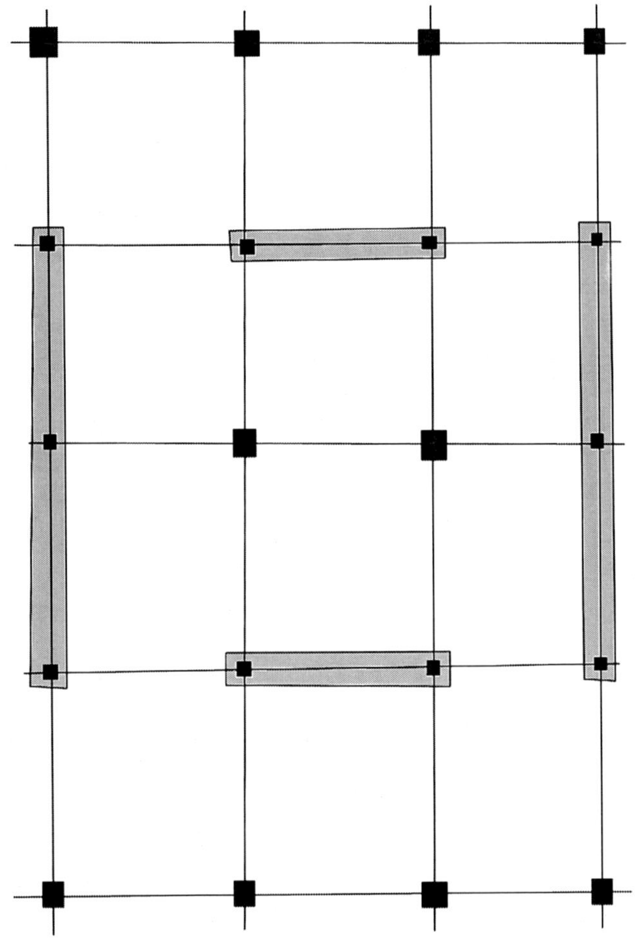

8.3 An example of balanced distribution in reinforced masonry walls within a hybrid system

There will be interconnecting forces between the two systems. According to the American building code concerning masonry structures: 'masonry walls shall not be connected to structural frames unless the connections and walls are designed to resist design interconnecting forces and to accomodate calculated deflections' (ACI 530-02/ASCE 5-02/TMS 402-02, 2002).

Another problem arising with the use of this structural system is twisting due to earthquake loads. In order to avoid this problem, the structure should be either symmetrical or close to symetrical. Thus, the reinforced masonry walls in both orthogonal directions in a masonry and frame hybrid system should be evenly distributed on the plan. This means that if there is only one reinforced masonry wall in this system, it should be close to the centre of gravity. If there are more reinforced masonry walls, they should be placed as couples balancing the effect of each other. Figure 8.3 shows an example of balanced distribution of reinforced masonry walls within a hybrid system. The subject of twisting instability due to earthquakes is studied further in chapter 9.

Structural guidelines for reinforced masonry and reinforced concrete frame hybrid systems are listed in Table 8.4.

A hybrid structure of reinforced masonry and reinforced concrete frame can have much larger openings in comparison to any other masonry structure. It can accommodate much larger spaces and it can be much higher than other masonry structures.

HYBRIDS OF REINFORCED MASONRY AND STEEL FRAME SYSTEMS

By replacing reinforced concrete vertical and horizontal tie-beams with steel frame elements, reinforced masonry can be integrated with steel frames. In this hybrid structure, the masonry walls are able to ensure strength against lateral loads by acting as bracing (Lashway & Throop, 2008).

An early example of a stone masonry and iron frame hybrid structure is W. Strutt and E. Darvin's Calico Mill, which was built in 1792 in Derby, the UK. Later, E.E. Viollet-le-Duc used iron frame

Table 8.4 Structural guidelines for reinforced masonry and reinforced concrete frame hybrid systems and the associated value system

Structural guidelines	*Value system*
Reinforced masonry parts of the structure should be designed according to the structural guidelines for the related reinforced masonry systems.	Safety
Reinforced concrete frame parts of the structure should be designed according to the structural guidelines for reinforced concrete frame systems.	Safety
Structural continuity between the two structures will be provided by connecting the reinforced concrete elements in both systems to each other.	Safety
Connecting parts between the two systems should be designed to resist interconnecting forces and to accommodate calculated deflections.	Safety
Reinforced concrete vertical tie-beams should be minimum 2cm × 20cm in plan.	Workmanship
Reinforced concrete horizontal tie-beams should be minimum 2cm deep.	Safety
Reinforced masonry walls in both orthogonal directions should be evenly distributed on the plan in order to avoid twisting instability problems due to earthquake loads.	Safety

elements with stone masonry in the 1850s (Turan, 2010). Steel frames started to be used together with masonry cladding during the 1880s and 1890s at the Chicago School of Architecture, USA. This approach provided fire protection for steel and enabled the presence of larger openings in masonry (Billington, 1985). The Empire State Building, which was built in New York, USA, in 1931, is 103 storeys and 443m high and has a steel frame structure combined with brick masonry walls (Nacheman, 2006). Today, there are many examples of combining steel frames with masonry structures in which the masonry parts play a structural role.

CASE STUDY 10: VILLA MÜLLER, CZECH REPUBLIC

The two case studies in this chapter are hybrid structures combining reinforced masonry systems and reinforced concrete frame systems. Together with reinforced concrete frame systems, the first case study has a reinforced brick masonry system and the second case study has a reinforced stone masonry system.

Adolf Loos' Villa Müller in Prague, as seen in Figures 8.4 and 8.5, combines brick masonry with a reinforced concrete frame system. The building was chosen as a case study because it looks like a brick masonry building from the outside, but it has large and

8.4 A sketch of case study 10: Villa Müller, Prague, Czech Republic, 1928 (drawn with the help of URL2, 2013)

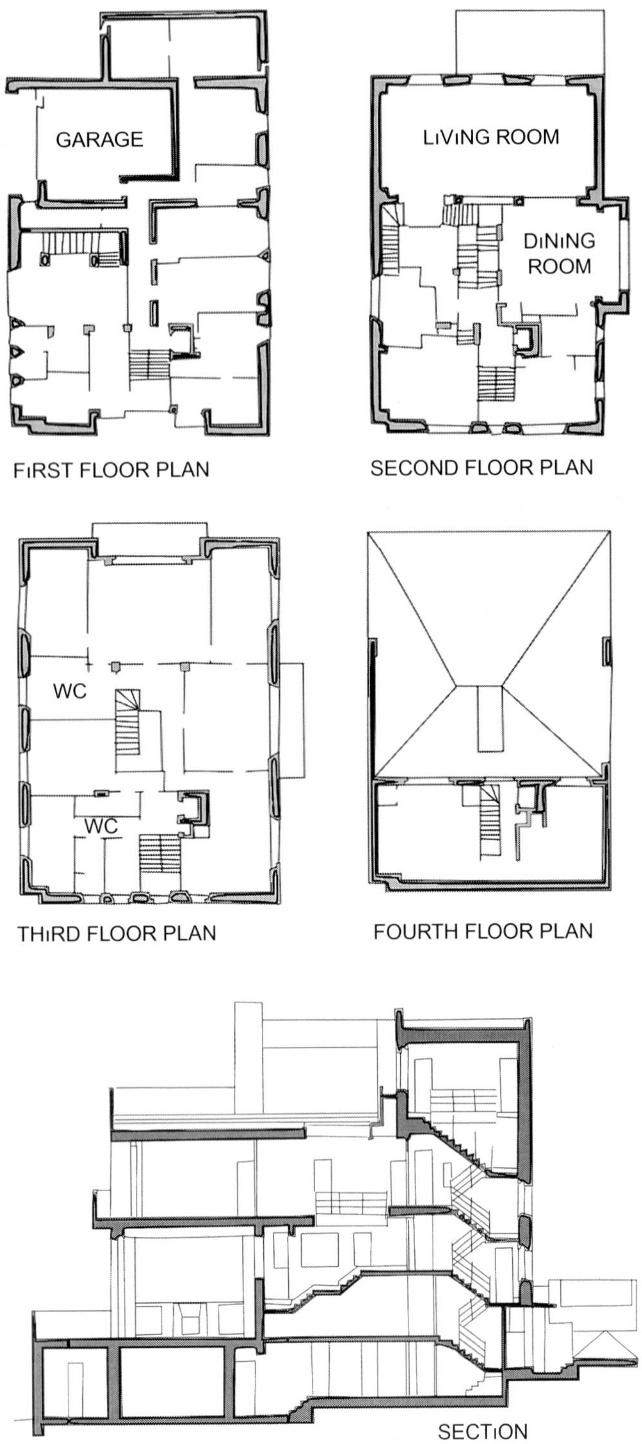

8.5 Plans and section of Villa Müller (drawn with the help of URL3, 2011)

continuous spaces inside. The plastered white brick masonry walls surround the building, whilst four reinforced concrete columns inside (one attached to the elevator shaft) help in the organisation of split levels. This is Loss' architectural idea of 'Raumplan'. However, it is also known that Loos used non-structural columns and beams for aesthetic purposes (Jara, 1995).

The main architectural idea of Villa Müller is to have a house in the form of a simple white box. Continuity between internal spaces is provided. Thus, the internal space reflects the tectonic characteristics of modern structural systems. The dominant physical entity which determines the tectonic characteristics of Villa Müller is structure.

When the masonry envelope around Villa Müller is analysed according to structural guidelines for reinforced brick masonry structures, it can be seen that the height of the building is above the limit of two storeys for reinforced brick masonry structures in earthquake zones. However, the building is in Prague, an intermediate earthquake zone (Solomos et al., 2008). The building is located on a steep topography and has three levels on one side and four on the other side. The length of the largest walls surrounding the indoor space is around 15m, which is not above the limit for reinforced brick masonry structures.

The reinforced concrete frame inside the building envelope is a simple regular frame, and thus does not present any problems according to structural guidelines for reinforced concrete frame systems. Although the author of this book thinks that there should be reinforced concrete tie-beams within the brick exterior walls, there is no information about this. The structure is very close to symmetrical in plan: thus, there is no danger of twisting due to earthquakes. There is an 'asymmetrical symmetry' both in the organisation of the facades and the plans of this building. This approach enabled the architect to combine the modern asymmetry with structural safety as regards twisting instability due to earthquakes. Since Villa Müller follows all structural guidelines whilst achieving its tectonic qualities, it has an affirmative relationship with structural guidelines.

CASE STUDY 11: FALLING WATER, USA

Frank Lloyd Wright's Falling Water in Pennsylvania, as seen in Figures 8.6 and 8.7, combines stone masonry with a reinforced concrete frame system. The building was chosen as a case study because, unlike Villa Müller, the hybrid nature of its structure is perceived both externally and internally. Both the thick stone walls and the frame elements, such as columns and cantilevers, are clearly seen. This is a result of Wright's wish to integrate indoor spaces with nature. The stone walls and the large cantilevers make a tectonic contrast to each other. The stone walls extend upwards from the rocks, whilst large cantilevers are parallel to the earth. Providing this contrast between horizontals and verticals is the main architectural concept of the building. The arrangement of verticals and horizontals provides continuity with the rocky natural environment and falling water. The dominant physical entity in achieving Falling Water's tectonic characteristics is topography.

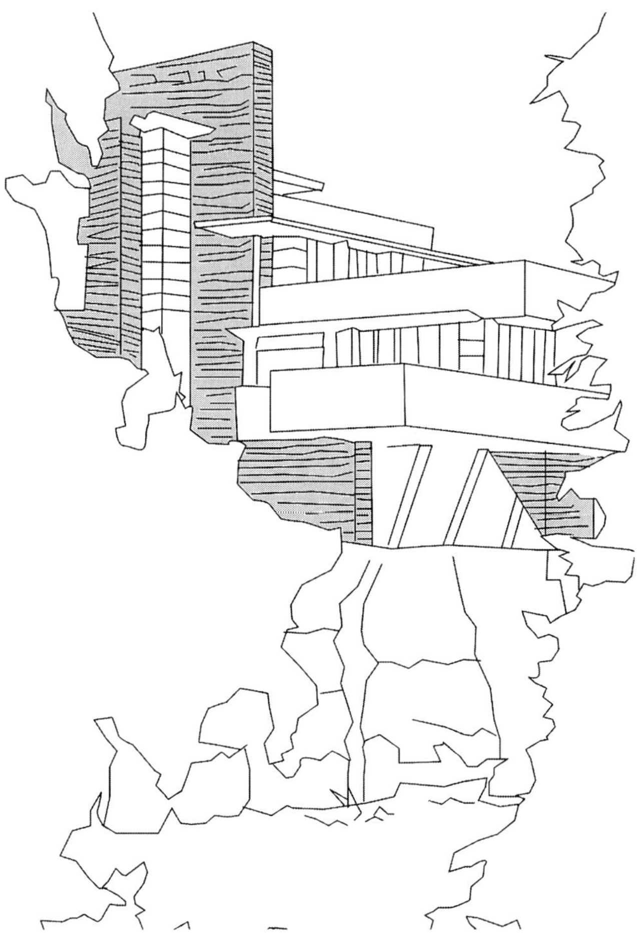

8.6 A sketch of case study 11: Falling Water, Pennsylvania, USA, 1935 (drawn with the help of URL4, 2014)

Falling Water is in Pennsylvania, a low seismic activity area (URL1, 1997), and this allows a design freedom not available in high seismic activity areas. The first striking feature of the structure is its asymmetrical organisation. Stone walls are in one area and cantilevers and reinforced concrete columns are in another. This organisation is weak against twisting instability, and so is unsuited to high seismic activity areas.

According to structural guidelines, the openings between the stone walls are not far enough from the corners of the walls in all facades of the building. The height of these openings is also not common for stone walls. Although it cannot be read from the plans, the author of this book believes that the walls have reinforced concrete tie-beams in them at strategic locations. Also the horizontal tie-beams, which connect the stone walls to the structure of slabs, should have been hidden for the tectonic purpose of not disturbing the verticality of stone walls.

When the reinforced concrete frame in Falling Water is analysed, the 5m and 6m long cantilevers are the most striking element. Many codes, such as the Turkish building code, limit the use of cantilevers to 1.5m (Ministry of Public Works and Settlement Government of the Republic of Turkey, 2007). This design freedom has resulted in large deflections at the cantilevers of Falling Water (Feldman, 2005). The design of Falling Water therefore has a contravening relationship with structural guidelines.

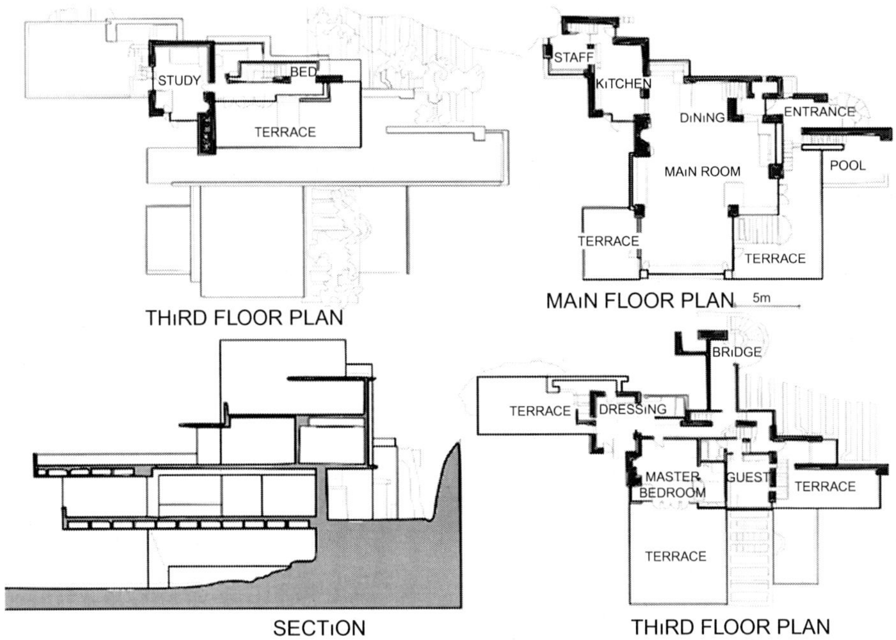

8.7 Plans and section of Falling Water (drawn with the help of Hernandez, 2014)

CONCLUSIONS

The case studies presented in this chapter illustrate two distinct approaches for when the design of the building structure is the main concern. Villa Müller conforms to structural guidelines, whilst the structure of Falling Water does not always follow them. Falling Water makes us think that the relationship between architecture and structural guidelines might depend upon the level of seismic risk.

Another discussion arises due to the presence of two different types of building codes. One is rather conservative by bringing physical limits to the design of contemporary masonry structures, and the other is more liberal by making any design, as long as it is mathematically analysed, possible. Since both of these two types of building codes exist for areas with high seismic risk, the difference between them cannot be related to this risk factor.

REFERENCES

ACI 530-02/ASCE 5-02/TMS 402-02 (2002) *Building Code Requirements for Masonry Structures*, Masonry Standards Joint Committee (viewed 5 September 2014: https://engineering.purdue.edu/~ramirez/CE479/FALL05/MasonryBuildingCode1-3-02.pdf)

Ambrose, J. (1991) *Simplified Design of Masonry Structures*, John Wiley and Sons: New York.

Billington, D.P. (1985) *The Tower and the Bridge: The New Art of Structural Engineering*, Princeton University Press: New Jersey.

Eurocodes Committee: Scientific and Technical Chamber of Cyprus under a Ministry of Interior's Program (2004) CYS Annex to CYS EN 1998-1. *Eurocode 8: Design of Structures for Earthquake Resistance* (viewed 20 October 2013: www.cys.org.cy/images/public/eurocodes)

Feldman, G.C. (2005) 'Fallingwater is No Longer Falling' *Structure Magazine*, September: pp.46–50.

Fodi, A., Bodi, I. (2011) 'Basics of Reinforced Masonry' *Concrete Structures*, Vol. 12 (viewed 26 October 2013: http://fib.bme.hu/cs2011/vb2011%20angol%20%20FODI-BODI%20page69-77.pdf)

Hernandez, J.M. (2014) *Falling Water, Mill Run* (viewed on 26 September 2014: www.jmhdezhdez.com/2013/05/fallingwater-house-frank-lloyd-wright.html)

Jara, C. (1995) 'Adolf Loos's Raumplan Theory' *Journal of Architectural Education*, Vol. 48, No. 3: pp.185–201.

Lashway, K., Throop, D. (2008) 'Masonry and Steel' *The Construction Specifier*, August: pp.76–81.

Ministry of Public Works and Settlement Government of the Republic of Turkey (2007) *Seismic Performance Evaluation of Dual Reinforced Concrete Systems Design According to Turkish Seismic Code*, trans. E.Y. Karcı (viewed 26 October 2014: www.belgeler.com/blg/22lc/seismic-performance-evaluation-of-dual-reinforced-concrete-systems-design-according-to-turkish-seismic-code)

Nacheman, R.J. (2006) 'The Empire State Building – Facade Evaluation and Repair of Engineering Landmark' *Structure Magazine*, January: pp.39–43.

Solomos, G., Pinto, A., Dimova, S. (2008) 'A Review of Seismic Hazard Zonation in National Building Codes in the Context of Eurocode 8' *JRC Scientific and Technical Reports*, EUR 23563–EN 2008.

Tsionis, G., Zhao, B., Taucer, F., Pinto, A. (2010) 'Seismic Design of Masonry Buildings According to Chinese Standards and Eurocode 8' *Codes in Structural Engineering: Developments and Needs for International Practice*, Croatian Society of Structural Engineers, Zagreb, Vol. 1: p.8.

Turan, M. (2010) 'Çatkı sanatının havarisi Viollet-le-Duc'un biçim arayışı ve uzak görüşü' (Apostle of the art of construction – Viollet-le-Duc's search for form) in eds. G. Pultar & Y. Hurol *Yapılar Fora – Mustafa Pultar'a Armağan Kitabı*, Tetragon İletişim: Istanbul.

URL1 (1997) *United States Seismic Zones Map* (viewed 16 February 2014: www.ivi-intl.com/pdfs/IVI_seismic_map_zones.pdf)

URL2 (2013) *Willa Mullerow w Pradze* (Villa Müller) (viewed 26 September 2014: http://pl.wikipedia.org/wiki/Willa_M%C3%BCller%C3%B3w_w_Pradze)

URL3 (2011) *Villa Müller Draftings* (viewed 26 September 2014: http://arch1201-samw.blogspot.com.tr/2011_03_01_archive.html)

URL4 (2014) *Laurel Highlands* (viewed 26 September 2014: www.visitpa.com/regions/laurel-highlands)

PART 3
THE TECTONICS OF FLEXURAL STRUCTURES

Flexural structures are the structures that respond to external loading by developing mainly bending stress in their elements. The elements of these structures are usually straight.

Reinforced concrete or steel frame systems and shear wall systems are studied in chapter 9. This chapter covers their elements, structural behaviour, the reasons for their use, and earthquake-resistant designs. Examples of these structures are given as case studies, and the chapter finishes with a discussion relating to the tectonics of flexural structures.

Most of the buildings that have flexural structures are modern. These buildings are frequently subjected to functional changes and so they are usually re-designed by interior architects at a later stage. During this re-design process, the structural elements of the building are also re-considered: they can be subtracted, changed or new elements can be added to the existing structure. Structural guidelines concerning the ways of handling flexural structures during the interior design process are collected in chapter 10.

Flexural structures are also used in the design of high-rise buildings. Frame and shear wall systems can be used up to a certain limit, but various types of tubular structures (which are also flexural structures) are used for the highest buildings. Strategies to increase the height of building structures and structural guidelines for high-rise building structures are studied in chapter 11.

9
The Tectonics of Frame and Shear Wall Systems

The main characteristics of frame systems can be better understood by comparing frame systems with post and lintel systems: Figure 9.1 shows these two systems.

The horizontal elements in post and lintel systems work with bending stress, and the vertical elements work with compression. However, both the horizontal and vertical elements in a frame system bend, as seen in Figure 9.2. The main reason for this is

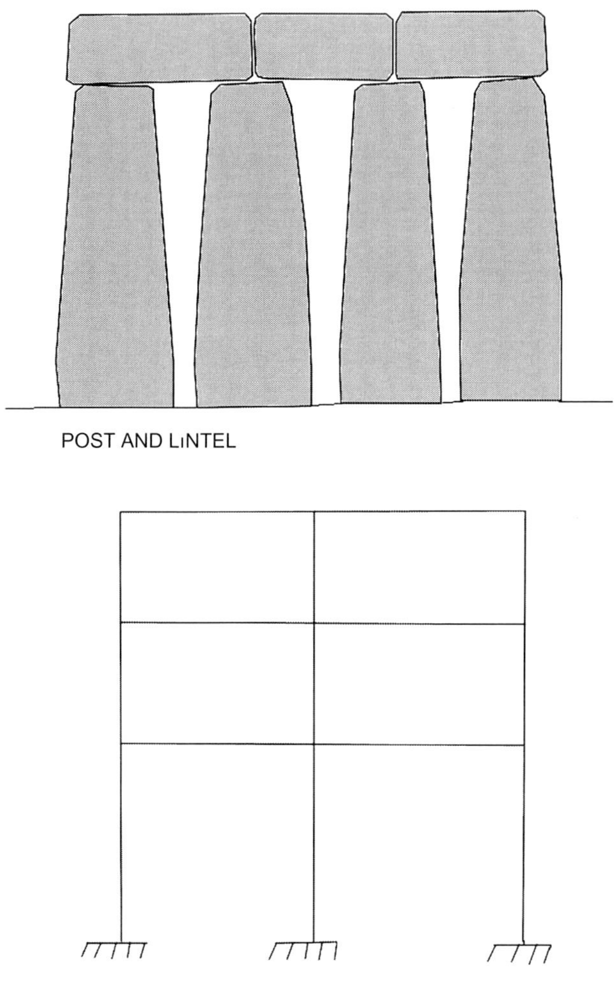

9.1 Frame system and post and lintel system

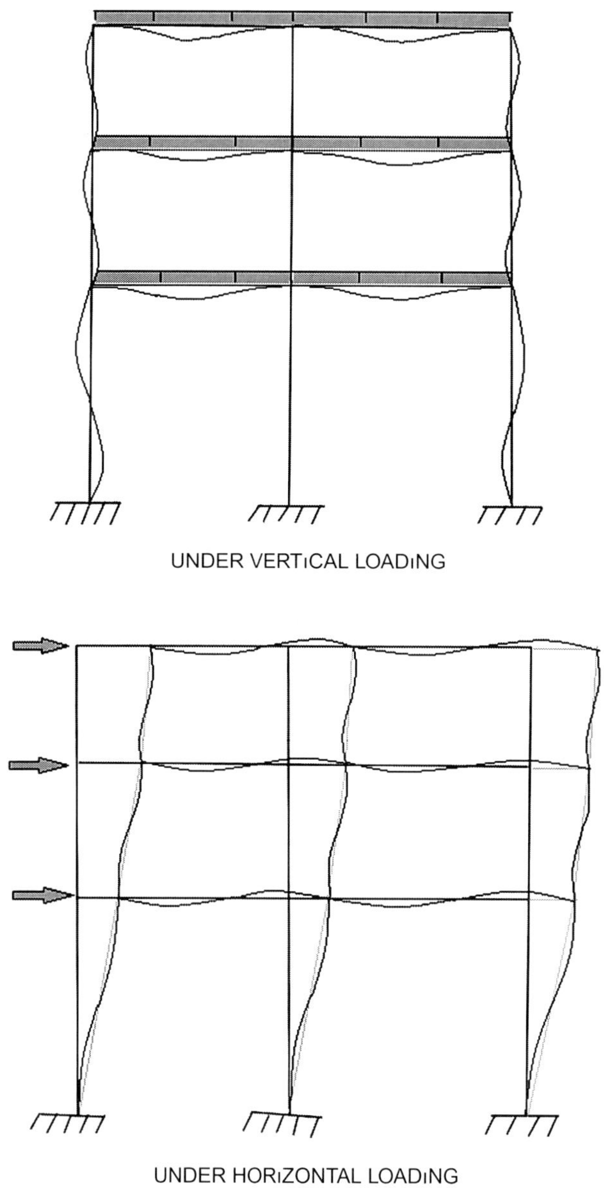

9.2 Deformation of frames under vertical and horizontal loads

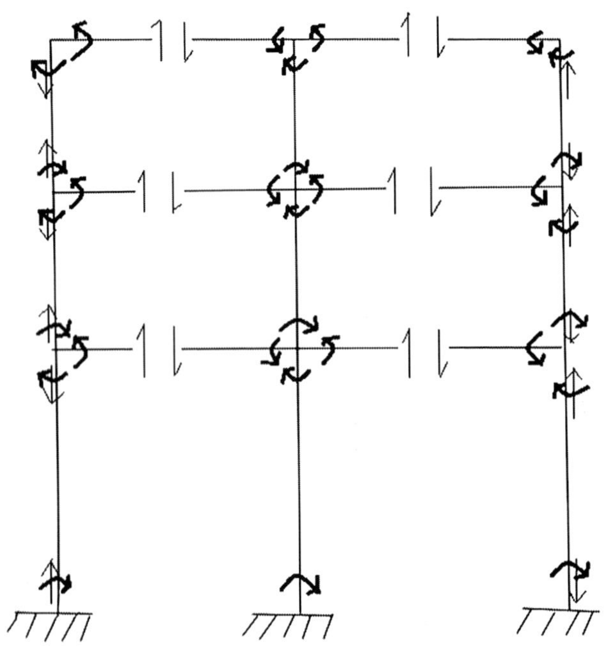

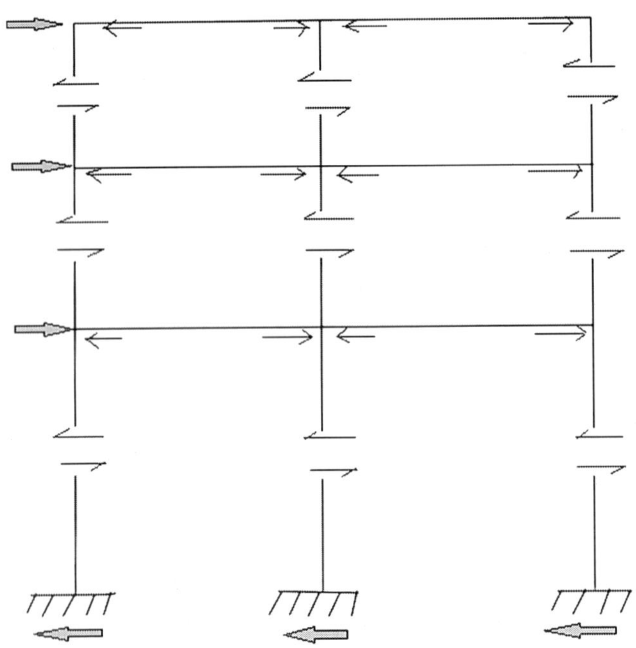

9.3 Axial force and shear force in the elements of frames

continuity between the elements. If a beam bends, the column adjacent to that beam will also bend due to continuity. Since the angle between the beam and the column is assumed to remain the same, the column also bends with the beam.

The compression in the columns of a frame affects the beams as shear force, as seen in Figure 9.3. Every joint should be in equilibrium in Figure 9.3. Similarly, the shear in the columns of a frame affects the beams as tension or compression. Axial force (tension or compression) and bending exist simultaneously in all frame elements.

Continuity makes frame systems stronger than post and lintel systems because all the frame elements help each other.

ELEMENTS OF FRAME SYSTEMS

As seen in Figure 9.4, three-dimensional frame systems can be analysed as two-dimensional frames in both orthogonal directions, with beams connecting them to create the whole. There are two-dimensional frames on axes A, B, C, 1 and 2.

Common elements which make up frame systems are as follows:

- Beams
- Columns
- Slabs
- Partition walls
- Stairs
- Foundations
- Expansion joints.

Beams

Beams in frames are usually horizontal elements that have bending, shear and axial stress simultaneously. They can be slightly curved in plan, but it is beneficial for them not to contain any corners. Corners and strong curves cause twisting and stress concentration in beams.

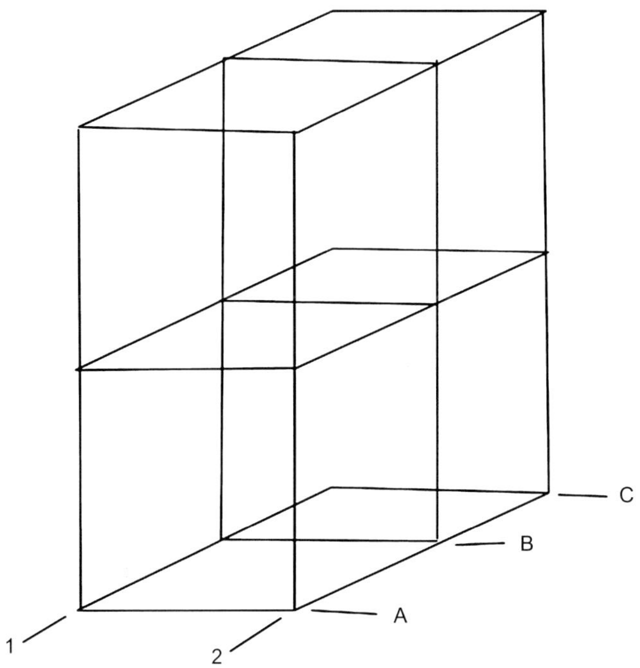

9.4 Two-dimensional frames within a three-dimensional frame

Approximate dimensions of beams can be determined by considering the type of material used. The approximate depth of reinforced concrete beams is calculated as follows:

d = length/10

where:

d is depth,
length is span (the distance between two supports/columns).

According to the Turkish building code, the minimum depth of a beam can be 30cm (Ministry of Public Works and Settlement Government of the Republic of Turkey, 2007).

The approximate depth of steel beams is calculated as follows:

d = length/20

The depth of beams is more critical than their width due to the direction of bending stress. Chapter 4 studies the subject of bending stress in detail. The minimum width of reinforced concrete beams is 20cm because this is the minimum required in order to physically place reinforcement into the formwork and work with it.

High-strength reinforced concrete beams can span up to approximately 15m (Engel, 1997). The moment created by the dead-weight of the beam might exceed the moment resisting capacity of it, depending on further increases in span. However, the optimum span for reinforced concrete beams is approximately 4.5–5m and following this guide reduces the cost of the structure.

Prefabricated, pre-stressed, reinforced concrete box girders can span up to 18m (Mieczyslaw & Zbigniew, 2014). However, the current research about UHPFRC (Ultra-High Performance Fibre Reinforced Concrete) and its applications show that the span of UHPFRC single-span beams can go up to 70m. The depth to span ratio of these beams can be 1/38. This means that a beam spanning 70m can be 1.8m deep, as in the Passerelle des Anges footbridge in Herault Gorges, France (Abrams, 2013; Resplendino & Toulemonde, 2010).

The optimum span of steel beams is around 7m, although specially designed steel beams can span up to 20m. Box girder bridges are examples for longer span steel structures, which can span 100–200m (Steel Construction Info, n.d.).

Columns

Columns in frames are usually vertical elements that have bending, shear and axial stress simultaneously. They can also be inclined. If the structure is located in a low-risk earthquake region, the approximate dimensions of columns can be determined by examining other existing structures that are similar in size and structural material.

The minimum practical plan dimensions of reinforced concrete columns are 20cm × 20cm: otherwise it would not be possible to place reinforcement into the formwork. The minimum for reinforced concrete columns is defined as 25–30cm by the Turkish building code (Ministry of Public Works and Settlement

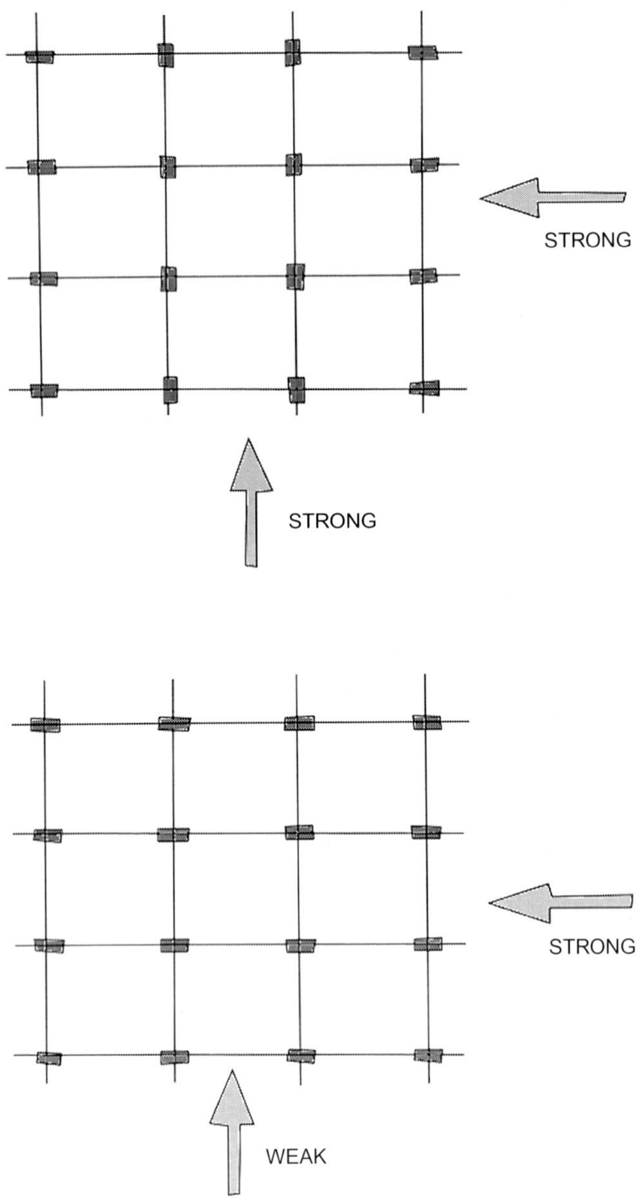

9.5 Directions of columns in plan

Government of the Republic of Turkey, 2007). The thickness of steel plates is around 4–5cm for steel columns. The proportions of these plates, in terms of the ratio between width and thickness, is an important consideration, especially to avoid buckling (Ministry of Public Works and Settlement Government of the Republic of Turkey, 2007). Columns usually resist bending and compression, although there may be buckling if the column is slender. The subject of buckling is studied in chapter 4.

It is better if all columns reach the foundations. The direction of the columns on a plan should be well distributed. It is better to use the outer columns perpendicular to the facade of the building, as seen in Figure 9.5, because these columns will resist against horizontal loads acting on the structure.

Slabs

Slabs can be classified according to their structural material. This book covers only reinforced concrete and steel slabs.

Reinforced concrete slabs

As seen in Figure 9.6, the most common reinforced concrete slab types are as follows:

- One-way slab
- Two-way slab
- Flat slab
- Ribbed slab
- Waffled slab.

If the slab is supported by beams in all directions and if the shorter dimension of the slab is under 7m, a one- or two-way slab is used. A one-way slab is preferred if the longer dimension of the slab is equal to, or more than, two times its shorter dimension. The structural plan and section of one-way and two-way slabs are seen in Figure 9.7. A two-way slab is preferred if the shape of the slab is closer to a square. The main reinforcement direction in one-way

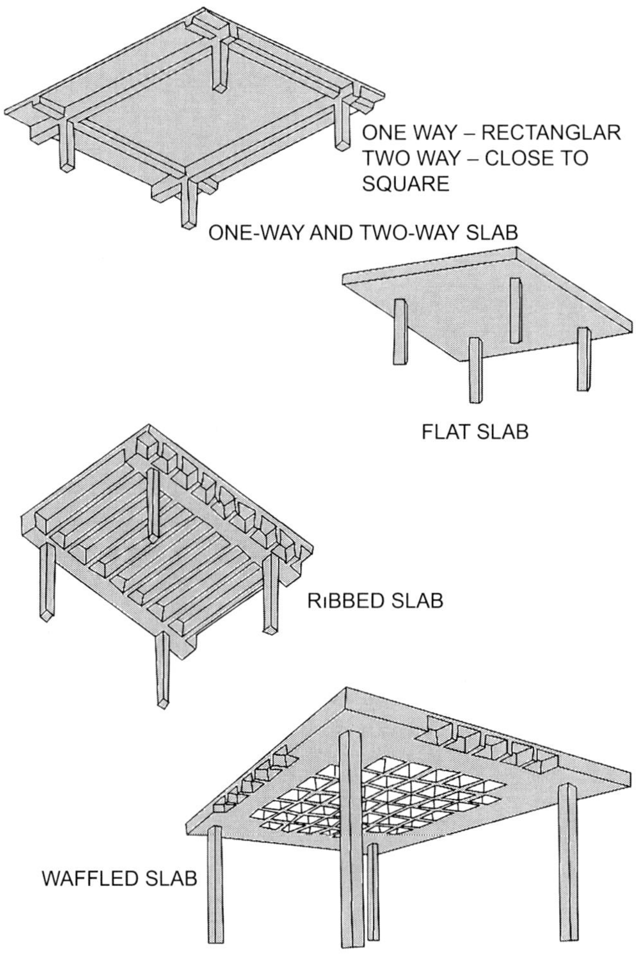

9.6 Reinforced concrete slab types

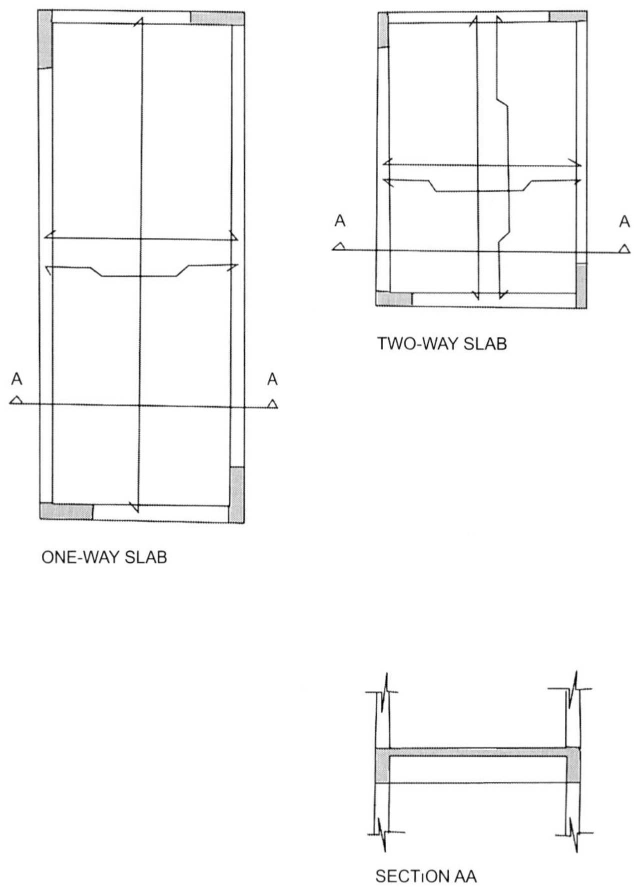

9.7 Structural plan (with reinforcement) and section of one-way and two-way slabs

slabs is in the shorter direction. Two-way slabs contain reinforcement in both orthogonal directions (see Figure 9.7).

The depth of one-way slabs can be between length/20 and length/30, while the depth of two-way slabs can be between length/30 and length/40, having length as the shorter span (ACI318-95, 1995). The thickness of these slabs is usually around 15cm. If the thickness exceeds 20cm, it is better to use either ribbed or waffled slabs in order to avoid extra dead-weight.

If a slab is not supported by beams, it is called a flat slab. There are no beams, but the reinforcement between the columns is more than the other parts of the slab. Figure 9.8 shows different applications of flat slabs. Since this type of slab is weaker than one-way and two-way slabs, its span is usually kept at around 4m in earthquake regions. However, according to the Cement and Concrete Association of Australia (2003), it is economic to span 6–8m with flat slabs. If they are pre-stressed, this span increases to 8–12m.

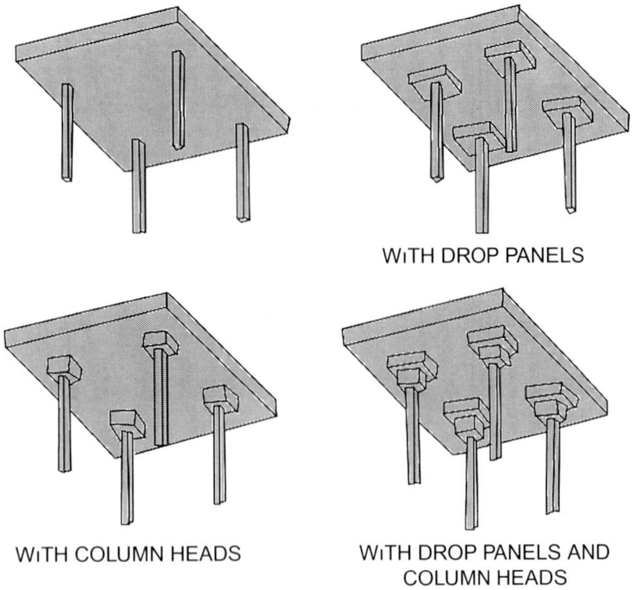

9.8 Different applications of a flat slab

Ribbed and waffled slabs are used if the shorter span of the slab is longer than 7m. A ribbed slab is preferred if the slab shape is rectangular and a waffled slab is preferred if the slab shape is close to square.

Concrete is weak against tension. Since the bottom parts of slabs are in tension under the effect of dead-weight, it is better to reduce the concrete and increase the steel at these locations. The concrete should be concentrated at the top part of the slab, where compression exists. Figure 9.9 shows a structural plan and section of a ribbed slab.

Ribs are not as thick as beams: they can be 15cm thick. The distance between two ribs is maximum 1m (see Figure 9.10). Since ribs are closely placed, ribbed slabs behave simultaneously: a force on one rib affects all ribs. If the slab size is large, another rib in the opposite direction is added to distribute the load to all ribs evenly. Ribs can be used in the longer direction in order to avoid heavy load on the longer beam. If an additional column supports the beam in the longer direction, then ribs can be used in the shorter direction.

Ribs are used in two directions in waffled slabs, as seen in Figure 9.11. The span of waffled slabs can be up to 15m (Cement and Concrete Association of Australia, 2003), and they can be longer if they are post-tensioned. It is also possible to have triangular waffled slabs, which can span longer distances. Waffled slabs can also be used for irregularly formed slabs. To avoid complication in the arrangement of formwork, the form of the beam can be designed as seen in Figure 9.12.

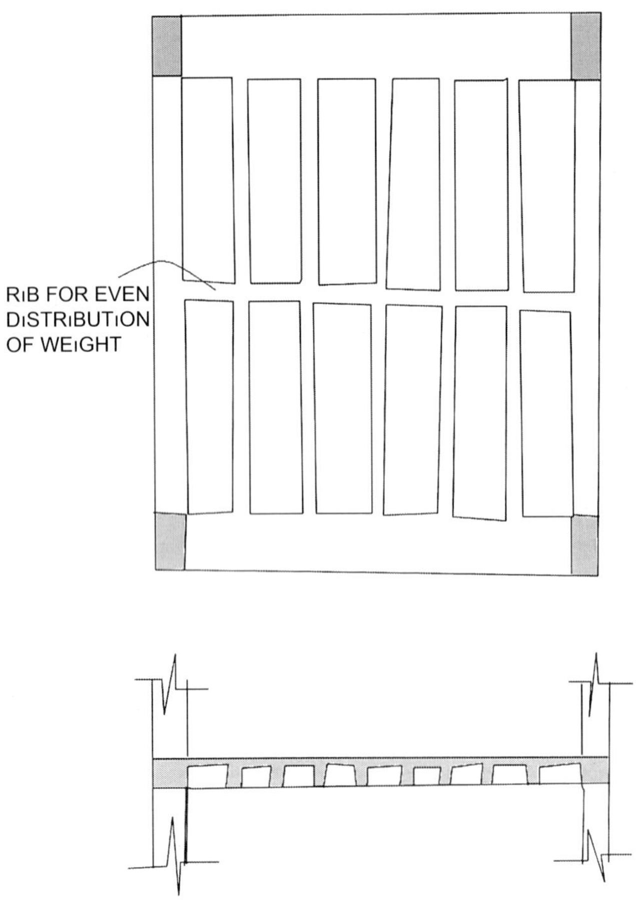

9.9 Structural plan and section of a ribbed slab

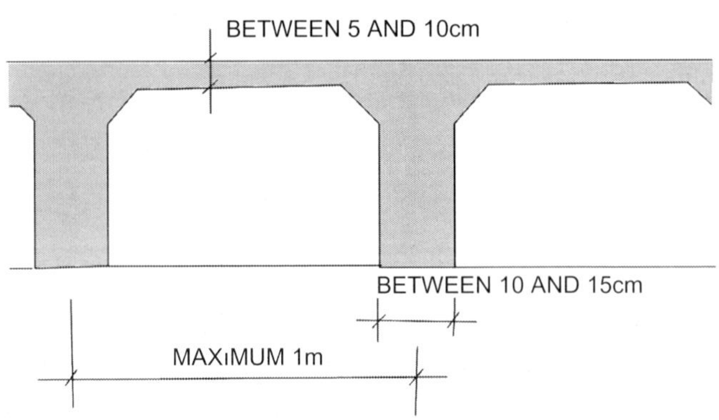

9.10 Dimensions of ribs

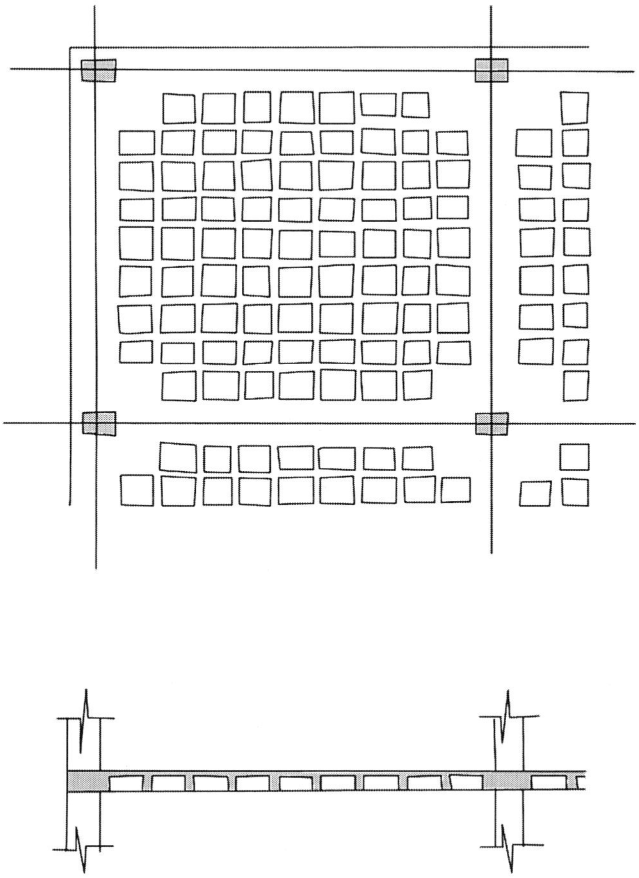

9.11 Structural plan and section of a waffled slab

9.12 An example of the arrangement of ribs and beams in irregularly formed slabs

The span to depth ratio of waffled slabs changes between 15 and 20 (Cement and Concrete Association of Australia, 2003). However, they are usually designed at the same depth as the beams surrounding them for aesthetic reasons. This can be done by increasing the width of the beam and reducing its depth. However, if the seismic risk is high, it is not good to reduce the depth of the beam too much. For example, the Turkish building code (Ministry of Public Works and Settlement Government of the Republic of Turkey, 2007) limits the depth to width ratio of reinforced concrete beams, by recommending that the width of the beam should not exceed the depth of the beam plus the width of the adjacent column. According to the New Zealand building code, ratio of length of the beam to its width should be equal to or below 25 (NZS3101.1, 2006).

Steel slabs

Steel slabs can be formed by using secondary steel beams or trusses. The secondary beams, which can be 7–20m long, can be placed in the shorter direction every 2–5m depending on their top cover, as seen in Figure 9.13. Similarly, secondary trusses might be spaced 1–3m apart. F.D.K. Ching's (1991) book shows various applications for such steel slabs.

Partition walls

One of the most important tectonic characteristics of frame systems is their lightness in comparison to masonry structures. Frames are lighter because the walls in the structure are non-load-bearing walls and so they can be replaced by large openings. Such walls are called partition walls and they are different to load-bearing walls. Non-load-bearing walls can be classified into two groups from a structural point of view. There can be lightweight partition walls, such as timber, gypsum and metal panels; and there can be rigid partition walls that can be built using various types of non-load-bearing bricks.

Unlike lightweight panels, rigid partition walls affect the structural behaviour of frame systems by not allowing deflection of

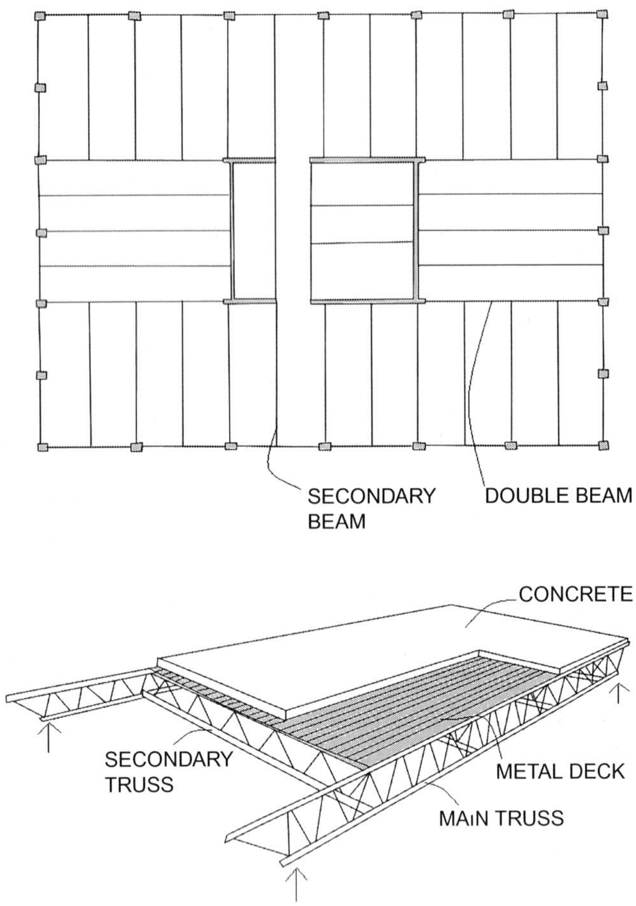

9.13 Use of secondary steel beams and trusses in steel slabs

adjacent beams and columns. Badly placed rigid partition walls can cause damage to structures during earthquakes. Yet correctly placed rigid partition walls can increase the earthquake resistance of structures. Thus, the arrangement of rigid partition walls should be carefully considered during architectural design. (The subject of earthquake-resistant design is studied later in this chapter.) It is better to place rigid partition walls over the beams (or very close to the beams) of a frame in order to be able to transfer their weight directly onto these beams.

Stairs

Stairs are special elements in frame structures. There can be various arrangements of staircase structures:

- Supporting flights and landings in a staircase with stringer beams.
- Design of landings and steps in a flight as cantilevers.
- Design of the whole staircase as a cantilever.

Supporting flights and landings in a staircase with stringer beams

The structure of these types of stairs can be understood by imagining flights (or steps) in a staircase as slabs, which can be carried by stringer beams in two different ways. As seen in Figure 9.14, there can be a stringer beam at the middle of the flight or there can be two stringer beams on both sides of the flight. These stringers can sit directly on beams, which take place on two sides of the stair, or they can be designed as bent stringers and sit on beams that are 1–1.5m away from the flight, as seen in Figure 9.15. These beams can be the beams of the frame system, which sit on columns on two sides.

However, if there are landings in a staircase, these landings should also be supported. Figure 9.16 shows some arrangements of beams to carry the flights and landings in different types of arrangements.

As seen in Figure 9.16, there is a hierarchy between the elements in staircases. The main beam of the frame is the first in the hierarchy; the beams around the landing are second in the hierarchy; and the beams carrying the flights (stringers) are third. These beams rest on each other, which is not seen to be a good solution for the beams of a frame. With the exception of the staircase beams, it is better not to let beams sit on each other. This is because, if beams sit on each other, this means that one of them applies concentrated load to the other and this increases the moment in the second one.

Different types of structural materials can be used for the stairs, such as steel, timber and reinforced concrete.

9.14 Structure of a flight with stringer beams

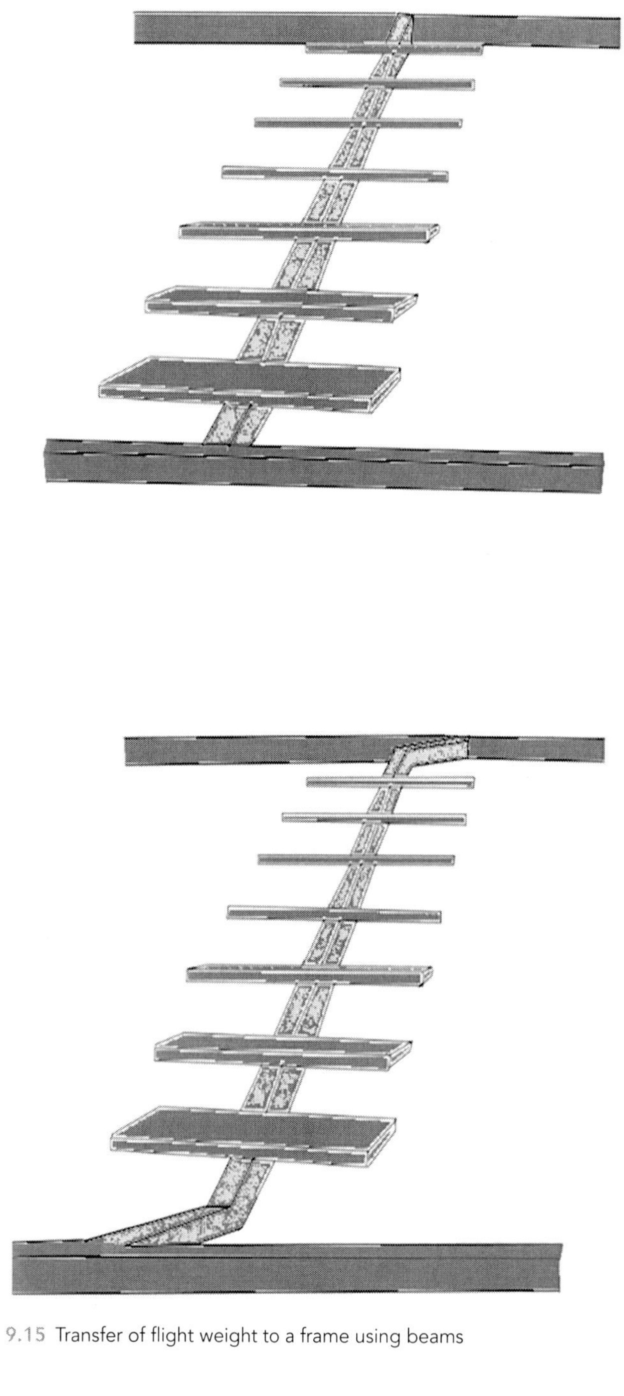

9.15 Transfer of flight weight to a frame using beams

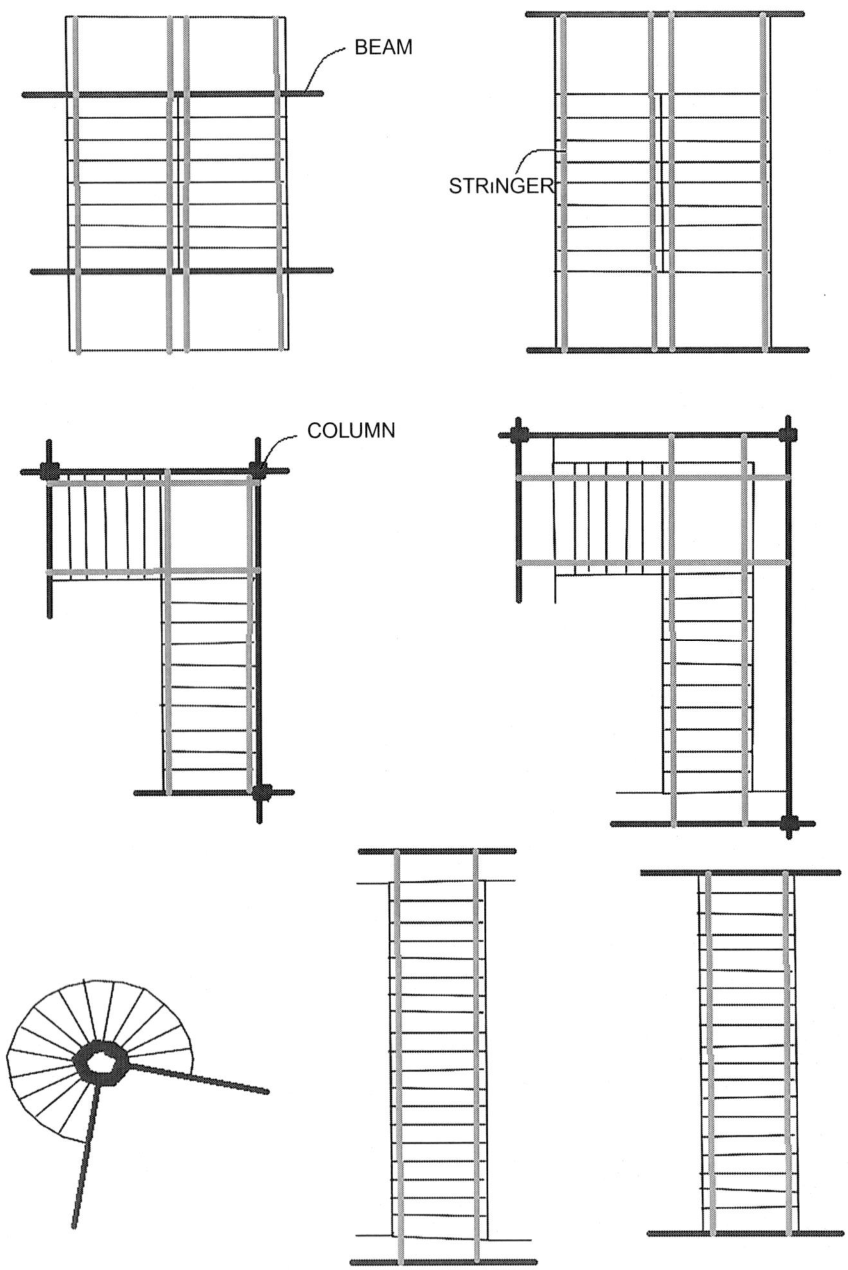

9.16 Use of beams to carry flights and landings in different arrangements

Design of landings and steps in a flight as cantilevers

Each step can be designed as a cantilever from a wall, a beam or a column. Figure 9.17 shows a staircase in which the steps are cantilevering from a beam. This type of stairs can also be achieved by using various structural materials, such as steel, timber and reinforced concrete. Many stone staircases are arranged with the help of stone steps cantilevering from stone walls.

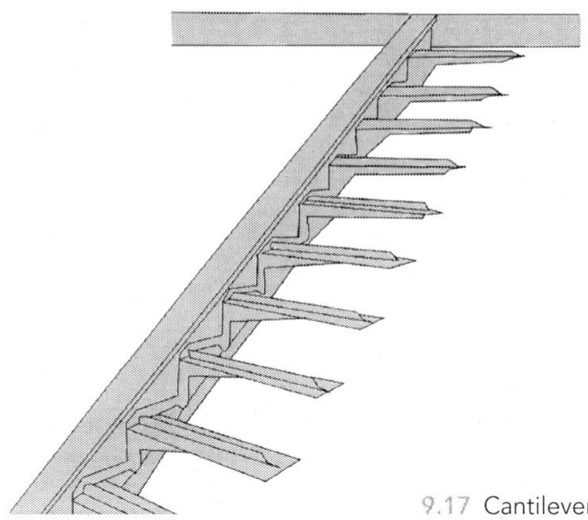

9.17 Cantilevering steps

Design of the whole staircase as a cantilever

If the structural material is monolithic, such as reinforced concrete, it becomes possible to design the whole staircase as a cantilever, as seen in Figure 9.18.

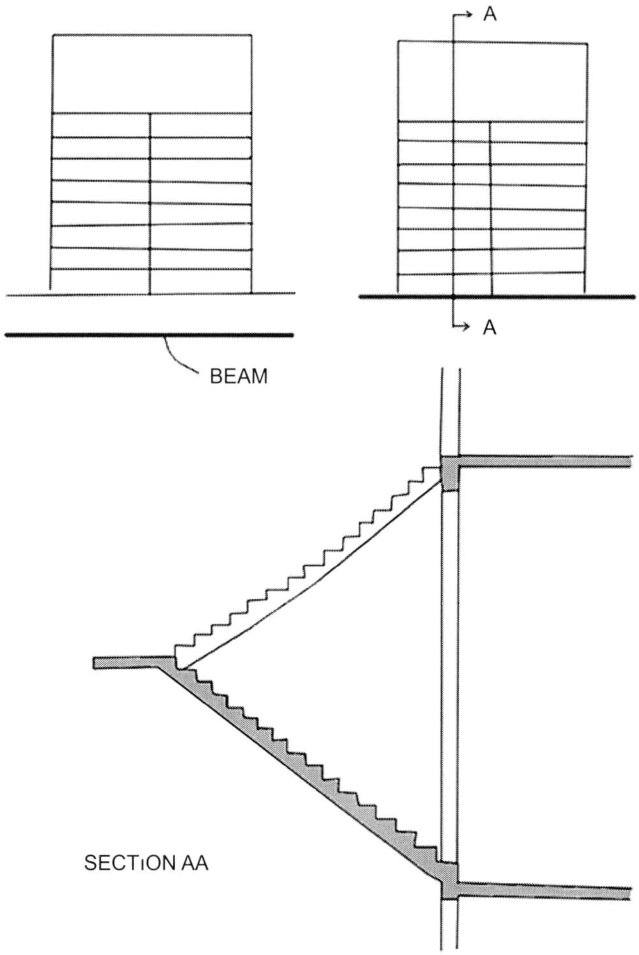

9.18 Whole staircase as a cantilever

Structure of slabs that are adjacent to stairs

Stairs are one of the first things to evaluate in a structural proposal for a student project. A common problem is not related to the structure of stairs, but to the structure of the adjacent slabs. It is better to have all of the slabs around a staircase surrounded by beams that sit directly on columns. Figure 9.19 shows a problematic and a correct solution for slab structures adjacent to staircases.

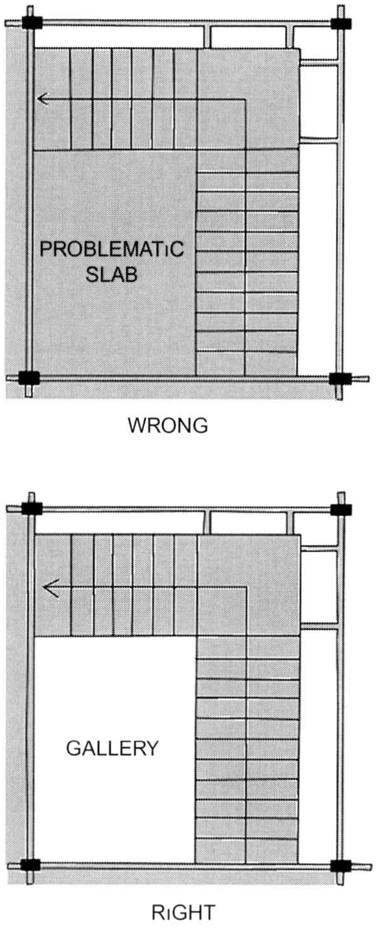

9.19 A problematic and a correct solution for slab structures adjacent to staircases

Foundations

The role of foundations is to transfer loads acting on the building structure to earth; and to avoid overturning instability, lateral sliding and uneven settlement. Types of foundation systems that can be used together with frame systems can be categorised into four groups:

- Individual footings
- Slab-on-ground foundations
- Raft (mat) foundations
- Pier and pile foundations.

The type of foundation is chosen according to soil quality and the loads affecting the building structure. If the soil is sufficiently strong to carry the weight of the building, individual footings might be preferred. Slab-on-ground foundations are used on expansive soil and for small buildings that are not higher than two storeys. Raft foundations distribute heavy loads to weak and expansive soil. If the firm soil is deep and difficult to reach, pier and pile foundations can be used.

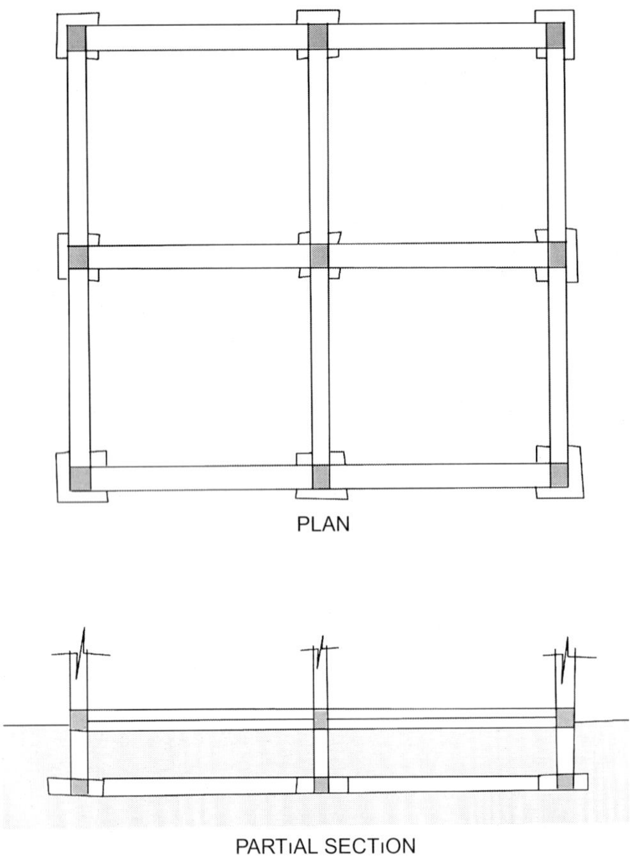

9.20 Plan and section of a structure with individual footings

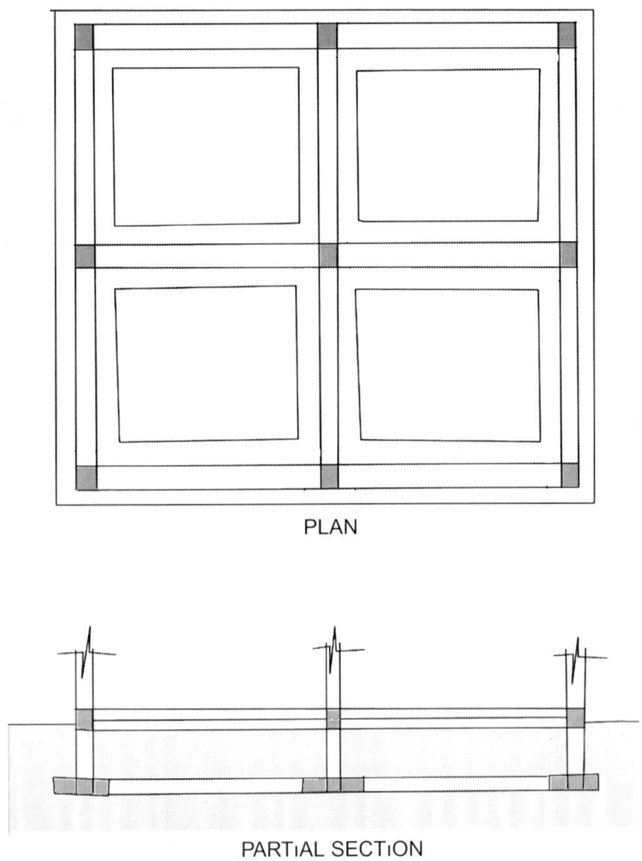

9.21 Connecting individual footings to each other

Individual footings are used under each column and they are connected to each other at footing level and earth level, as seen in Figure 9.20. The depth of these foundations is determined according to the depth of firm soil and frost level. However, they are often around 80cm deep. The area of the footings is determined according to the bearing capacity of the soil and the weight of the building. If individual footings get too close to each other, they are connected as shown in Figure 9.21.

Slab-on-ground foundations have to be a minimum 50cm deep in the ground and 10cm outside the ground: a total of 60cm deep, as seen in Figure 9.22. They are used for buildings of less than two storeys in order to avoid overturning in areas of high earthquake risk. If the building structure sits on rock, the reinforced concrete foundation surface should be connected to the rock with the help of borings.

Raft or mat foundations can be imagined as inverted slabs that can be in various forms, as seen in Figure 9.23. The simplest type of raft foundation has similarities to a boat in the sea. If the weight of the removed soil from the building site is equal to the weight of the building, then the soil at that level can easily carry the weight of the building. This structure is similar to an inverted flat slab. By adding inverted beams, ribs and waffles, the resistance of the inverted slab can be increased.

Pile foundations are more slender than pier foundations and they can reach deeper levels. End-bearing pile foundations can reach the firm soil at deep levels. If the firm soil is unreachable, friction piles are used (see Figure 9.24). The surfaces of friction piles are designed in such a way that the piles can carry the weight of the building with their surface friction. Structural materials for piles can be concrete, wood and steel. Encased in concrete with a circular steel shell can also form a pile.

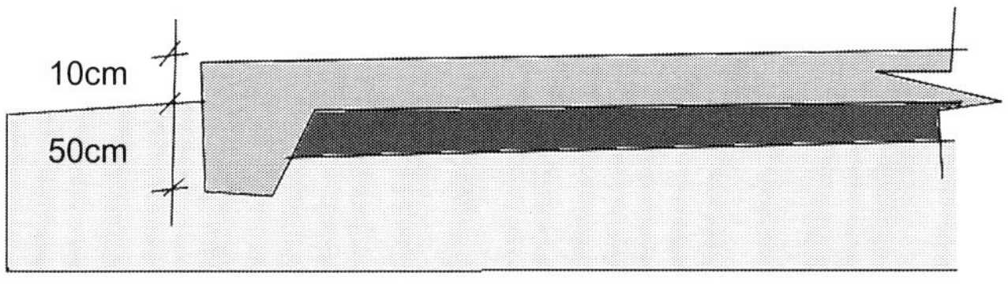

9.22 Section of a slab-on-ground foundation

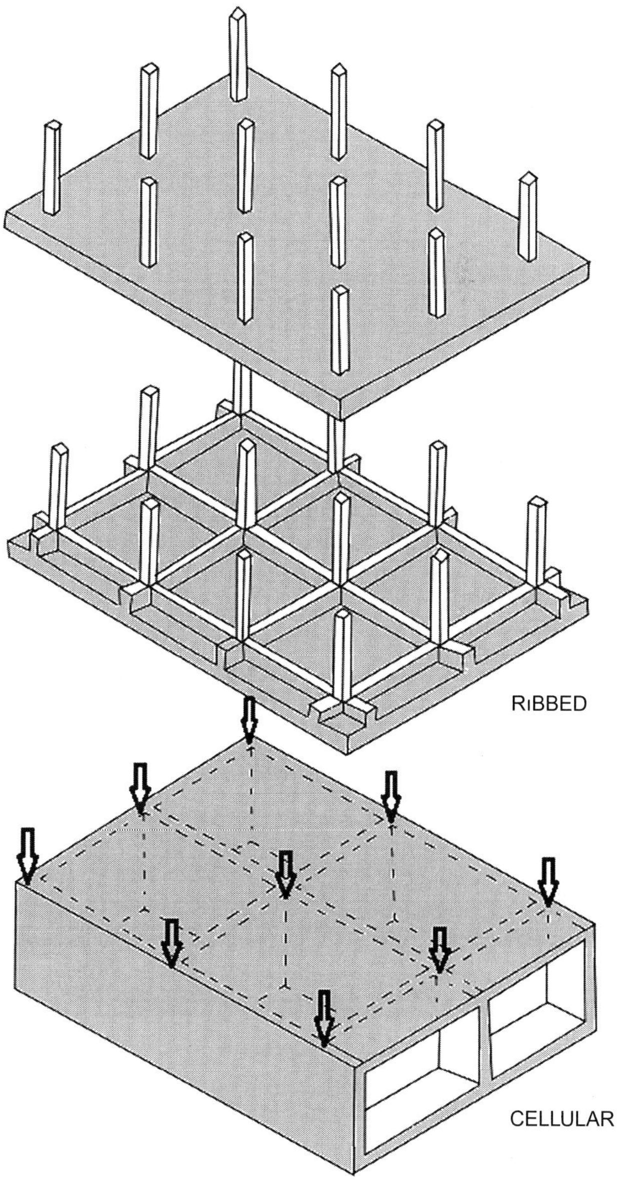

9.23 Various applications of a raft foundation

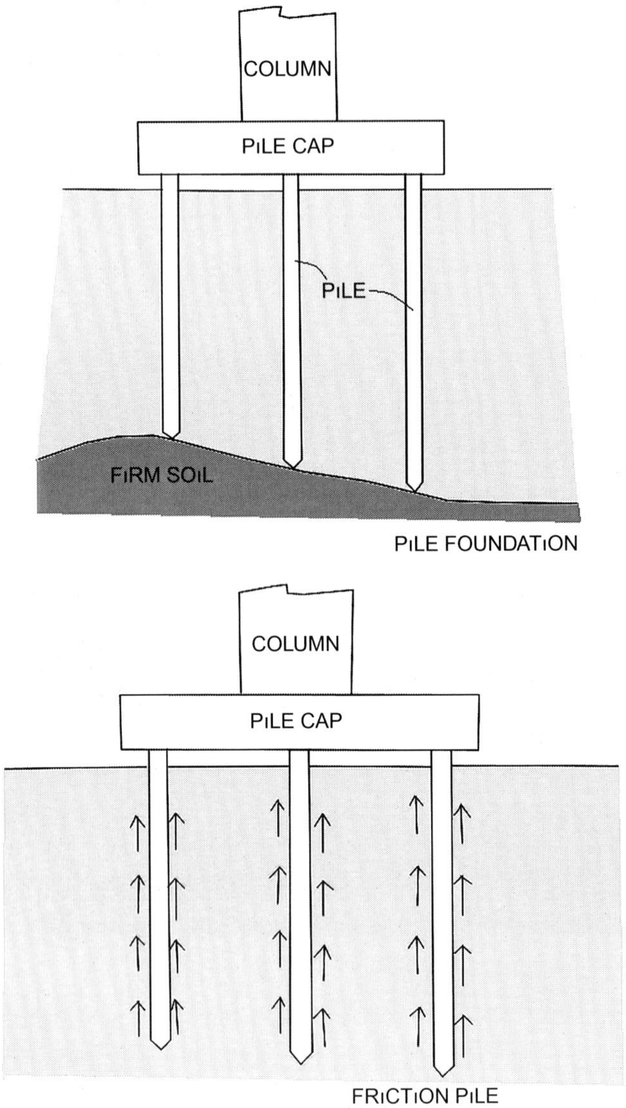

9.24 Pile foundations and the use of friction piles

Expansion joints

Expansion joints separate building structures into parts for various reasons. Expansion joints might be needed in the following conditions, which are shown in Figure 9.25:

a. If one part of the building is much higher than the other parts, expansion joints are used to separate the structures of those parts with different heights.
b. If one part of the building has another structural material/structural system, expansion joints are used to separate those parts with different structural materials/systems.
c. If the plan of the building contains deep recesses, expansion joints are used to separate the recesses from the main body in areas of high earthquake risk.
d. If the plan of the building is larger than 30m × 30m, expansion joints are used to separate the structure of the building into smaller pieces in order to decrease temperature load (see chapter 4 for more information regarding loads).
e. If the building sits partially on strong soil and partially on weak soil, expansion joints are used to separate the part of the building on strong soil from the part on weak soil.

If the expansion joint is needed to avoid temperature load, then it is not necessary to continue the expansion joint at foundation level. However, if the expansion joint is needed for any of the other reasons, then it is better to continue the expansion joint at foundation level. If the expansion joint continues at foundation level, it simplifies the structural design process because the different parts of the structure become totally independent from each other.

The minimum thickness of an expansion joint is approximately 3cm. According to the Turkish building code, if the building is higher than 6m, then 1cm is added to the thickness of expansion joints for every additional 3m of building height (Ministry of Public Works and Settlement Government of the Republic of Turkey, 2007). For example, for a building of five storeys, the expansion joint will be approximately 6cm.

Expansion joints can be arranged in framed structures in three different ways, as shown in Figure 9.26. Similar applications can

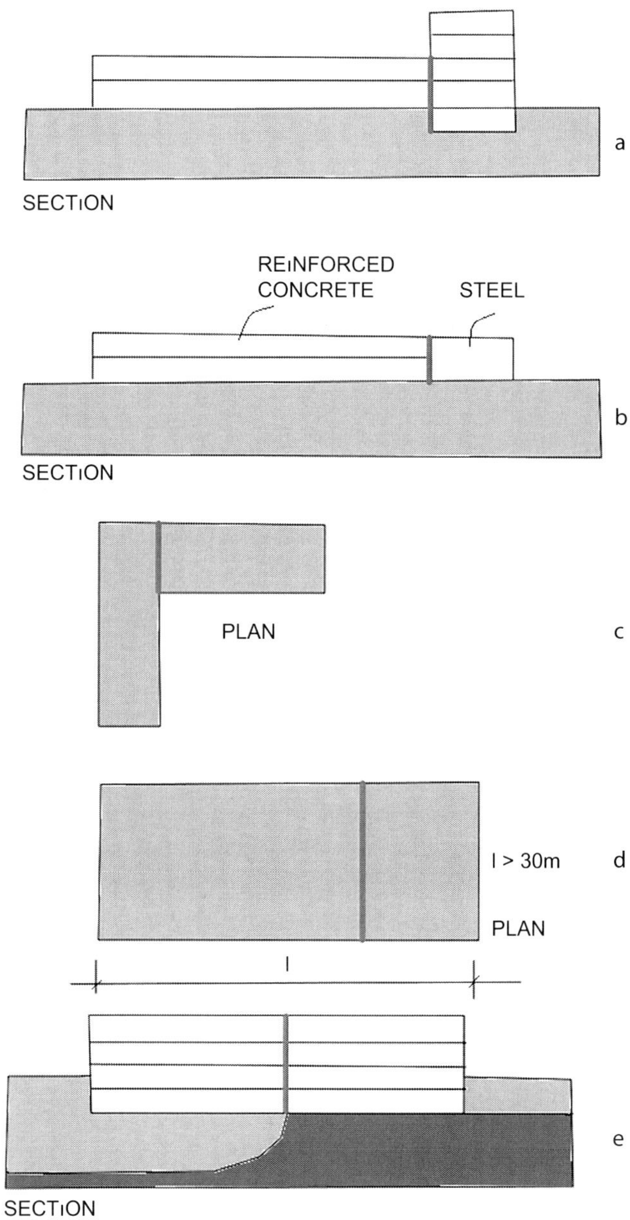

9.25 Places to use expansion joints

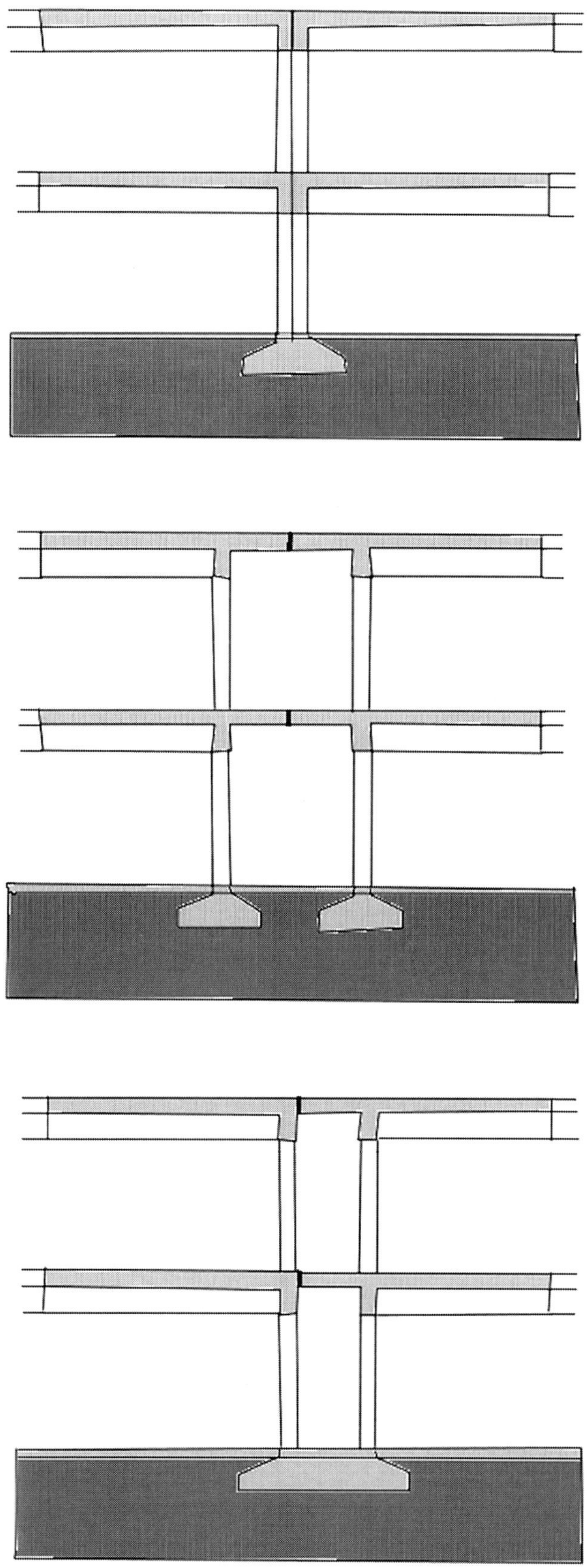

9.26 Different applications of expansion joints

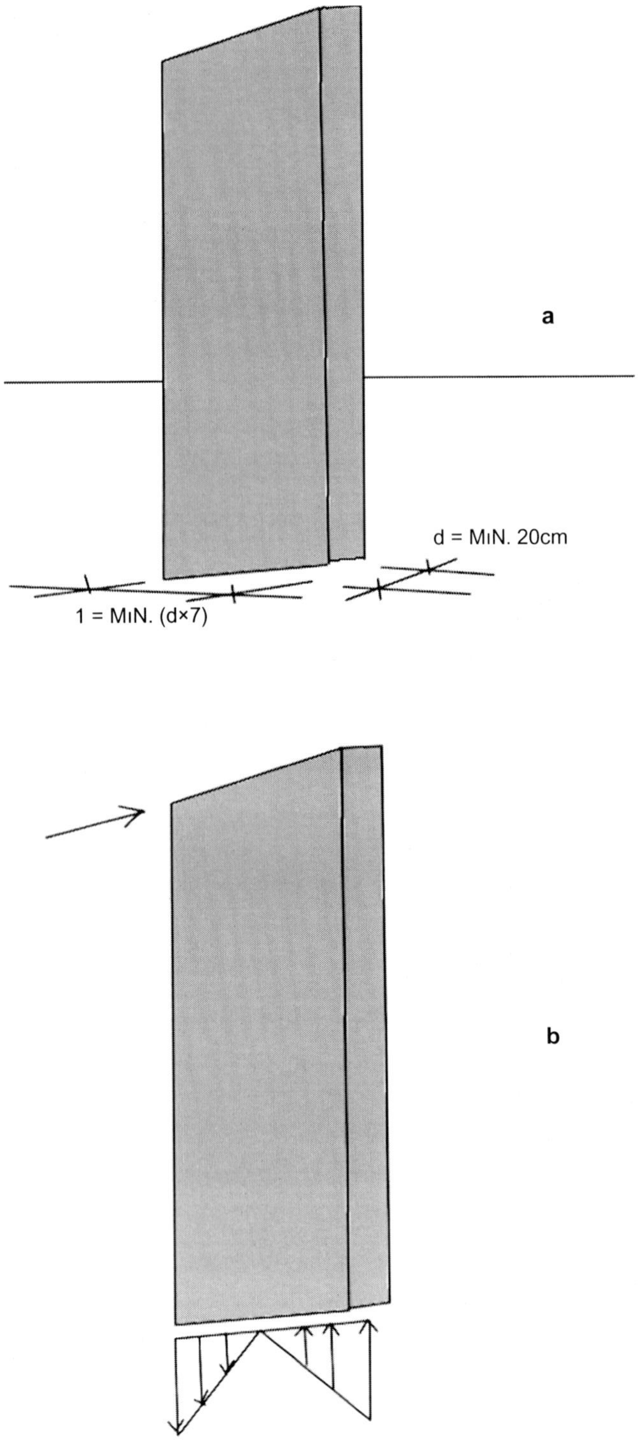

9.27 **(a)** Minimum dimensions of reinforced concrete shear walls; **(b)** Transfer of horizontal load by reinforced concrete shear walls

be carried out by separating the foundations. Whether or not the foundations are separated, the expansion joint can be located between the columns or cantilevers.

SHEAR WALL SYSTEMS AND USE OF SHEAR WALLS WITH FRAME SYSTEMS

The Turkish building code defines the minimum dimensions of reinforced concrete shear walls as 20cm in thickness and seven times the thickness in length, as seen in Figure 9.27a (Ministry of Public Works and Settlement Government of the Republic of Turkey, 2007). The American Concrete Institute defines the thickness of shear walls as length/25, having 'length' as either height or length, whichever is the shorter (ACI318-95, 1995).

A shear wall is very useful in resisting horizontal (lateral) loads such as earthquake and wind loads. It can transfer horizontal loads to foundations, as seen in Figure 9.27b. Thus, all shear walls should reach the foundations.

The ratio of wall area to floor area can be between 5% and 10%, according to the NIST (National Institute of Standards and Technology, 2012). There can also be steel shear walls with various applications for bracing, as seen in Figure 9.28. Shear force in a steel shear wall is taken by the diagonal elements, as seen in Figure 9.29. One diagonal is in tension and the other is in compression. Similarly, one column is in tension, while the other is in compression.

Buildings can be formed by using shear walls as the only structural system. However, the most common application for shear walls is with frame systems. Shear wall structures are very frequently used together with frames in order to increase the resistance of frames against horizontal loads. The Japanese building code suggests the use of frames together with shear walls as one of the main strategies against earthquakes (Paz, 1994). These shear walls should be evenly distributed in both orthogonal directions, as seen in Figure 9.30, in order to avoid a twisting instability problem, which is described later in this chapter.

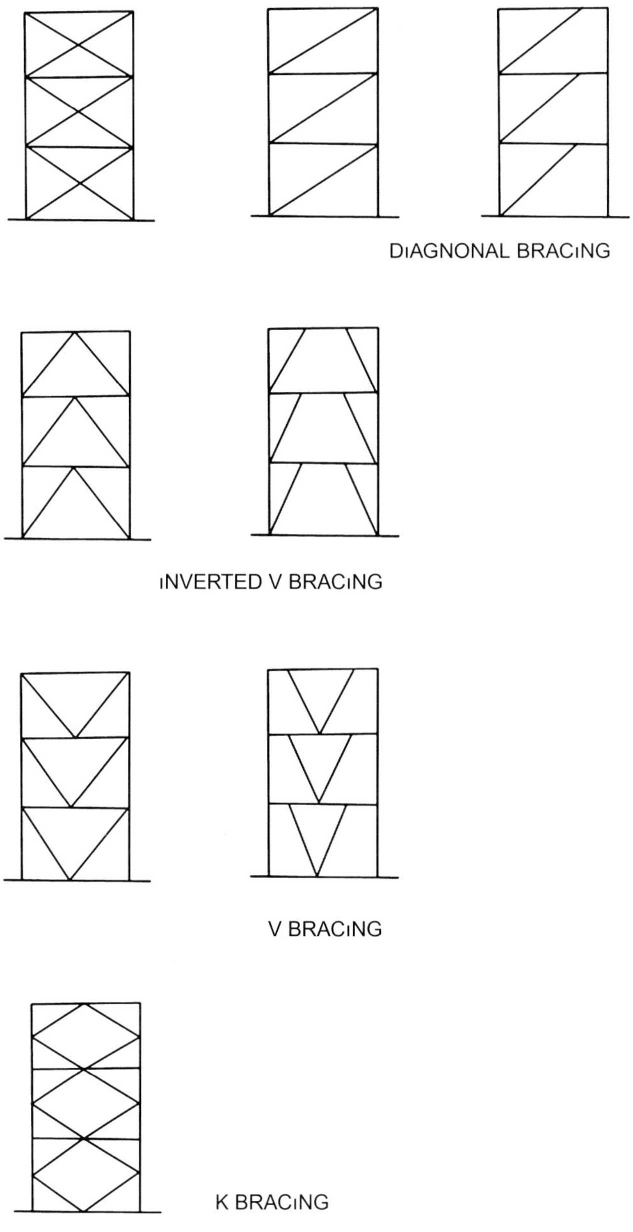

9.28 Bracing in steel shear walls

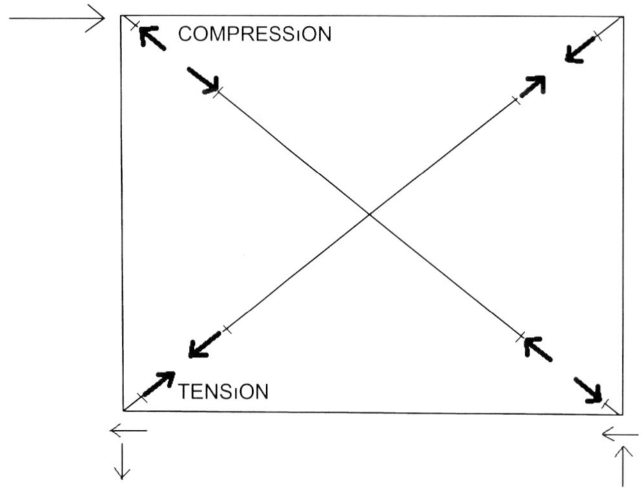

9.29 Internal forces in a steel shear wall

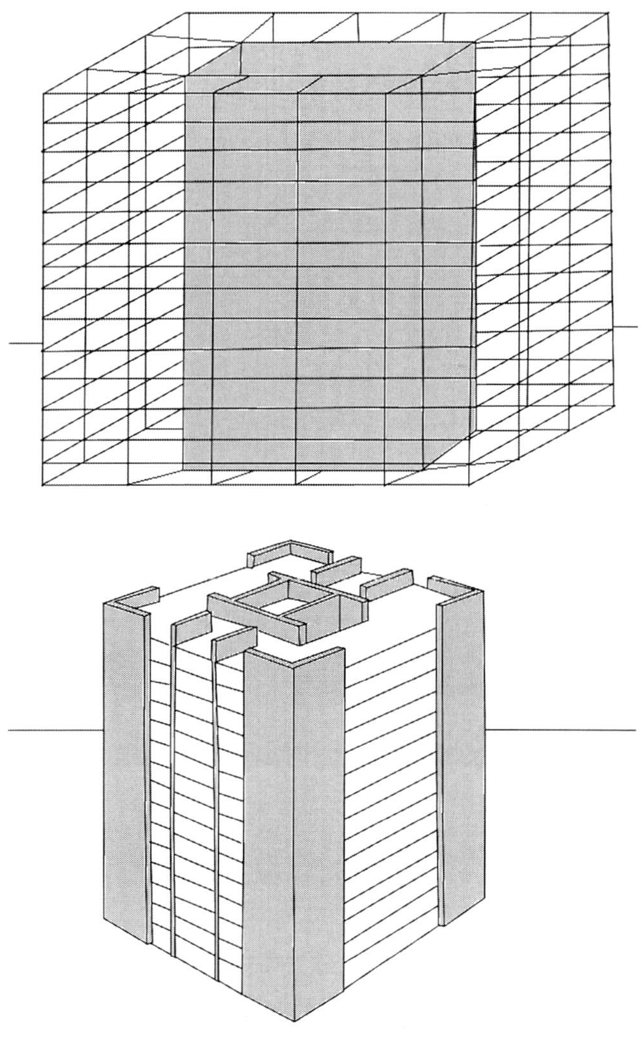

9.30 Distribution of shear walls within a structure

FINDING INTERNAL FORCES IN FRAMES

Finding internal forces in frames is not one of the common practices in an architect's professional life. However, it is still taught in schools of architecture for the following reasons:

- To be able to communicate at the level of scientific knowledge regarding building structures.
- To solve problems teaches a lot about the structural issues, which are difficult to teach in other ways.
- To be able to communicate with structural engineers.
- To be able to make some simple calculations during architectural design.

The author of this book believes that the way in which architects are educated regarding internal forces should be different to the education of structural engineers. Graphical and approximate methods are useful for teaching students of architecture. Knowledge of deformed or deflected shapes of structures should be included so that students can imagine what might happen to their structures in the future. Hence this book teaches the following methods:

- Drawing N (axial force), V (shear force), M (moment) diagrams and the deflected shape of determinate systems (small systems with maximum three unknown support reactions).
- Drawing N, V, M diagrams and the deflected shape of indeterminate frames (larger systems with more than three unknown support reactions), by using the Portal Method.

Drawing N, V and M diagrams and the deflected shape of determinate systems

The first step in drawing internal force diagrams N (axial), V (shear), M (moment) and deflected shape is to find the unknown reactions. Reactions of determinate systems, which have only three unknown reactions, can be found by using three equations of equilibrium, as described in chapter 4. Then, the diagrams can be drawn. The N diagram is not related to V and M diagrams and deflected shape, but M diagrams can be drawn with the help of V diagrams, and deflected shape can be drawn with the help of M diagrams.

N diagrams show axial forces at every point of the structural member. It is drawn by considering the effects of forces that are parallel to the axis of the structural member. The four steps of drawing N diagrams are shown in Figure 9.31. The first step is to draw the system without any forces and reactions. The second step is to show the forces that are parallel to the axis of the member on the system. The third step is to sign the application points of these forces. There is an axial force between these points: if the forces are towards each other, there is compression; and if the forces are going away from each other, there is tension. Compression is shown as minus and tension is shown as plus in the axial force diagram. The fourth step is to draw the diagram.

V diagrams show shear forces at every point of the structural member. It is drawn by considering the effects of forces that are perpendicular to the axis of the structural member. To be able to draw the diagram, one can start from the left side of the diagram and move his/her pencil together with the forces that are perpendicular to the axis of the member. Figure 9.32 shows the steps of drawing a shear diagram of the same system. If the V diagram is not closed, this means that there is something wrong in the ΣF_y equation.

M diagrams show the moments at every point of the structural member. It is drawn by considering the effects of all forces, including vertical and horizontal forces, and all moments. Figure 9.33 shows the steps of drawing M diagrams. The first step is to find the moment values at the end points of the system. The moment arrows which press the top of the member can be accepted as positive, and the moment arrows which press the bottom of the member can be accepted as negative. However, the opposite of these signs can also be used.

The rest of the diagram is drawn by considering that: change in the moment diagram is equal to the corresponding area in the shear diagram. For this purpose, areas in the shear diagram are calculated as the second step. Then, starting from the left of the M diagram one can move his/her pencil diagonally upwards with the positive shear areas and diagonally downwards with the negative shear areas. This forms the third step in drawing

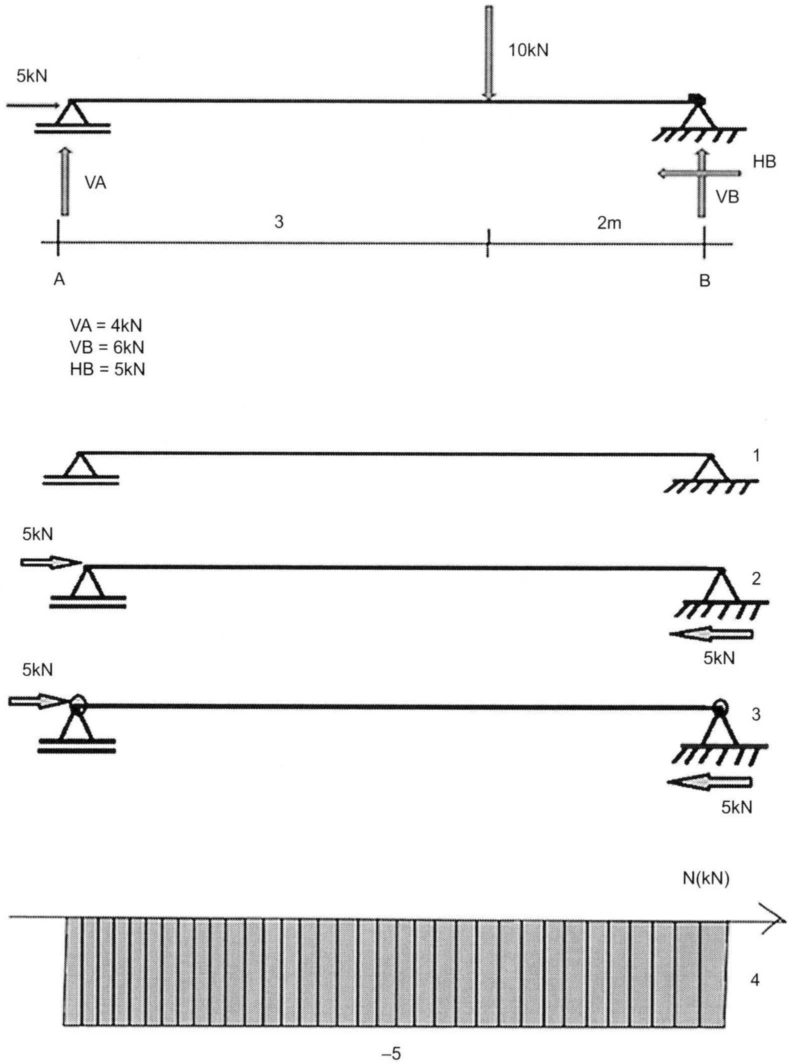

9.31 Steps to draw an N diagram

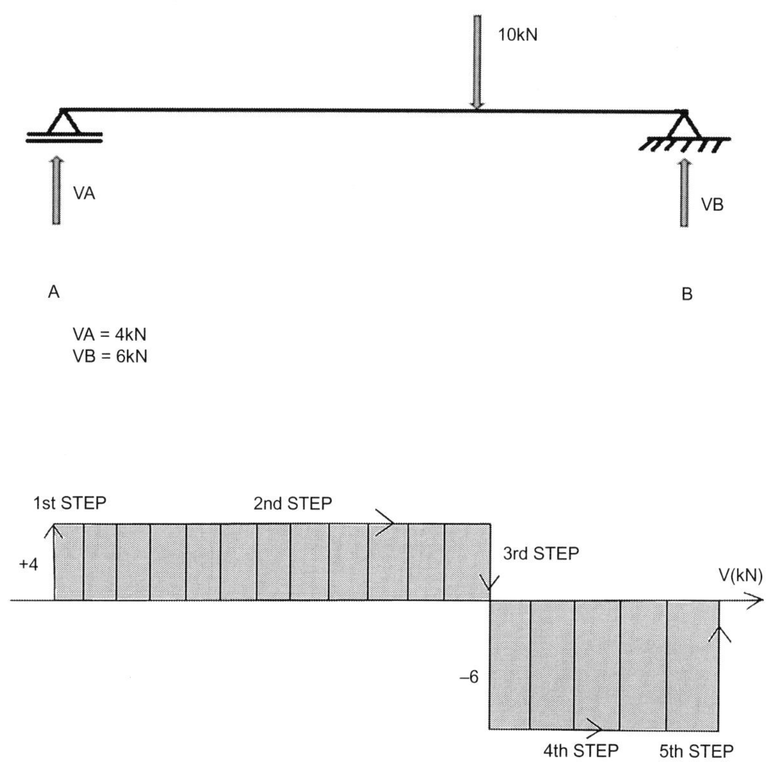

9.32 Steps to draw a V diagram

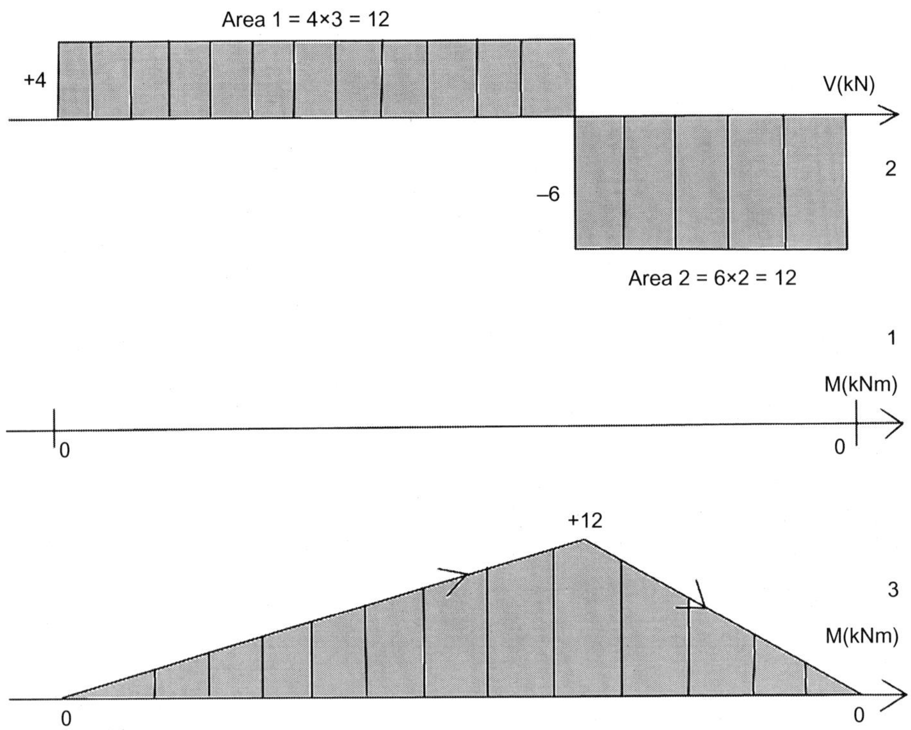

9.33 Steps to draw an M diagram

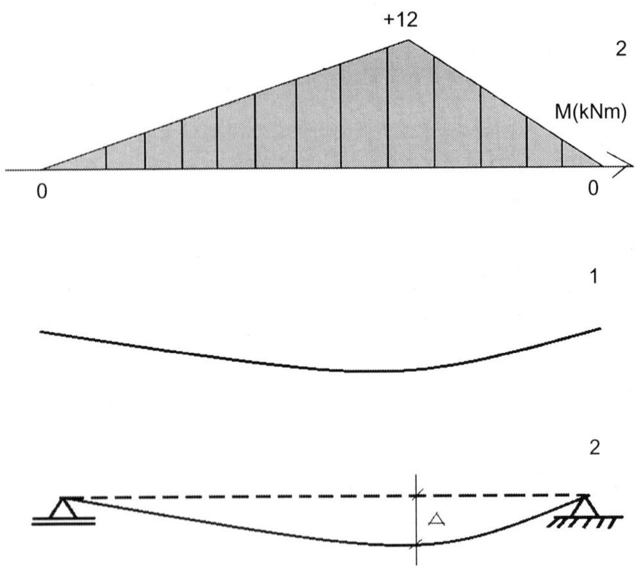

9.34 Steps to draw deflected shape

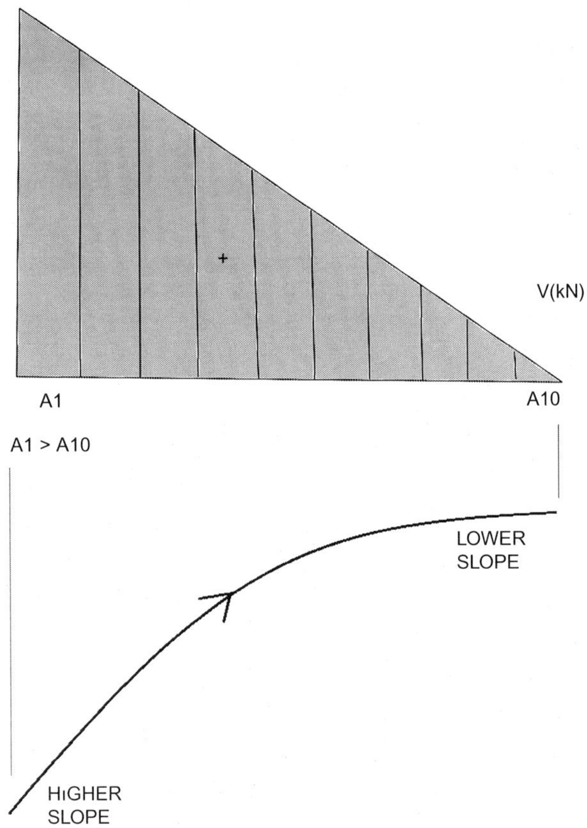

9.35 Drawing a parabolic curve in the M diagram

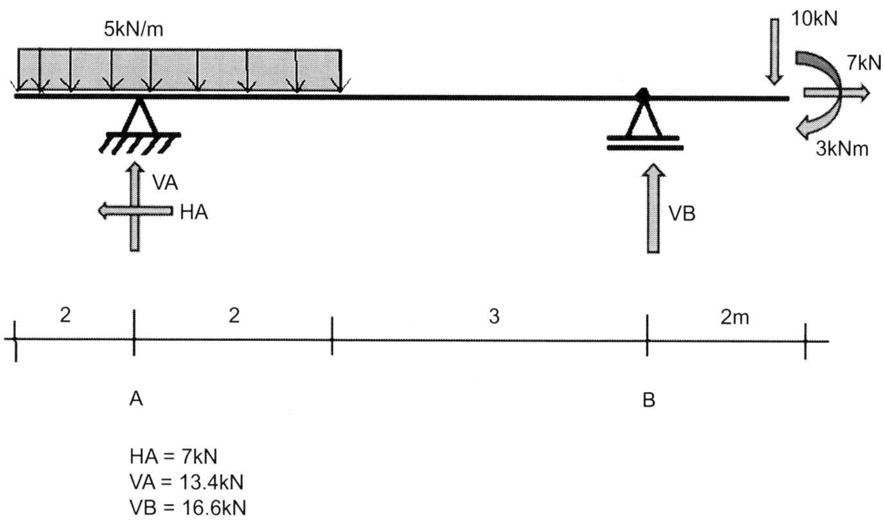

9.36 A system to draw N, V and M diagrams and deflected shape

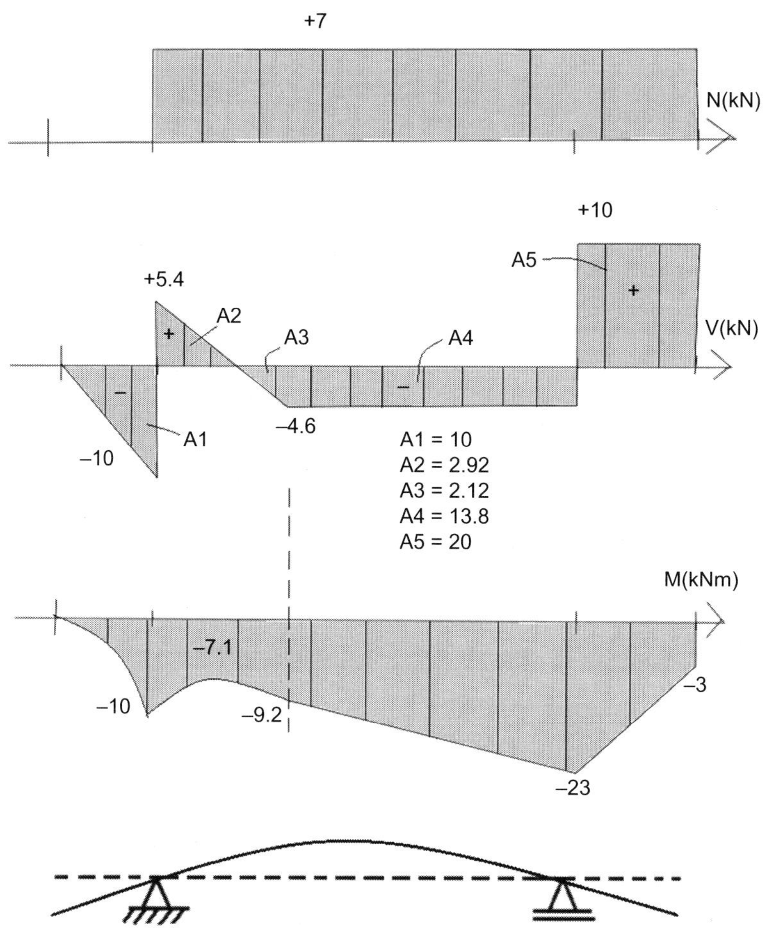

9.37 N, V and M diagrams and deflected shape of the system in Figure 9.36

a moment diagram. If the M diagram is not closed, this means that there is something wrong in the application of the three equations of equilibrium.

Deflected shape can be drawn with the help of the M diagram. If there is a positive moment in the member, this means that the deflected shape will be negatively curved. If there is negative moment, this means that the deflected shape will be positively curved, as seen in Figure 9.34. Determining the shape of the curvature forms the first step to drawing the deflected shape. Considering that there cannot be any deflection at the supports of the system, the deflected shape should therefore pass from the support points. This is the second step to drawing the deflected shape.

Concentrated forces cause rectangular areas in the V diagram and triangular areas in the M diagram. Distributed forces cause triangular areas in the V diagram and parabolic areas in the M diagram. To be able to draw the parabolic curves in the M diagram, one should divide the corresponding triangular area in the V diagram into small pieces and consider that the smaller areas will create less change, whilst the larger areas create more change, as seen in Figure 9.35.

EXAMPLE 9.1: Draw the N, V and M diagrams and the deflected shape of the system shown in Figure 9.36.

Figure 9.37 shows the N, V and M diagrams and the deflected shape of the system shown in Figure 9.36.

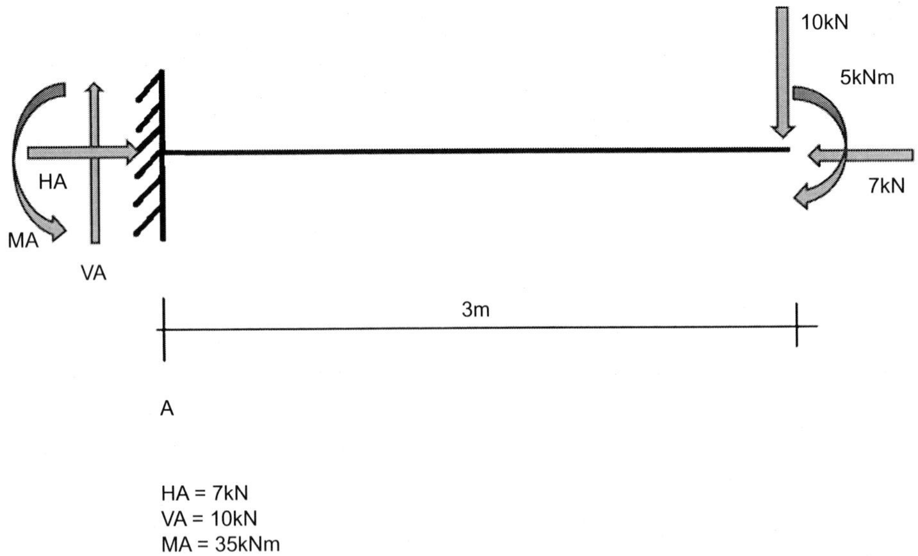

9.38 An alternative system to draw N, V and M diagrams and deflected shape

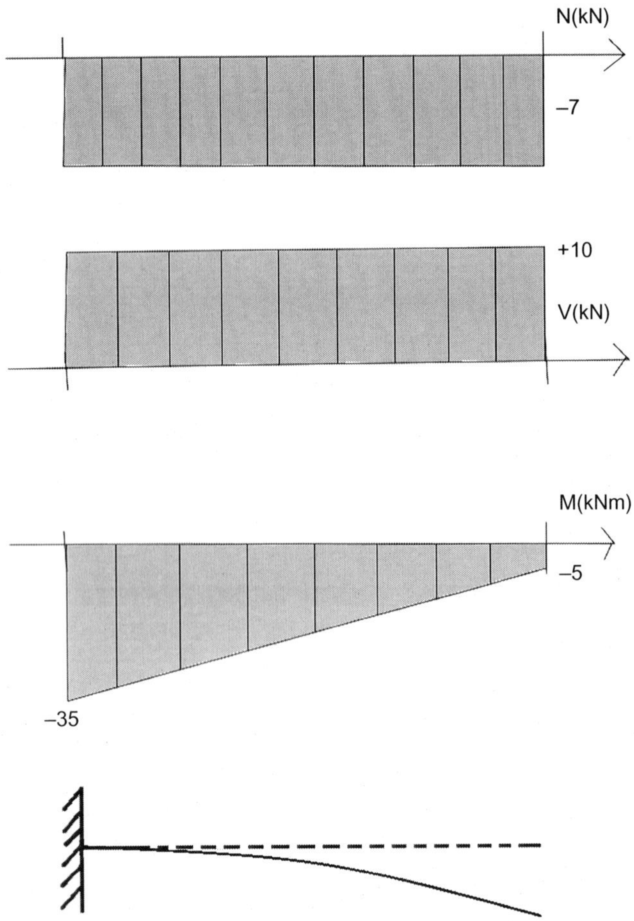

9.39 N, V and M diagrams and deflected shape of the system in Figure 9.38

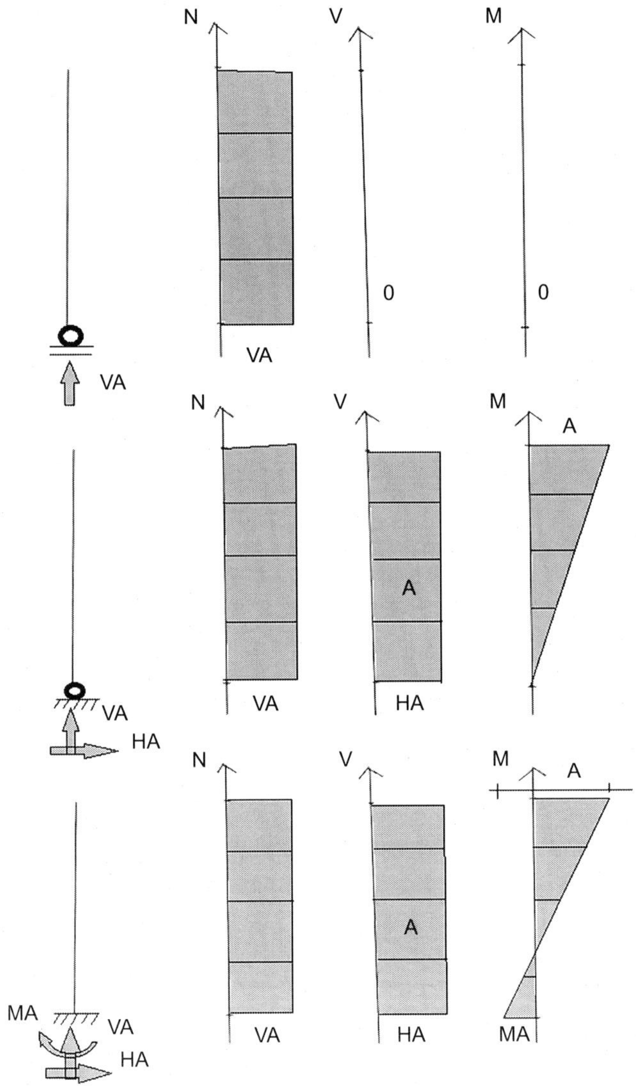

9.40 Types of columns in determinate systems

EXAMPLE 9.2: Draw the N, V and M diagrams and the deflected shape of the system shown in Figure 9.38.

Figure 9.39 shows the N, V and M diagrams and the deflected shape of the system shown in Figure 9.38.

The determinate systems can also have columns. As seen in Figure 9.40, there can be three types of columns according to the support type. The N, V and M diagrams and deflected shape of these columns are also drawn in the same way. The left side of the column is accepted as the positive side and the right side is accepted as the negative side. The axial force in columns is seen as shear in beams, and the shear force in columns is seen as an axial force in beams. Thus, there is no relationship between the columns and beams when drawing N and V diagrams. However, this is not true for the M diagram: the M value in columns affects the M value in beams. When drawing M diagrams, one should consider the equilibrium of joints.

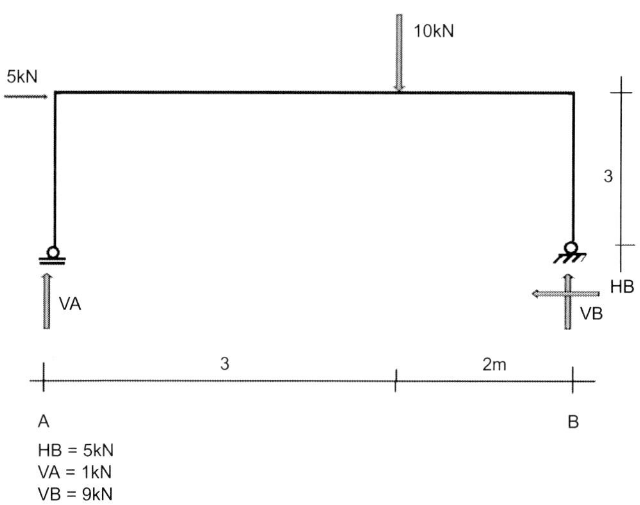

9.41 A system with columns to draw N, V and M diagrams and deflected shape

EXAMPLE 9.3: Draw the N, V and M diagrams and the deflected shape of the system shown in Figure 9.41.

Figure 9.42 shows the N, V and M diagrams and the deflected shape of the system shown in Figure 9.41.

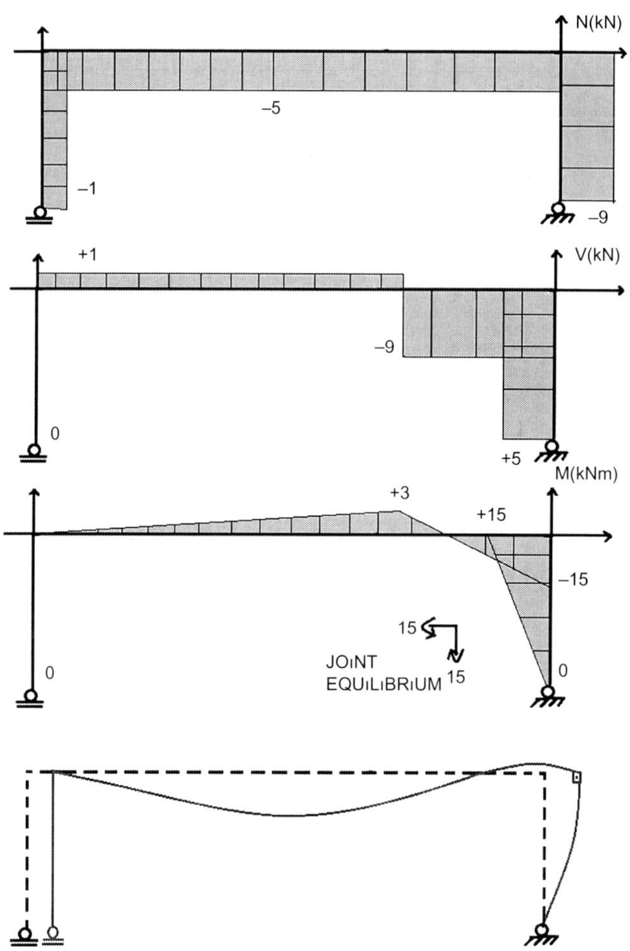

9.42 N, V and M diagrams and deflected shape of the system in Figure 9.41

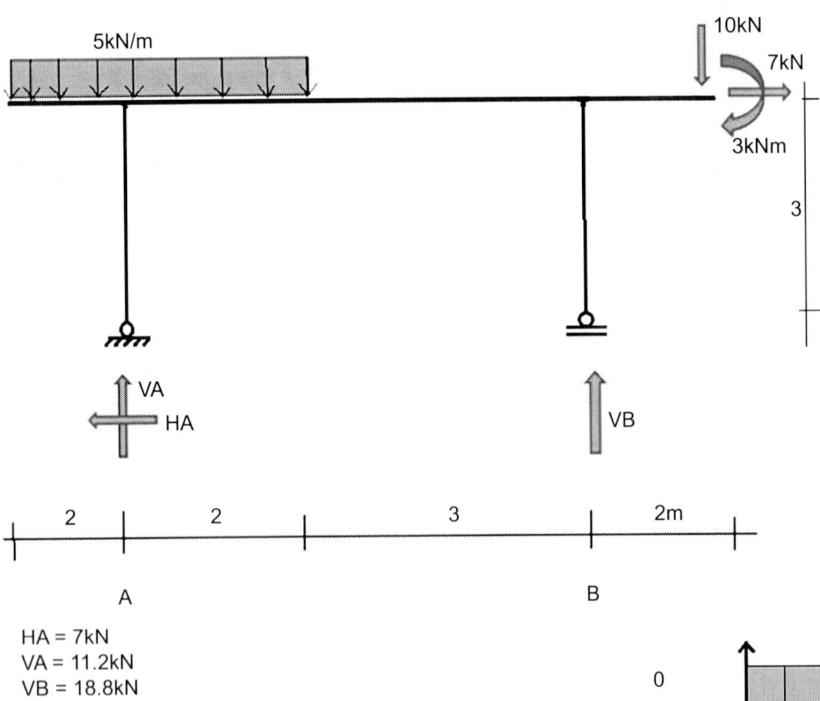

9.43 An alternative system with columns to draw N, V and M diagrams and deflected shape

EXAMPLE 9.4: Draw the N, V and M diagrams and the deflected shape of the system shown in Figure 9.43.

Figure 9.44 shows the N, V and M diagrams and the deflected shape of the system shown in Figure 9.43.

The Portal Method

The Portal Method is an approximate method of analysis of the effect of horizontal forces on frame systems that are not slender. It is used if the height of the structure is less than three times the width of the structure (h < 3 × w). According to Lin and Stotesbury's (1981: p.224) description of the method, the major assumptions of the Portal Method contain the following items:

- The moment at the top and bottom of each column are equal to each other, and the moment values at the middle of the columns are equal to zero.
- The moment at the left and right side of each beam are equal to each other, and the moment values at the middle of the beams are equal to zero.
- Shear values in interior columns are twice the shear values of exterior columns.

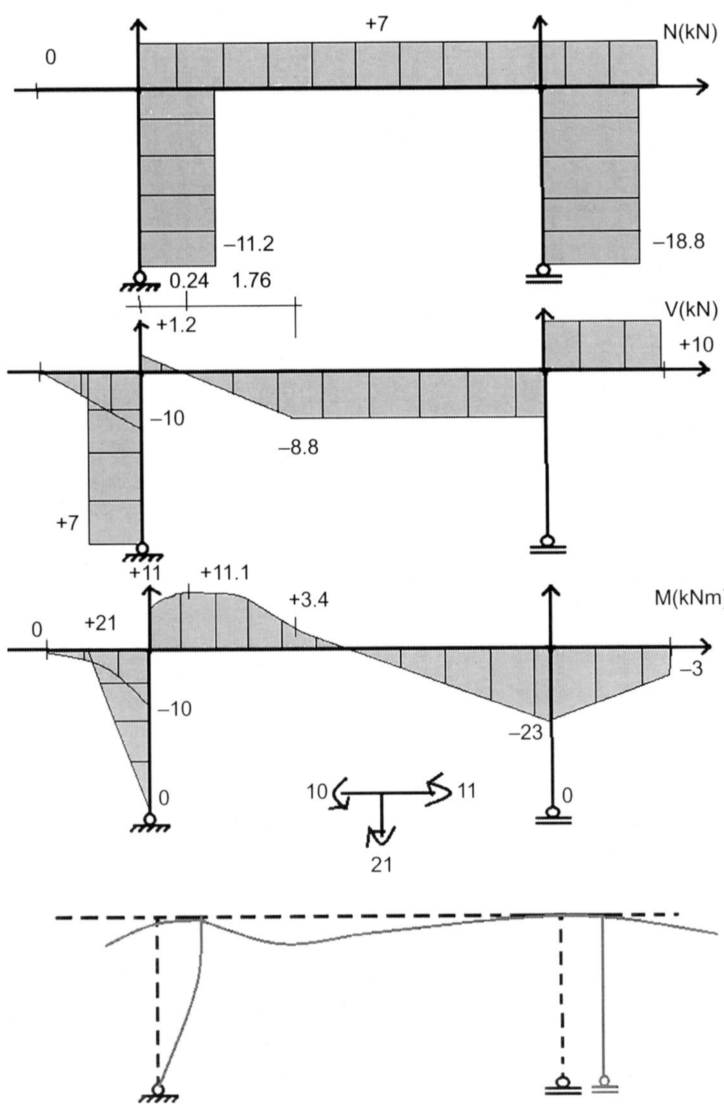

9.44 N, V and M diagrams and deflected shape of the system in Figure 9.43

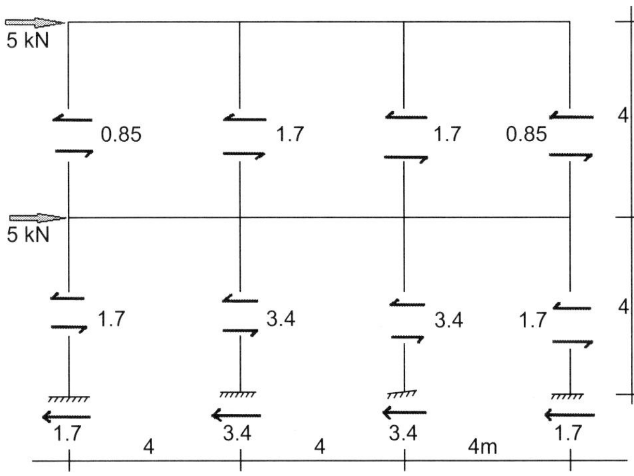

9.45 Finding shear in columns

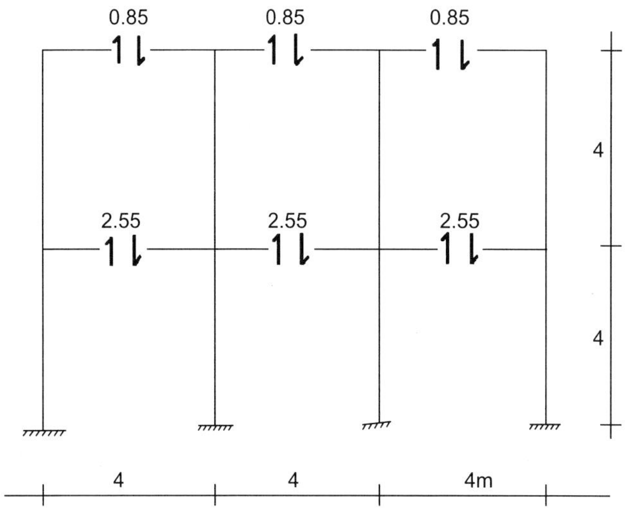

9.47 Finding shear in beams

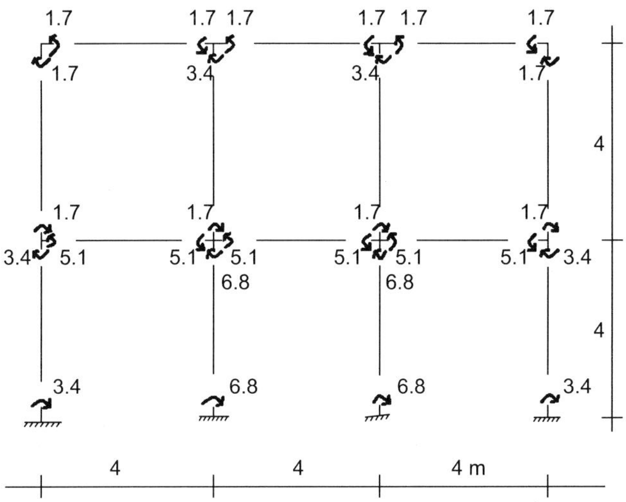

9.46 Finding moments in columns and beams

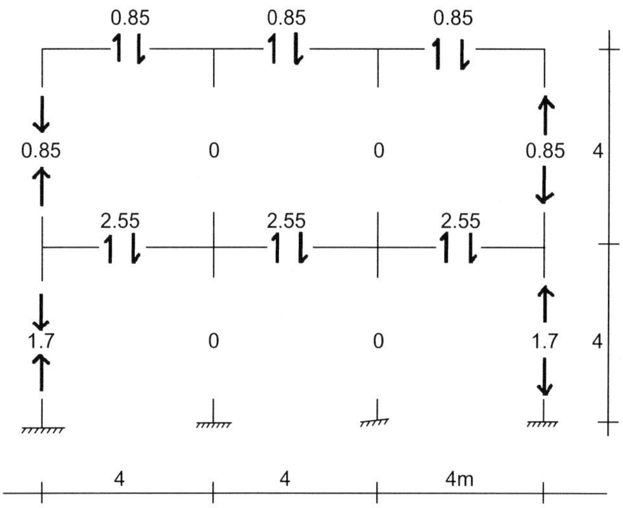

9.48 Finding axial force in columns

The steps for the Portal Method are as follows:

1 Determine the shear in the columns by considering that the shear in inside columns will be twice the shear in outer columns (see Figure 9.45). Consider that these shear forces should balance the horizontal loads.
2 Find the moment in the columns by considering that M = V × (l/2), where V is shear in the column and l is the length of the column (see Figure 9.46).
3 Find the moment in the beams by considering the joint equilibrium. Consider that positive moment presses the top of the element, whilst negative moment presses the bottom (see Figure 9.46).
4 Determine the shear in the beams by considering that V = M / (l/2), where M is moment in beams and l is the length of the beam (see Figure 9.47).
5 Find the axial force in the columns by considering the joint equilibrium (see Figure 9.48).
6 Find the axial force in the beams by considering the joint equilibrium (see Figure 9.49).

The N, V and M diagrams and the deflected shape of a frame are shown in Figure 9.50.

EXAMPLE 9.5: Draw the N, V and M diagrams and the deflected shape of the system shown in Figure 9.51.

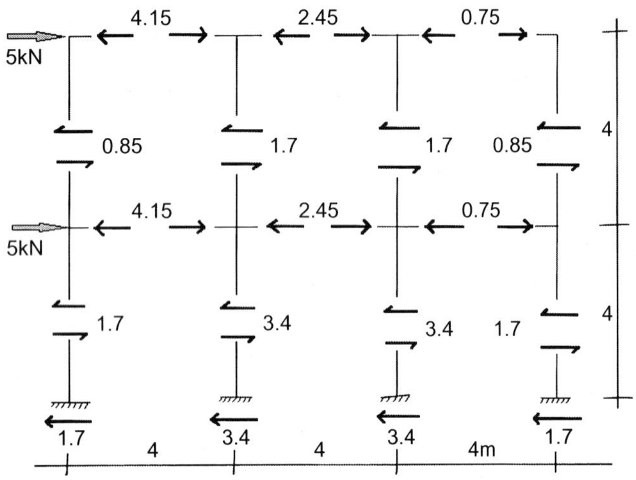

9.49 Finding axial force in beams

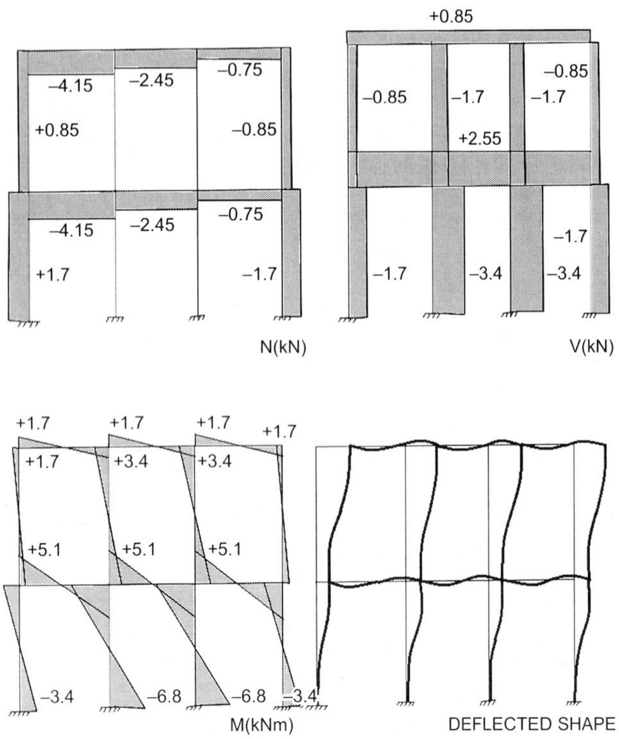

9.50 N, V and M diagrams and deflected shape of a frame

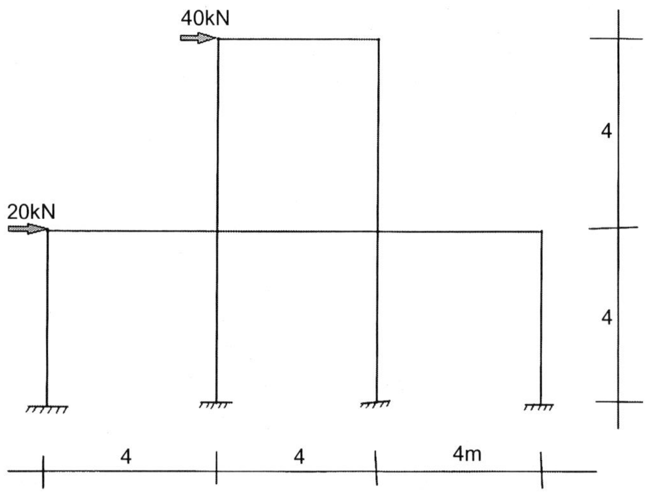

9.51 A frame to draw N, V and M diagrams and deflected shape

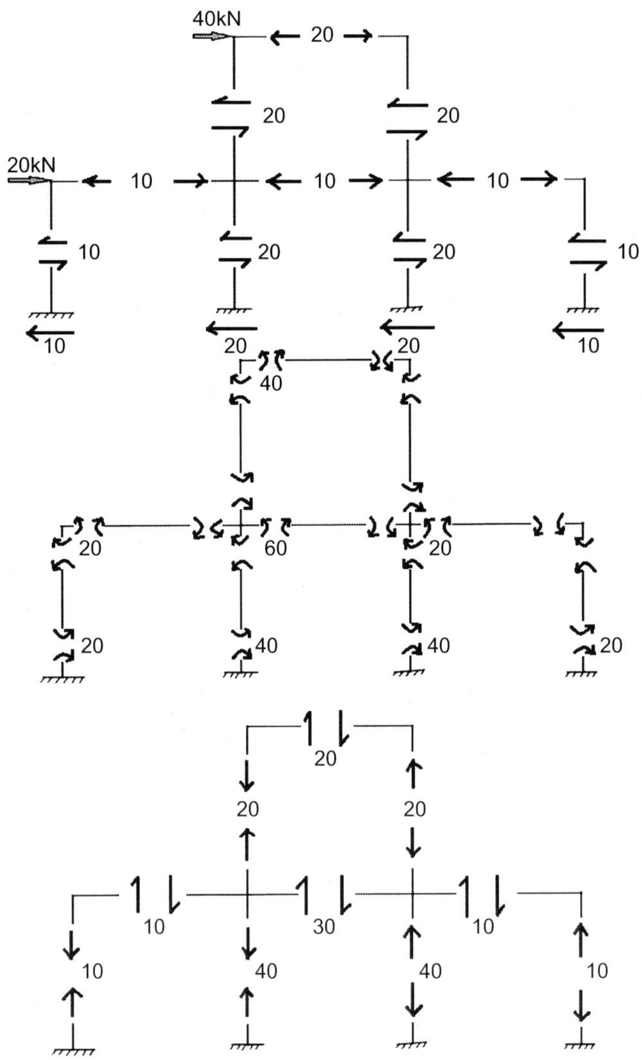

9.52 Finding internal forces in the elements of a frame

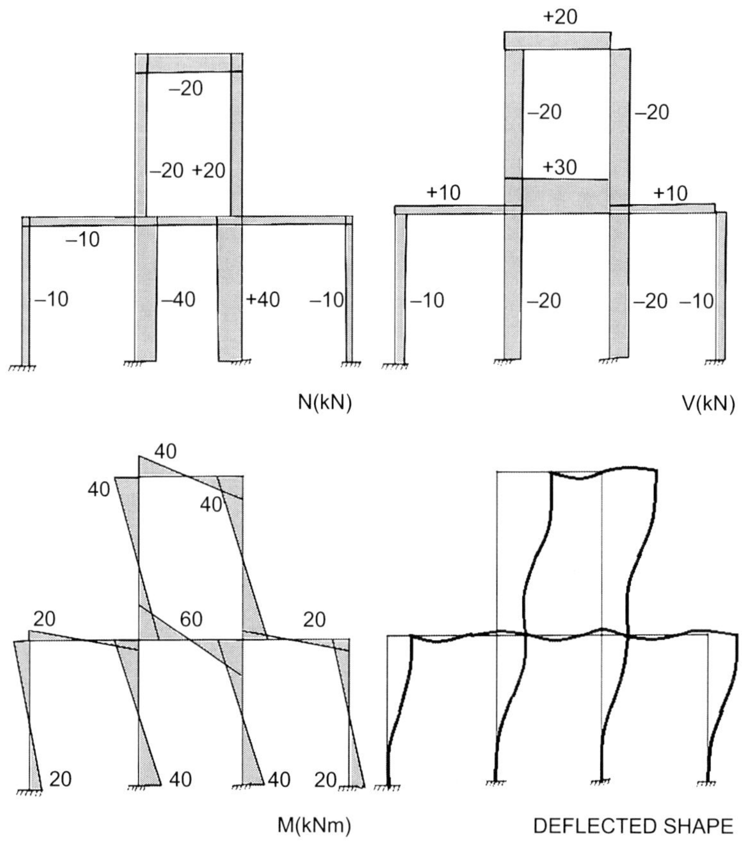

9.53 N, V and M diagrams and deflected shape of the frame in Figure 9.51

Internal forces in the elements of the frame can be found as shown in Figure 9.52.

Figure 9.53 shows the N, V and M diagrams and the deflected shape of the frame shown in Figure 9.51.

WHEN TO USE FRAME SYSTEMS

The preference for frame systems can be based on economy and/or limitations of certain spans of slabs/beams and certain height ranges of the building. The limits of reinforced concrete and steel beams and slabs are listed in Table 9.1.

Table 9.1 Limits of reinforced concrete and steel beams and slabs

Material	*Structural element*	*Span*
Reinforced concrete	Beam	Up to 15m (if high-strength reinforced concrete is used)
	Simple slab	Up to 7 × 7m
	Ribbed/waffled slab	Up to 15 × 15m (can reach 25m with triangular waffled slab)
Steel	Beam	Up to 20m
	Box girder	Up to 200m

Reinforced concrete frame systems are economical in buildings up to 20 storeys. If they are used together with shear walls, they become economical up to 50 storeys (Mir, 2001). Steel frames are economical in buildings up to 30 storeys. If they are used with steel shear walls, they become economical up to 40 storeys. If they are used with steel shear walls and belt trusses, they become economical up to 55 storeys (Dallaire, 1983). Economic limits of various applications for frame systems are shown in Figure 9.54.

ORGANISATION OF ELEMENTS

Beams, columns, shear walls, slabs, stairs, foundations and partition walls take place within a system in order to form a structure for a building. They are not arbitrarily placed: they have to transfer load to each other and act in unity against loads. Thus, their continuity should be provided. Load is transferred from slabs to beams, from beams to columns, and finally from columns to foundations, as shown in Figure 9.55.

Beams, columns and foundations come together to form frames, and frames come together to form three-dimensional frame systems, as seen in Figure 9.56. Each two-dimensional frame within this three-dimensional frame is analysed separately by structural engineers.

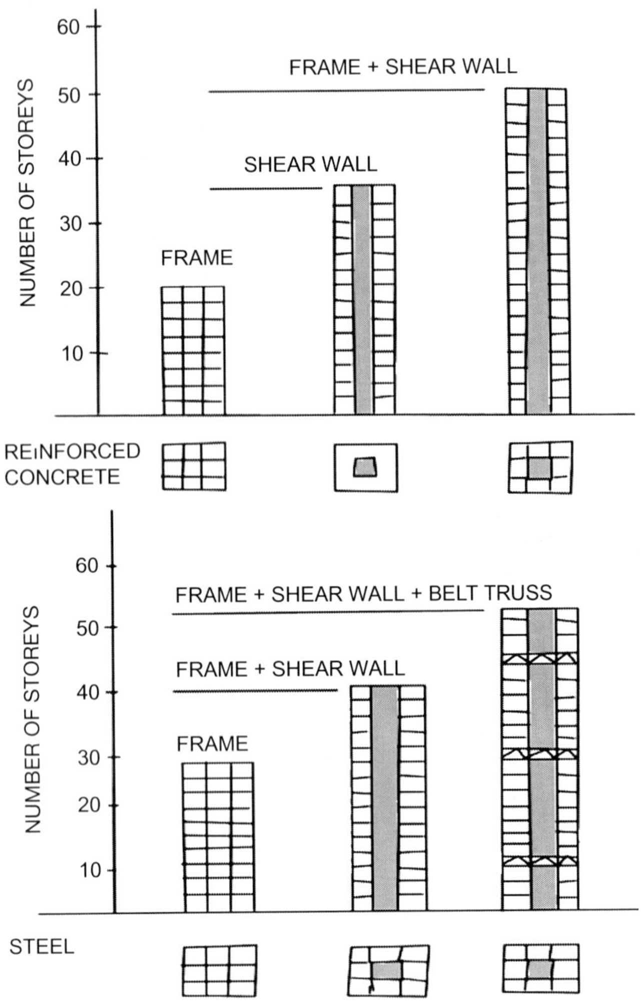

9.54 Economic height limits for various frame applications

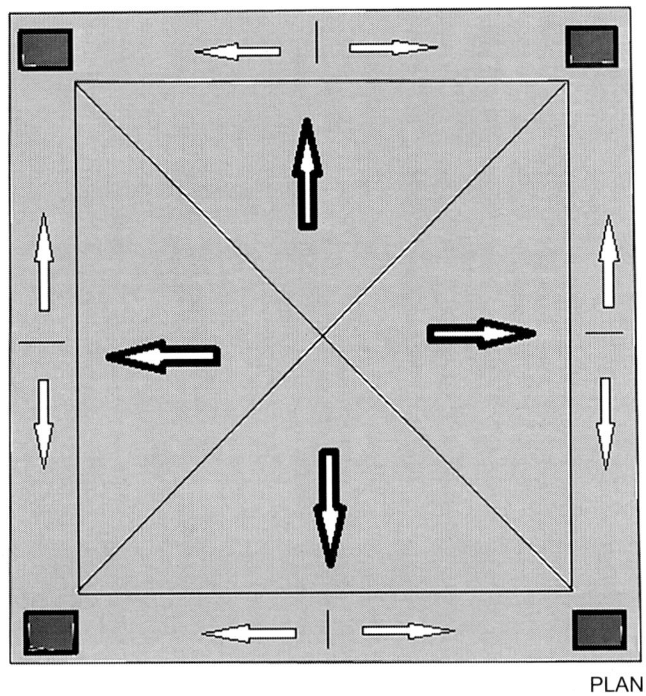

9.55 Transfer of load to foundations

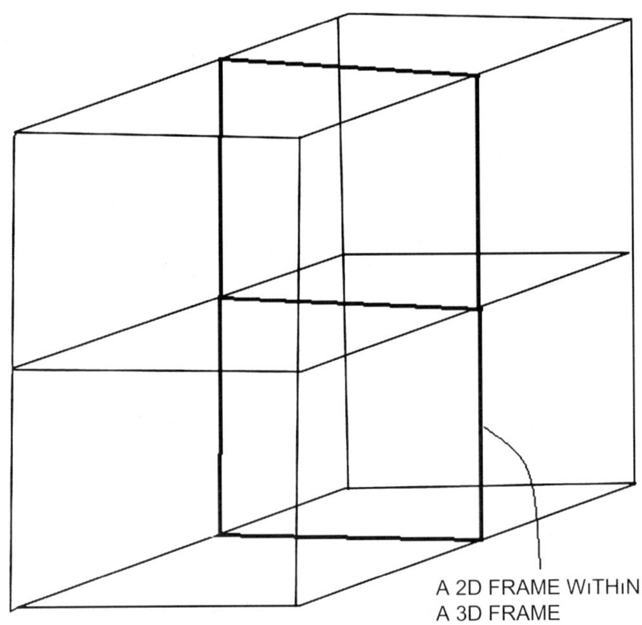

9.56 Two-dimensional frame systems taking place within a three-dimensional frame

Each two-dimensional frame forms a column axis in plan. There can be straight, curved and broken axes, as seen in Figure 9.57. Column axes might also have different angles with respect to each other. However, the structures that contain column axes with various angles will be more expensive than the structures with regular orthogonal axes, because building codes, such as the Turkish building code, may require more loading for the structural analysis of irregular axes (Ministry of Public Works and Settlement Government of the Republic of Turkey, 2007).

In a well-designed frame system, all column axes start from one end of the structure and end at the other end, as seen in Figure 9.58. It is not good design to have intersecting beams

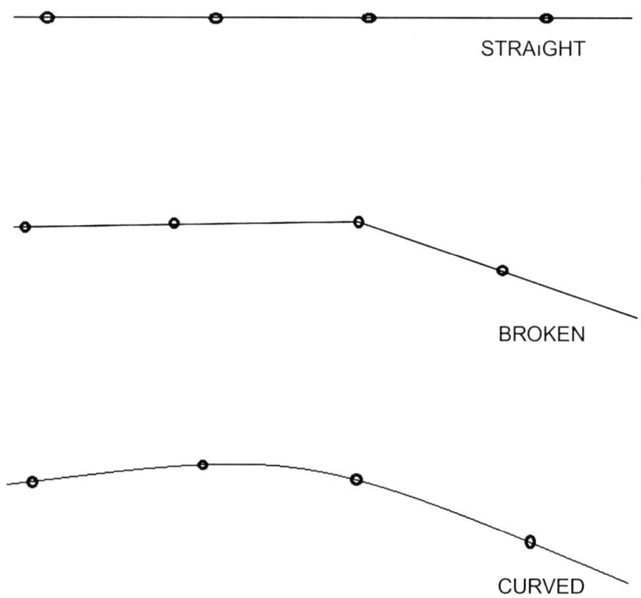

9.57 Types of column axes

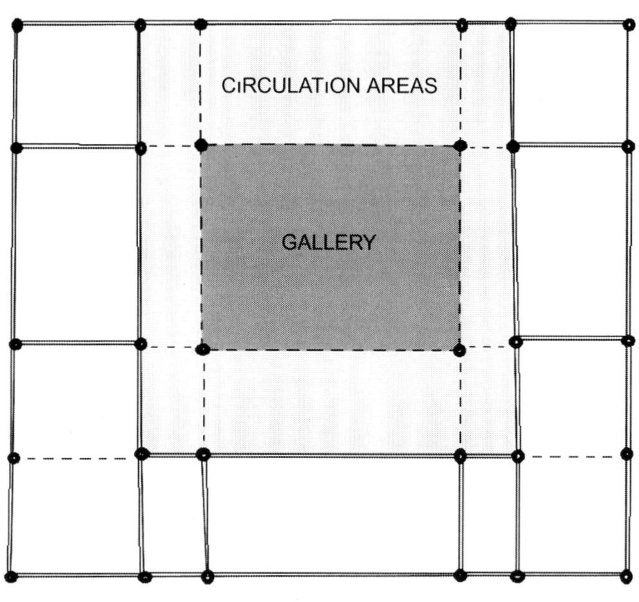

9.58 Column axes starting at one end of the structure and ending at the other, with intersecting beams and unconnected frame pieces

and unconnected frame pieces within a frame system. Intersecting beams cannot directly transfer horizontal loads to the columns; and unconnected frame pieces cannot behave in unity against horizontal loads.

EARTHQUAKE-RESISTANT DESIGN OF FRAME SYSTEMS

Frames should be resistant against dead-load as well as against horizontal loads, such as wind loads and earthquake loads. Building codes suggest different applications for high and low earthquake risks. Places of the world with a high earthquake risk are shown in Figure 9.59. As seen from this figure, the shores of the Pacific Ocean, southern Europe, the Middle East, mid-Asia and Japan form the high-risk earthquake regions (University of California, Seismological Laboratory, 2008).

The earthquake load is proportional to the mass of the building. The Peru building code suggests reducing weight in order to combat earthquake load (NTE E.030, 2003). There are many structural engineering books concerning earthquake-resistance of buildings, such as Charleson's (2008) *Seismic Design for Architects: Outwitting the Quake*. These books and reports written after earthquakes cover many problems in buildings that have been affected by an earthquake.

This book describes the problems that can be eliminated or reduced through architectural design. The following problems are addressed and categorised with the help of the International Building Code (ICC, 2000) and the Turkish building code (Ministry of Public Works and Settlement Government of the Republic of Turkey, 2007):

- Dimensions of columns in relation to dimensions of beams.
- Plan irregularities (including: general shape of plan; total area of galleries; all discontinuities in a horizontal force resistance path; and twisting instability).
- Vertical irregularities (including: soft and weak storey; short column; and weight irregularity).

If earthquake risk is high and the building's height is over 80m, the International Building Code does not allow any irregularities to

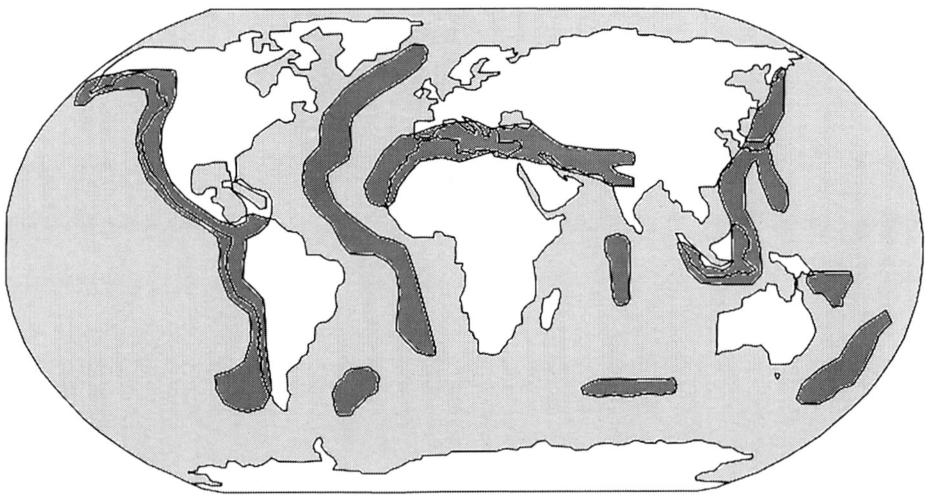

9.59 Earthquake map of the world (drawn with the help of URL1, n.d.)

exist in an architectural project. If the building is under 80m high, extreme twisting instability, and extreme soft- and weak-storey problems should be avoided in regions with high earthquake risk. The other problems are considered during the structural analysis by increasing the load acting on the building.

However, the International Building Code of 2012 (ICC IBC, 2012) does not contain these categories, and structural analysis is preferred as a tool to eliminate potential problems.

Dimensions of columns in relation to dimensions of beams

Horizontal loads directly cause shear and moment forces in columns. Thus, the architectural dream of having very few and slender columns might not be realised if there is a high earthquake risk. However, if the earthquake risk is low, having slender columns is possible for low-rise structures: 25cm × 25cm columns can be sufficiently strong for this type of structure.

There are two strategies to follow in high-risk earthquake regions:

- Using shear walls in a systematic way, together with slender columns.
- Making the columns thicker.

Different countries' building codes suggest these solutions depending on the economic condition of the country. For example, Japanese building code suggests the first strategy (Paz, 1994), while Turkish building code only suggests the use of shear walls if the building is higher than 13m. For lower buildings, the second strategy is suggested (Ministry of Public Works and Settlement Government of the Republic of Turkey, 2007).

According to the Turkish building code, columns should be able to carry 20% more moment in comparison to the moment carried by beams. The author of this book suggests that architects may translate this into physical terms by saying that column dimensions should be similar to beam dimensions. For example, if two columns carry a 5m long reinforced concrete beam, which is 50cm deep, then the dimensions of these columns can be 25cm × 50cm, having the 50cm similar to the depth of the beam. It is possible to have one or two slender columns within a structure. However, the majority of columns should have similar dimensions to the beams.

Plan irregularities

Plan shapes that contain deep recesses are not recommended if there is a high earthquake risk. It is better to separate these deep

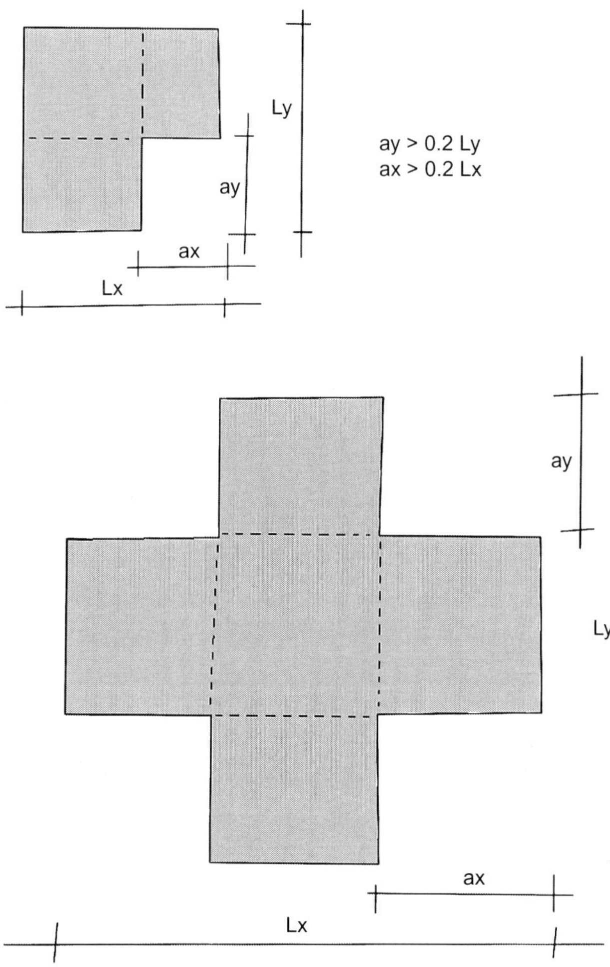

9.60 Separating deep recesses with the help of expansion joints

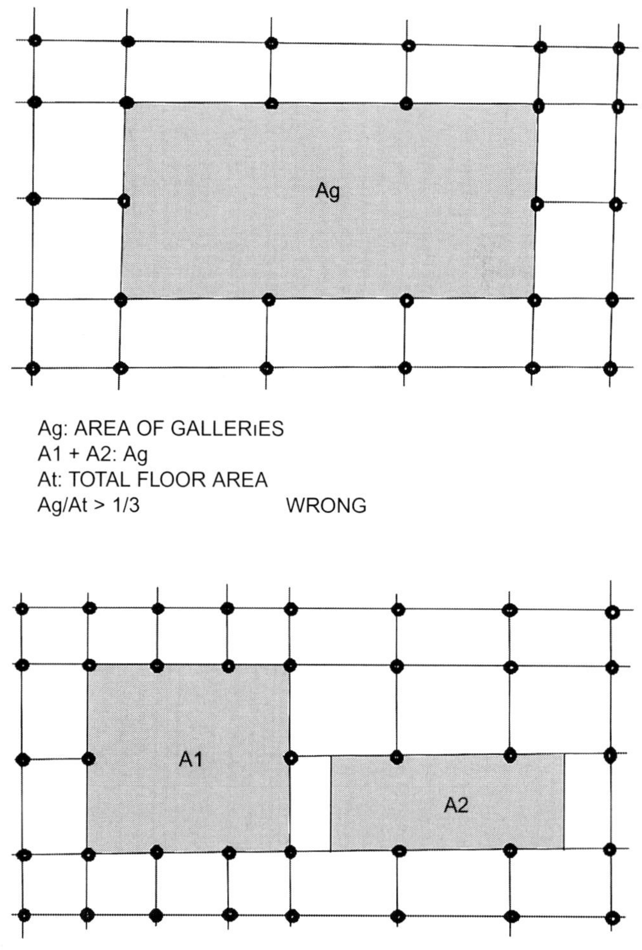

9.61 Examples of problematic arrangements for galleries

recesses from the rest of the structure with the help of expansion joints, as seen in Figure 9.60. According to the Turkish building code, these recesses should not exceed 20% of the whole length of the building (Ministry of Public Works and Settlement Government of the Republic of Turkey, 2007).

Large galleries are not recommended in slab structures. The area of galleries in a slab should not exceed 1/3 of the slab area. Abrupt discontinuities in these slabs eliminate the diaphragm behaviour of the slabs. Figure 9.61 shows some problematic organisations for galleries.

All discontinuities in a horizontal force resistance path, such as having discontinuous axes, beams intersecting each other, and having non-parallel axes, are not recommended, as discussed earlier in this chapter.

Twisting instability is one of the major irregularities that frequently cause problems in earthquakes. Buildings can be twisted

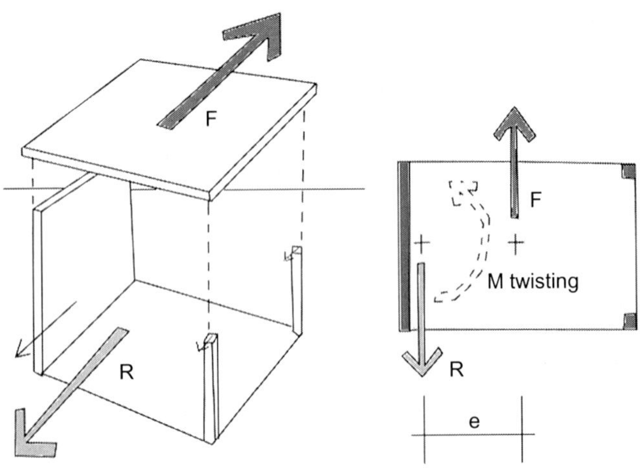

9.62 Earthquake force (F) and resistance of structure (R)

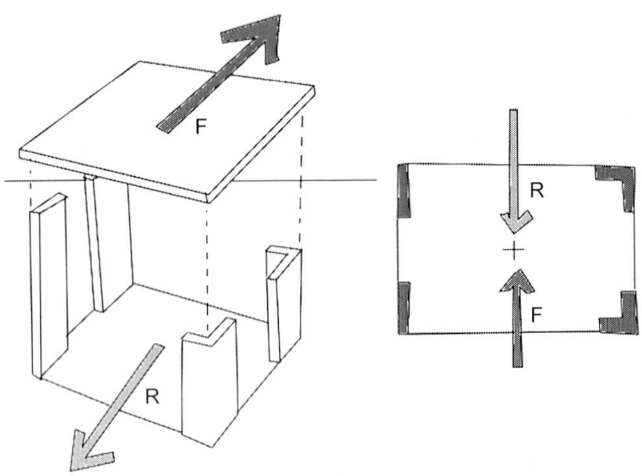

9.63 Reducing eccentricity

around themselves during earthquakes. To understand twisting instability, it is necessary to know where on plan the earthquake force is applied, and the source of the structure's resistance. The earthquake force is applied to the centre of the geometry of the plan, whilst resistance of the structure comes from the centre of gravity of the structural plan, as seen in Figure 9.62.

The distance between the two forces is called eccentricity (e). If eccentricity is high, the two forces act as a couple to create a twisting moment. This problem can be solved in two different ways:

- Designing the structural plan as symmetrical.
- Reducing eccentricity by balancing the structural plan as seen in Figure 9.63.

Vertical irregularities

The soft-storey problem is one of the major vertical irregularities that cause problems in earthquake. Frames are flexible structures, but the use of rigid partition walls (such as brick walls) eliminates the flexible movement of frames. Thus, the placement of rigid partition walls in frames can cause significant problems. Figure 9.64 shows some acceptable and unacceptable arrangements of rigid partition walls and windows within frame systems.

If the bays of the frame are filled with windows, this will be a flexible but acceptable structure. If the bays are filled with rigid partition walls, this will be a rigid and acceptable structure. However, if one of the lower storey's bays is filled with windows and the bays of the rest of the structure are filled with rigid partition walls, then the level with windows is called soft storey.

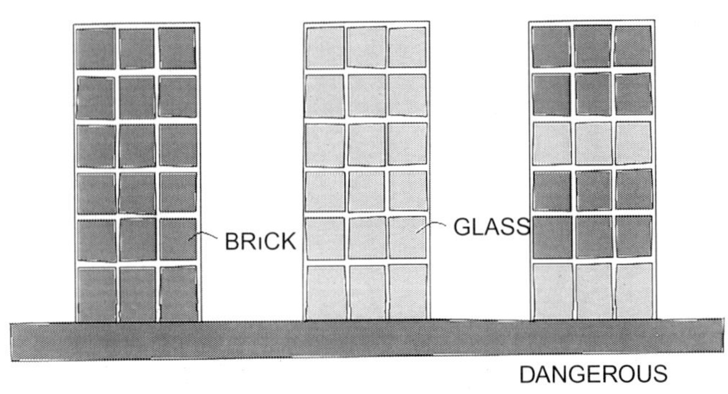

9.64 Acceptable and unacceptable arrangements of rigid partition walls and windows within frame systems

Stiffness of this level is considerably less than the stiffness of the other storeys. Columns of the storeys with rigid walls cannot bend, while the columns of the soft storey are bent too much, and so these columns might collapse due to excessive bending.

The soft-storey problem can be solved in the following ways:

- Replacing rigid partition walls with more flexible walls, such as metal panels.
- Using bracing for the soft storey (see Figure 9.65a).
- Increasing the thickness of columns at the soft-storey level (see Figure 9.65b).
- Increasing the number of columns at the soft-storey level (see Figure 9.65c).
- Adding shear walls to the system (see Figure 9.65d).
- Decreasing the size of openings at the soft-storey level (see Figure 9.65e).
- Placing elastic material between the structural elements and the rigid walls, to free the movement of elements (see Figure 9.65f).

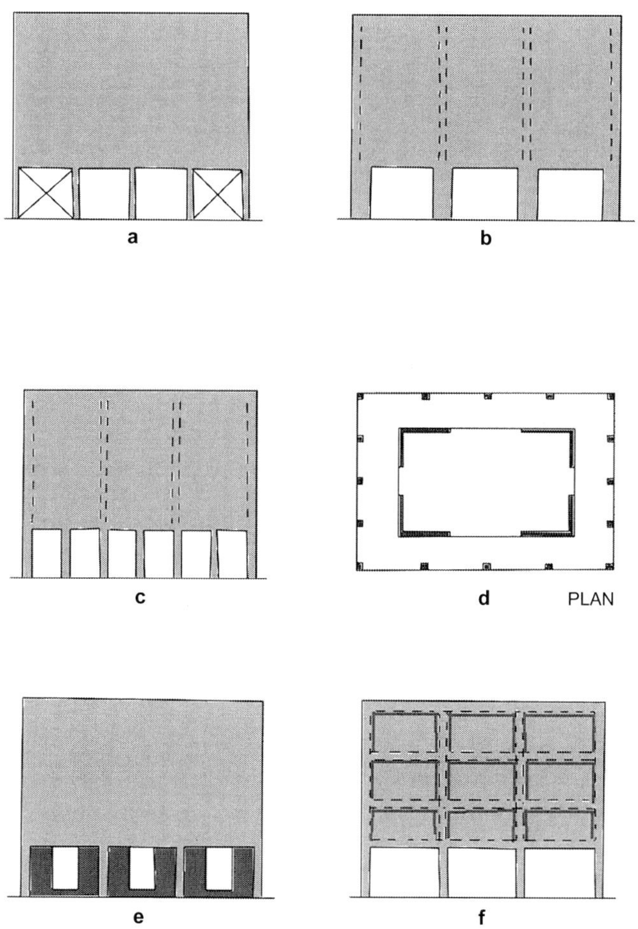

9.65 Examples to solve soft-storey problems

Most of the solutions are architectural solutions, except for increasing the thickness of columns, which can be achieved under the control of a structural engineer.

Weak-storey problem occurs if the strength of a storey against horizontal loads is less than 80% of the strength of the storey above (ICC, 2000). This problem can occur if the dimensions of columns and shear walls are reduced at an intermediate level. Another cause of the problem can be due to the removal of rigid partition walls at one intermediate level.

Short-column problem is also due to the incorrect use of rigid partition walls within a flexible frame. Another cause of the short-column problem is the form of openings within these rigid partition walls. As seen in Figure 9.66, most forms of openings are acceptable: however, ribbon windows can cause short-column problems. Since the rigid walls eliminate deflection of the columns' lower parts, there occurs a concentration of shear within the columns' upper parts beside the ribbon windows. These columns can be cut due to excessive shear. It is possible to make the toilet paper analogy here: because of the holes organised over lines, it is easier to tear the toilet paper through these lines.

If the short-column problem exists due to the form of openings, it can be eliminated by taking the following precautions:

- Replacing rigid partition walls with more flexible walls, such as metal panels.
- Changing the form of the opening.
- Increasing the thickness of short columns.
- Adding shear walls to the system.

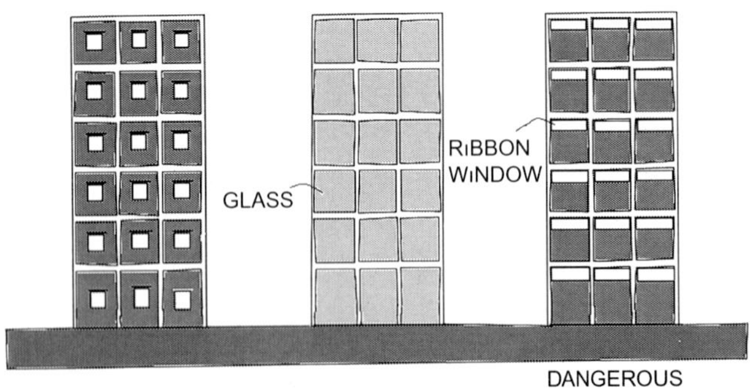

9.66 Form of openings and short-column problems

- Placing elastic material between the structural elements and the rigid walls, to free the movement of elements.

Short-column problems can also occur at various different situations where the height of the column is reduced for some reason. Figure 9.67 shows that short-column problems might occur at staircases and inclined sites.

Weight irregularity problems occur if the weight of one storey is more than 150% of the adjacent storeys.

STRUCTURAL GUIDELINES FOR FRAME AND SHEAR WALL SYSTEMS

Structural guidelines for frame and shear wall systems can be presented under the following three categories:

- General structural guidelines for all structural materials.
- Structural guidelines for reinforced concrete frame (and shear wall) systems.
- Structural guidelines for steel frame (and shear wall) systems.

These structural guidelines are listed in Tables 9.2, 9.3 and 9.4.

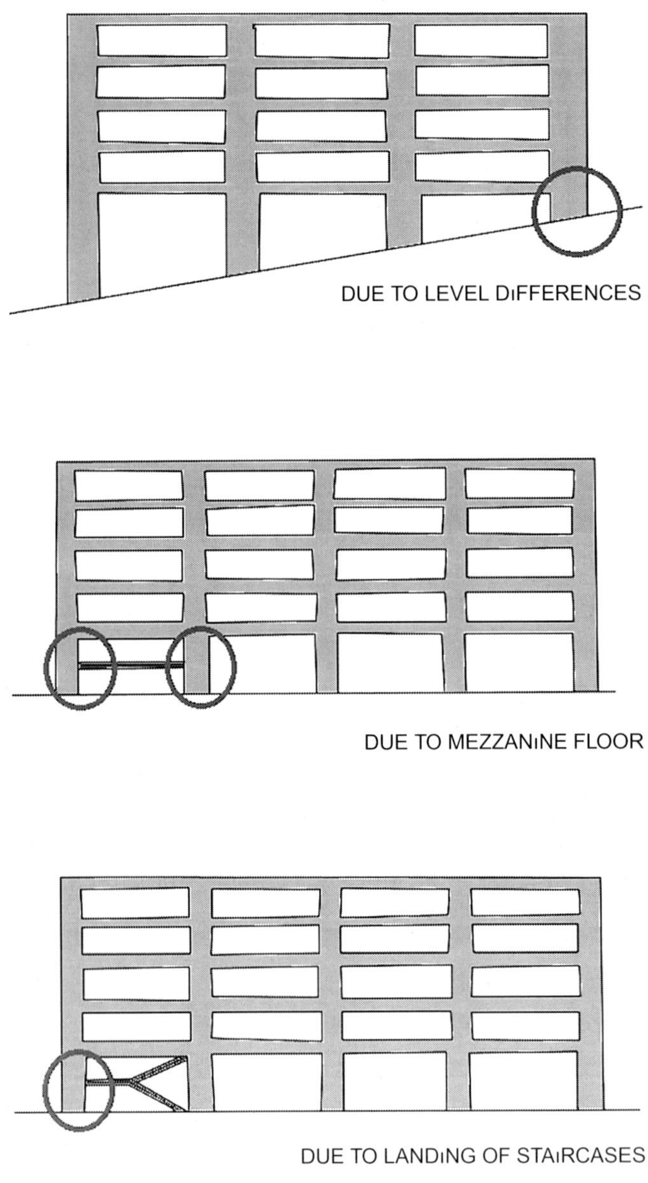

9.67 Examples of causes of short-column problems

Table 9.2 General structural guidelines for frame (and shear wall) systems with any structural material and the associated value system

Structural guidelines	*Value system*
Elements of frame systems should be continuous. The loads from slabs should be directly transferred to beams; the loads from beams should be directly transferred to columns; and the loads from columns should be directly transferred to foundations.	Safety
Form of the elements of frame systems should be designed to reduce bending moment.	Economy
It is less problematic to avoid strongly curved beams or beams with corners in order to prevent twisting.	Economy
Columns should not be slender in order to avoid buckling.	Safety
Columns and shear walls should reach foundations.	Safety
Directions of columns and shear walls should be well distributed on plan.	Safety
Rigid partition walls should be placed over beams.	Safety
Building parts with considerable height differences should be structurally separated by expansion joints.	Economy
It is better to separate building parts with different structural materials/structural systems with the help of expansion joints.	Economy
If there is high earthquake risk and if the plan shape contains deep recesses, these parts should be separated from the main body with the help of expansion joints.	Economy
If the plan is longer than 30m × 30m, the structure should be divided into different parts with the help of expansion joints.	Economy
Minimum width of an expansion joint is 3cm. If the building is higher than 6m, 1cm should be added to this value for each 3m of height.	Safety
Slabs adjacent to stairs should be surrounded by beams.	Safety
Column axes should start from one end of the building and end at the other end to be able to resist horizontal loads. It is better not to have any intersecting beams and unconnected frame pieces if there is a high earthquake risk.	Safety
Building weight can be reduced to decrease earthquake load.	Economy
If there is high earthquake risk and if the building height is over 80m, irregularity problems in the structure should be eliminated.	Safety
Either shear walls should be used systematically, or columns should have similar dimensions to beams, in order to resist earthquake loads. If the building height is over 13m, there should be shear walls.	Safety
If there is high earthquake risk, area of galleries should not be over 1/3 of the slab area. There should not be abrupt discontinuities in the slabs in order to distribute earthquake load evenly to the vertical elements of the structure.	Safety
If there is high earthquake risk, twisting instability problems should be solved either by designing a symmetrical structure or by balancing it to reduce eccentricity.	Safety
If there is high earthquake risk, soft-storey problems should be eliminated.	Safety
If there is high earthquake risk, weak-storey problems should be eliminated.	Safety
If there is high earthquake risk, it is better to eliminate short-column problems during the architectural design stage.	Economy
If there is high earthquake risk, it is better to solve weight irregularity problems during the architectural design stage.	Safety

Table 9.3 Structural guidelines for reinforced concrete frame (and shear wall) systems and the associated value system

Structural guidelines	*Value system*
Depth of a reinforced concrete beam can be length/15. The minimum depth is 30cm.	Safety
Minimum width for a reinforced concrete beam should be 20cm.	Practicality
High-strength reinforced concrete beams can span up to 15m.	Safety
Optimum span for a reinforced concrete beam is 4.5–5m.	Economy
Minimum dimensions of a reinforced concrete column can be 25cm × 30cm.	Safety
Shorter span of one-way and two-way slabs can be maximum 7m.	Economy
One-way slabs are used for rectangular slabs and two-way slabs are used for forms close to square.	Economy
Depth of a one-way slab can be between length/20 and length/30; and depth of a two-way slab can be between length/30 and length/40. Minimum depth for these types of slabs is 9cm.	Economy
Pre-stressed flat slabs can span up to 8–12m. However, it is better not to exceed 4m if there is high earthquake risk.	Safety
Ribbed and waffled slabs are used for spans longer than 7m. Waffled slabs can span up to 15m.	Economy
Span to depth ratio of waffled slab can change between 15 and 20.	Economy
Very deep and very shallow (but wide) beams should be avoided.	Safety
Minimum thickness of reinforced concrete shear walls is 20cm. Minimum length of shear walls is 7 times their thickness.	Safety
In a shear wall structure the ratio of shear wall area to floor area can be between 5% and 10%.	Safety
Reinforced concrete frame systems can be economic up to 20 storeys. If they are used together with shear walls, they can be economic up to 50 storeys.	Economy

Table 9.4 Structural guidelines for steel frame (and shear wall) systems and the associated value system

Structural guidelines	*Value system*
Depth of a steel beam can be length/20.	Safety
Optimum span of steel beams is 7m. However, they can span up to 20m.	Economy
Box girders can span up to 200m.	Economy
Secondary beams of steel slabs can be 7–20m long and they can be placed 2–5m apart. Secondary trusses can be spaced 1–3m apart.	Economy
Steel frames can be economic up to 30 storeys. If they are used with shear walls, they can be economic up to 55 storeys.	Economy

CASE STUDY 12: VILLA SAVOYE, FRANCE

The most important architectural characteristics of Le Corbusier's Villa Savoye are its simple geometric form, the presence of pilotis that raise the building mass over columns, and the use of ribbon windows (see Figures 9.68 and 9.69). Le Corbusier said that it is possible to see the horizon without any break through these windows (Leatherbarrow & Mostafavi, 2005: pp.42–43). Since there were few buildings with frame systems in 1928, this was a strikingly different architectural characteristic. Le Corbusier was

9.68 A sketch of case study 12: Villa Savoye, Poissy, France, 1928 (drawn with the help of URL2, 2011)

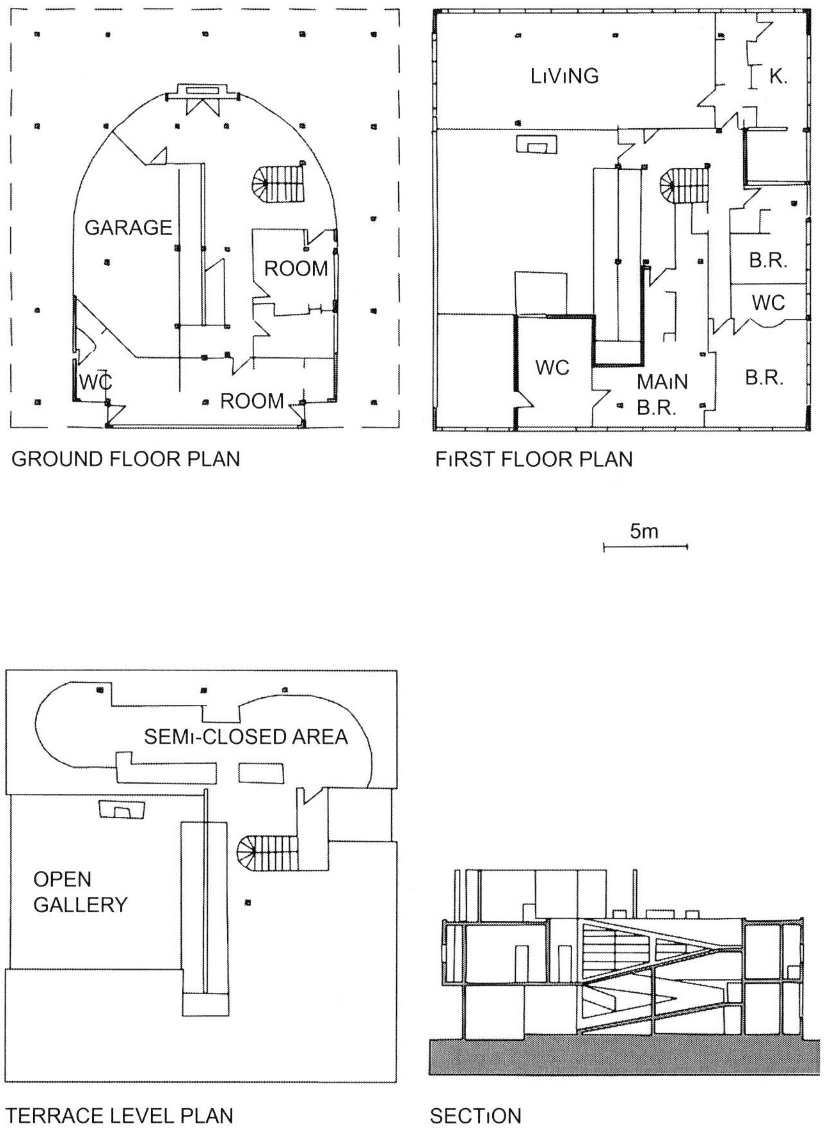

9.69 Plans and section of Villa Savoye (drawn with the help of URL3, 2011)

providing continuity between indoor and outdoor spaces. He was also providing continuity between the different indoor spaces. Such continuity was becoming possible because of the tectonic characteristics of frame systems. Since it is possible to remove the partition walls; continuity can be achieved. Buildings with frame systems are much lighter than buildings with masonry systems. The dominant physical entity in achieving the tectonic qualities of Villa Savoye is structure.

When the characteristics of Villa Savoye are examined with the help of the structural guidelines listed in Tables 9.2 and 9.3, it can easily be seen that such a building could not be built if there was high earthquake risk. Pilotis cause soft-storey and weak-storey problems and ribbon windows cause short-column problems. Furthermore, the columns are slender and thus not as strong as the beams. Although the building is nearly symmetrical, these problems would not allow the same style to be applied in high-risk earthquake regions. However, since Villa Savoye is not in a high-risk earthquake region, it can be stated that it has an affirmative relationship with structural guidelines, because it achieves its tectonic qualities without contravening them.

CASE STUDY 13: NATIONAL ASSEMBLY IN DACCA, BANGLADESH

For the National Assembly in Dacca, Louis Kahn did not try to achieve simple geometric forms. He was dividing the mass into pieces by separating different types of functions from each other (Leatherbarrow & Mostafavi, 2005: pp.215–226). He was also surrounding these masses with a wall, which forms an envelope as part of a climatic response (see Figures 9.70 and 9.71). Kahn used frames and shear walls together to form the structure of

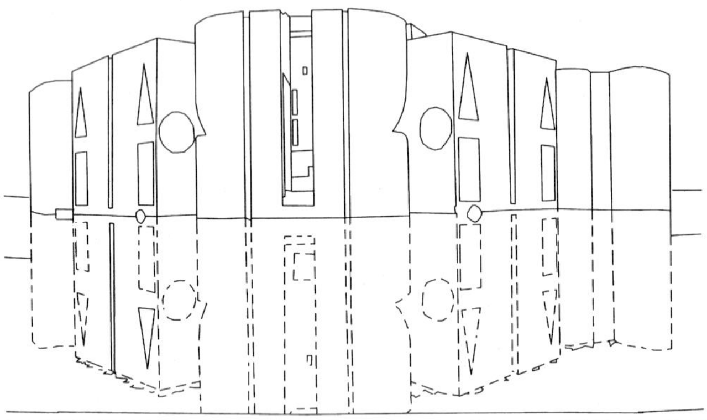

9.70 A sketch of case study 13: National Assembly in Dacca, Bangladesh, 1962–1974 (drawn with the help of URL4, 2004)

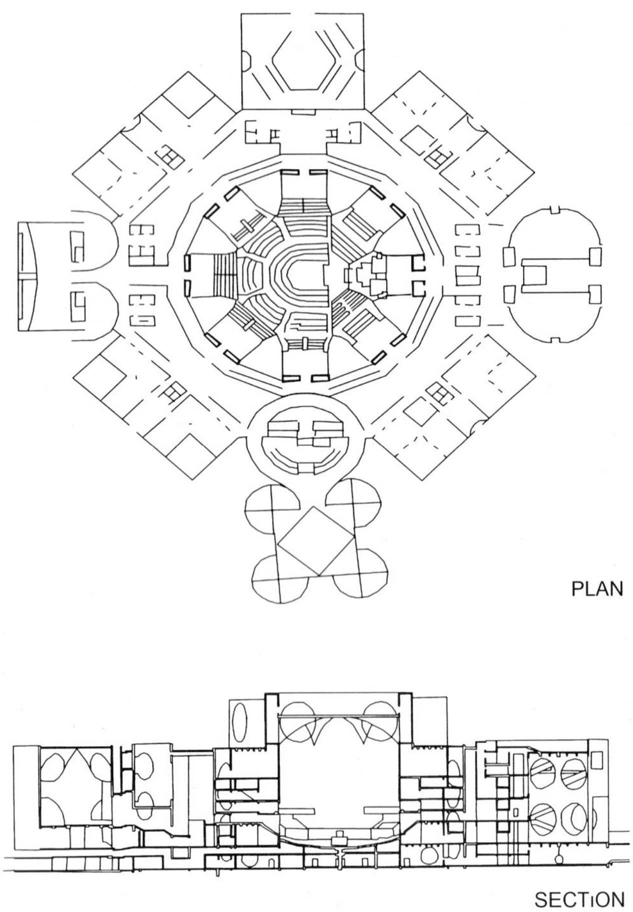

9.71 Plan and section of Dacca National Assembly (drawn with the help of URL5, 2014)

his buildings. Kahn's structures were not economic because some structure axes were too close to each other (Frampton, 2001: pp.232–233).

The indoor spaces between the masses of the National Assembly building in Dacca have the characteristics of outdoor spaces. Street furniture is used in these spaces and natural light penetrates from the roof. The walls surrounding the National Assembly building are stone and they have a haptic effect. However, the openings on these walls have simple geometric forms, which are optic. Thus the optic openings on haptic walls make a contrast. The dominant physical entities in achieving the tectonic characteristics of the National Assembly in Dacca are materials, details (such as the details of openings) and structure.

When the plan and section of National Assembly are analysed with the help of the structural guidelines as listed in Tables 9.2 and 9.3, it is understood that there should be some expansion joints separating the structure into various parts because this is a large building and the hall in the middle has a different roof structure.

There are large galleries between the masses that form the building. Since the building is in a seismic zone, it could have been difficult to achieve such large galleries. However, one can guess that there are expansion joints between the masses. There are galleries between these masses, but they do not work as galleries dividing horizontal diaphragms into pieces. The problem created as a result of the design was compensated with the help of expansion joints. The relationship of the National Assembly in Dacca to structural guidelines is therefore contravening because it achieves the interior spaces despite the limitations of earthquake regions.

CASE STUDY 14: CHURCH OF LIGHT AND SUNDAY SCHOOL, JAPAN

Tadao Ando's projects, especially the Church of Light and Sunday School, form successful tectonic examples of reinforced concrete shear walls. The most striking tectonic characteristic of these buildings is the form of its windows (see Figures 9.72 and 9.73). One is in the form of a cross and the other is between the two shear walls with different angles. The mullions of these windows

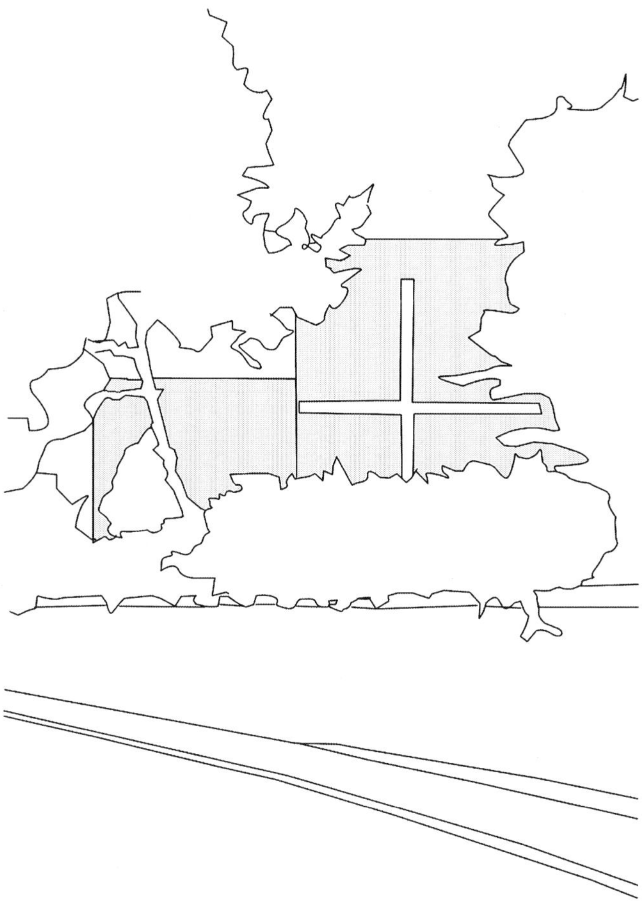

9.72 A sketch of case study 14: Church of Light and Sunday School, Osaka, Japan, 1999 (drawn with the help of URL6, 2006)

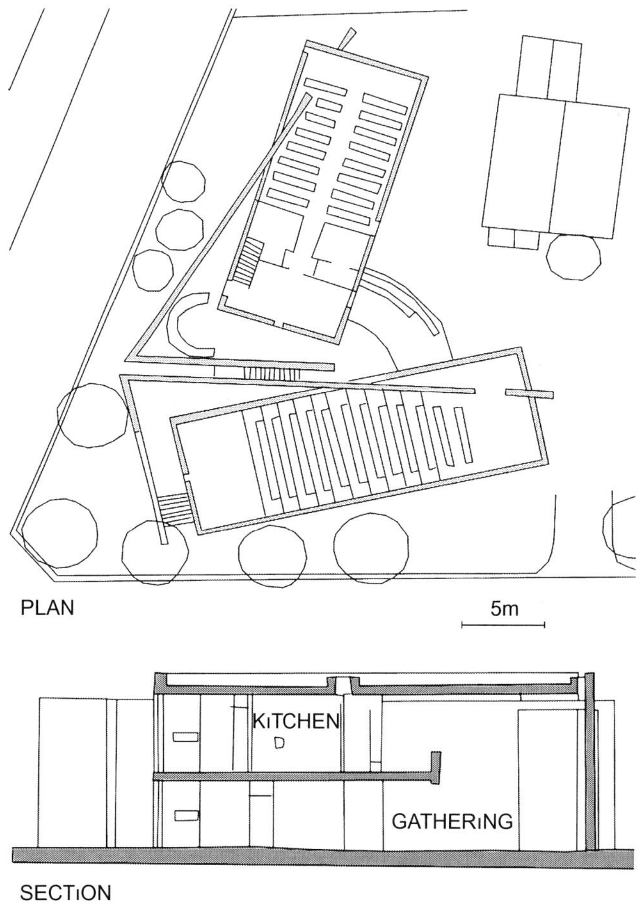

9.73 Plan and section of Church of Light and Sunday School (drawn with the help of URL7, 2012 and URL11, 2014)

are also designed not to be seen from inside or outside. If one searches the internet for interior photographs of this building, s/he sees that these windows create interesting light effects. Thus, it can be stated that the dominant physical entities in achieving the building's tectonic characteristics are details and materials. The window details contribute to the tectonic qualities of the building, as well as the continuous concrete surfaces.

However, the window in the form of a cross cancels a shear wall. The shear walls with different angles should be connected to each other through horizontal elements in order to avoid their movement in different directions. Since there is a high seismic risk, the structure of the building should have been designed in unity. Another tectonic characteristic of the Sunday School is the brutalist expression of shear walls that still have construction marks on them. Thus, the Church of Light and Sunday School has a contravening relationship with structural guidelines for shear wall structures.

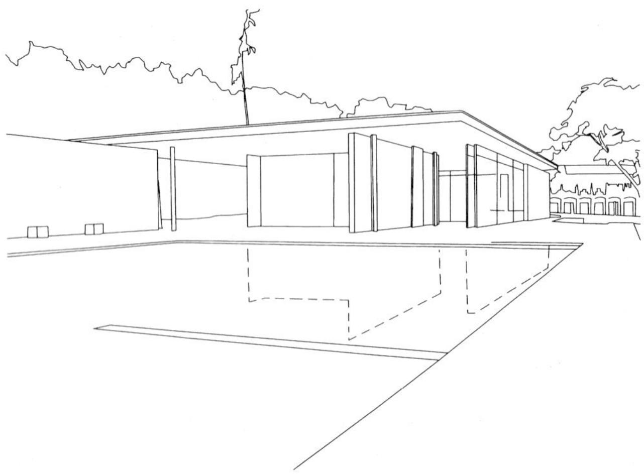

9.74 A sketch of case study 15: Barcelona Pavilion, Barcelona, Spain, 1928–1929 (drawn with the help of URL8, 2012)

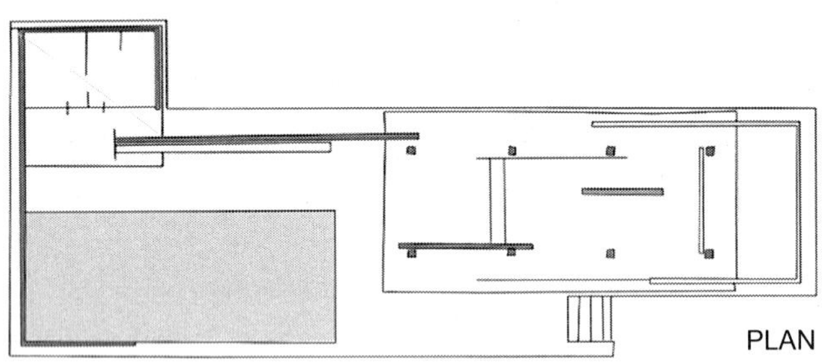

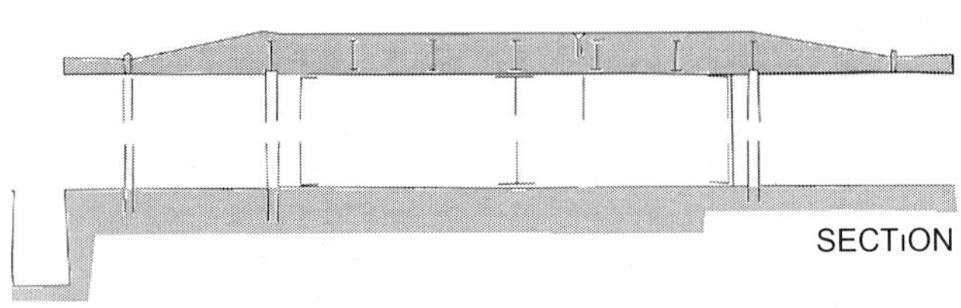

9.75 Plan and section of Barcelona Pavilion (drawn with the help of URL9, 2003 and URL10, 2011)

CASE STUDY 15: BARCELONA PAVILION, SPAIN

The architectural concept of Ludwig Mies van der Rohe's Barcelona Pavilion is based on the idea of a steel frame (see Figures 9.74 and 9.75). Mies van der Rohe wanted to express the tectonic qualities of steel frames, despite covering the steel columns with chrome and hiding the steel beams of the slab. He thought that the best way of expressing the steel frame was to hide its elements and show them in a different way (Frampton, 2001: p.177). Thus, the dominant physical entities in this building are its structural system and details.

There are false stone walls inside and outside the building. These walls are not structural, and they do not have any functional role. Despite their lack of reason, Mies van der Rohe thought that these walls gave presence (Hartoonian, 1994: pp.68–80). These walls are used very close to columns, and at two places the columns are placed symmetrically in front of the walls (Frampton, 2001: p.175).

When Barcelona Pavilion's plan and section are evaluated according to the structural guidelines as listed in Tables 9.2 and 9.4, it can be seen that the columns are very slender and they do not have similar dimensions to the beams. However, Barcelona Pavilion does not take place within an area of high seismic risk. These columns should have been checked against buckling only. Thus, it can be stated that Barcelona Pavilion has an affirmative relationship with the structural guidelines because it follows the structural guidelines applicable for no earthquake risk.

DISCUSSION ON THE CASE STUDIES

When the four case studies in this chapter are studied, it can be understood that the structure of the buildings in regions of high seismic risk can be very different to the ones in regions of low seismic risk. Thus, the architectural styles used in areas with low seismic risk should be used very carefully in areas with high seismic risk because problems such as soft storey, short column and slender columns have to be solved without disturbing the architectural quality.

When the two case studies built in high-risk seismic zones (the National Assembly and the Church of Light and Sunday School) are examined, it can be seen that their tectonic qualities have been achieved with careful thought. The National Assembly's large galleries have been achieved with the help of a detail – expansion joints – which separate the different masses of the building. The tectonic effect of the Church of Light and Sunday School is not based on structural form. Instead, tectonic qualities are achieved with the help of space quality and details, such as the window details and construction marks on the reinforced concrete shear walls. Thus, it can be stated that tectonic qualities can also be achieved with the help of details.

CONCLUSIONS

The tectonic qualities of frame systems are very different to those of masonry systems. Continuous surfaces with small openings formed the basis of the tectonic characteristics of masonry structures, together with the symmetry and sculptural effects of cross-walls and buttresses. Also the texture of stone, brick, adobe or timber surfaces were very effective in determining the tectonic quality of these buildings.

The most dominant tectonic quality of frame systems is their lightness in comparison to masonry structures. All the partition walls can be replaced by glass surfaces to achieve maximum transparency. However, this lightness can also be compared with the strong desire of structural engineers to create light structures. This is achieved through using the minimum amount of structural materials, while lightness in architecture can be achieved by reducing the number of walls within the building. Continuity between indoor and outdoor spaces as well as continuity between the interior spaces also became possible with the use of frame systems.

Form is also used as a tectonic characteristic in buildings with frame systems. The early buildings with frame systems have simple geometric forms (for example, Villa Savoye). Later, form was divided into pieces to form assemblages of forms (for example, the National Assembly in Dacca). The use of natural forms is also seen as desirable by many architects (for example, the roof structure of Le Corbusier's Ronchamp Chapelle in France). The architecture of the 1990s used the form in a different way, by creating a kind of formlessness (for example, Frank Gehry's Guggenheim Museum in Spain). The tectonics of contemporary architecture (which is stamped by the tectonics of frame systems) can be more playful than both the tectonics of traditional architecture (which is stamped by the tectonics of masonry structures), and the tectonics of engineering structures.

Thus, the general tectonic characteristics of frame systems can be listed as:

- Lightness
- Continuity of spaces
- Importance of form.

When these characteristics are brought together with structural guidelines (as listed in Tables 9.2, 9.3 and 9.4), it can be seen that they have a very critical relationship with earthquake-resistant architectural design. The demand for lightness results in the use of slender columns and the removal of partition walls. The demand for continuity might result in the presence of large galleries to connect different levels to each other. Being playful with the form can easily cause twisting instability problems due to the lack of symmetry.

It can even be stated that the relationship between designers and structures has been changed considerably after the invention of frame systems. Previously, there was a tendency to design small-scale traditional structures in line with structural guidelines concerning earthquake risk. Designers contravened these guidelines only if there was a strong reason or demand for it. However, contemporary small structures – it might even be better to say contemporary styles – run against structural guidelines concerning earthquakes. Symmetry is disliked; slender columns are in favour; and large galleries are spacious and desirable. Thus, the architects who design within high-risk earthquake regions should be very careful with contemporary architectural styles.

PROBLEMS TO SOLVE

9.1: Draw N, V and M diagrams and the deflected shape of the following two systems.

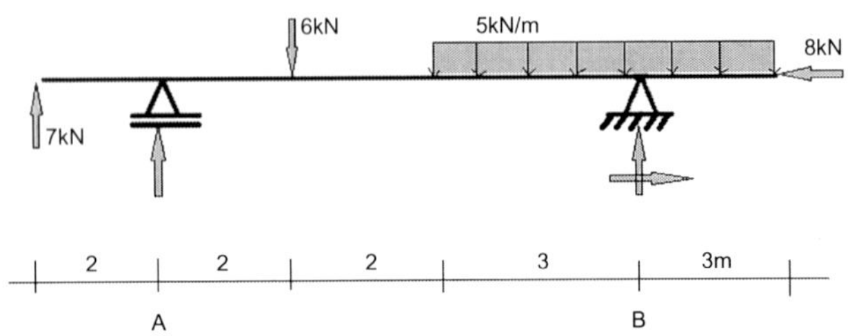

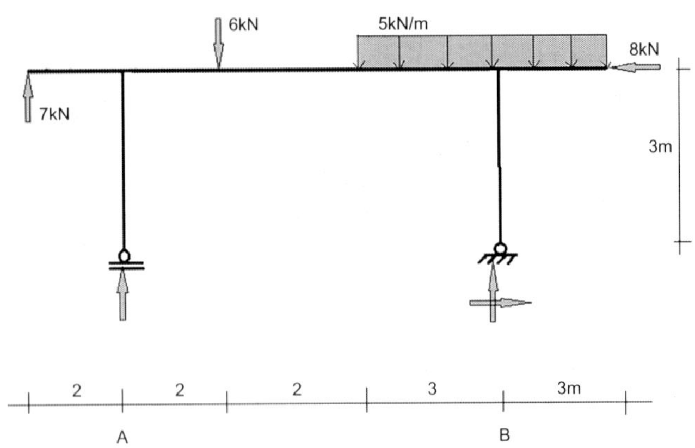

9.2: Draw N, V and M diagrams and the deflected shape of the following two frames.

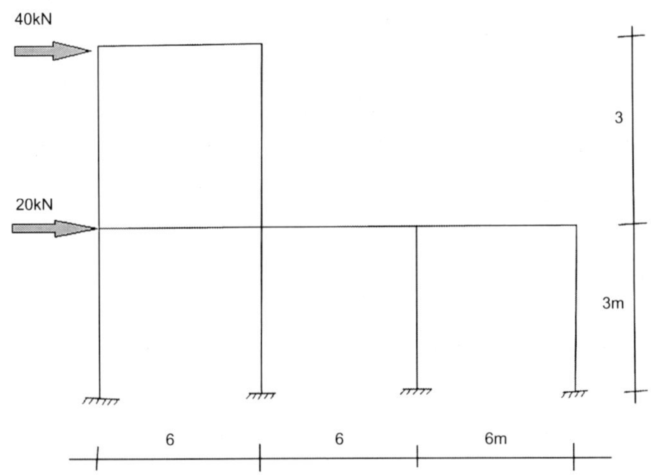

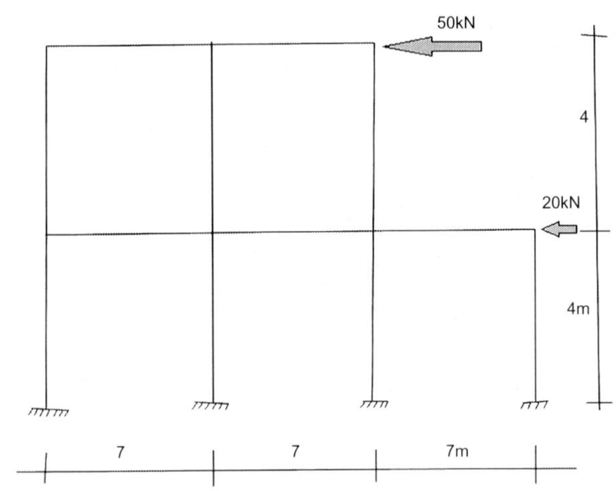

REFERENCES

Abrams, D.S. (2013) *Practical Limitations of Single Span Ultra-high Performance Concrete Beams*, unpublished Master thesis, Massachusetts Institute of Technology: USA.

ACI 318-95 (1995) *Building Code Requirements for Reinforced Concrete*, American Concrete Institute: Michigan.

Cement and Concrete Association of Australia (2003) *Guide to Long-span Concrete Floors*, 2nd edition, Cement and Concrete Association of Australia: Sydney.

Charleson, A.W. (2008) *Seismic Design for Architects: Outwitting the Quake*, Architectural Press: New York.

Ching, F.D.K. (1991) *Building Construction Illustrated*, 2nd edition, Van Nostrand Reinhold: New York.

Dallaire, E.E. (1983) 'The Quiet Revolution in Skyscraper Design' *Civil Engineering*, Vol. 53, No. 5: pp.54–59.

Engel, H. (1997) *Structure Systems*, 3rd edition, Gerd Hatje Publishers: Ostfildern, Germany.

Frampton, K. (2001) *Studies in Tectonic Culture: The Poetics of Construction in Nineteenth and Twentieth Century Architecture*, ed. J. Cava, The MIT Press: Cambridge, Massachusetts.

Hartoonian, G. (1994) *Ontology of Construction*, Cambridge University Press: New York.

ICC (2000) *International Building Code*, International Code Council: Illinois.

ICC IBC (2012) *International Building Code*, 2nd edition, International Code Council: Illinois.

Leatherbarrow, D., Mostafavi, M. (2005) *Surface Architecture*, The MIT Press: Cambridge, Massachusetts.

Lin, T.Y., Stotesbury, S.D. (1981) *Structural Concepts and Systems for Architects and Engineers*, John Wiley and Sons: New York.

Mieczyslaw, W., Zbigniew, M. (2014) 'Demountable Bridge Spans made of Prefabricated Box Beams' *IABSE Reports* (viewed 23 August 2014: http://dx.doi.org/10.5169/seals-42768)

Ministry of Public Works and Settlement Government of the Republic of Turkey (2007) *Seismic Performance Evaluation of Dual Reinforced Concrete Systems Design According to Turkish Seismic Code*, trans. E.Y. Karcı (viewed 26 October 2013: www.belgeler.com/blg/22lc/seismic-performance-evaluation-of-dual-reinforced-concrete-systems-design-according-to-turkish-seismic-code)

Mir, M.A. (2001) 'Evolution of Concrete Skyscrapers: From Ingalls To Jin Mao' *EJSE Electronic Journal of Structural Engineering*, Vol. 1, No. 1: pp.2–14.

NIST (National Institute of Standards and Technology) (2012) *Comparison of US and Chilean Building Code Requirements and Seismic Design Practice*, US Department of Commerce: Washington.

NTE E.030 (2003) *Technical Standard of Building E.030 Earthquake Resistant Design*, Peru National Building Code, Peru Japan Center of Seismic Research and Disaster Mitigation: Lima.

NZS3101.1 (2006) *Concrete Structures Standard. Part 1: The Design of Concrete Structures*, Cement and Concrete Association of New Zealand: Wellington.

Paz, M. (1994) *International Handbook of Earthquake Engineering*, Chapman & Hall: New York.

Resplendino, J., Toulemonde, F. (2010) *Designing and Building with UHPFRC*, Wiley-ISTE: New Jersey.

Steel Construction Info (n.d.) *Box Girder Bridges* (viewed 14 April 2014: www.steelconstruction.info/Box_girder_bridges)

University of California, Berkeley: Seismological Laboratory (2008) *Where Earthquakes Occur* (viewed 15 April 2014: http://seismo.berkeley.edu/blog/seismoblog.php/2008/09/29/where-earthquakes-occur)

URL1 (n.d.) *Earthquakes* (viewed 25 September 2014: http://lab-space.open.ac.uk/mod/resource/view.php?id=416333)

URL2 (2011) *Villa Savoye* (viewed 27 September 2014: www.gal-insky.com/buildings/savoye/)

URL3 (2011) *5 Projects* (viewed 27 September 2014: http://archinect.com/features/article/2673501/5-projects-interview-3-matthew-persinger)

URL4 (2004) *My Architect* (viewed 27 September 2014: www.smh.com.au/articles/2004/10/06/1096949579883.html)

URL5 (2014) *National Assembly* (viewed 27 September 2014: http://cpoh1.workflow.arts.ac.uk/personal-project-information-file)

URL6 (2006) *Church of Light, Osaka, Japan* (viewed 27 September 2014: www.galinsky.com/buildings/churchoflight/)

URL7 (2012) *Architecture As Aesthetics* (viewed 27 September

2014: http://architectureassociate.blogspot.com.tr/2012/12/church-of-light.html)

URL8 (2012) *A Bit of Bauhaus in Barcelona* (viewed 27 September 2014: http://vmcinteriordesign.edublogs.org/tag/barcelona-chair/)

URL9 (2003) *Another* (viewed 27 September 2014: http://another29.exblog.jp/m2003-09-01/)

URL10 (2011) *Barcelona Pavilion by Ludwig Mies van der Rohe* (viewed 27 September 2014: http://simoncyho.blogspot.com.tr/2011/03/case-study-1.html)

URL11 (2014) *Iglesia de la Ruz – Tadao Ando* (Church of Tadoa Ando) (viewed 27 December 2014: http://sobrearquitecturas.wordpress.com/2014/03/31/iglesia-de-la-luz-tadao-ando/)

10
The Tectonics of Frame Systems in Interior Architecture

Traditional and vernacular architecture were based on the use of masonry structures. The spatial qualities of masonry structures can be evaluated with the help of the concept of stereotomics ('stereos' meaning 'solid' and 'tomia' meaning 'to cut') (Frampton, 2001: p.5). As discussed in Part 2, structural guidelines for masonry structures are more restrictive for designers than structural guidelines for frame systems. With frame systems, it is possible to have many forms, higher buildings, and larger openings: frame systems are known to be much lighter structures in comparison to masonry structures. Modernity is generally known for the freedom it gives to individuals and this is also valid for the freedom of designers. It must also be stated that this freedom would not be possible without the opportunities given by modern structures, and especially by frame systems.

Another interesting characteristic of frame systems is their potential for change. Since the partition walls of frames can be removed, it becomes possible to change the positions of all walls, thus replacing the old walls with totally different arrangements. This type of change was not possible with masonry structures. Thus, flexibility of frame systems is more than that of masonry systems.

Although one group of architects believe that buildings should not be changed without the permission of architects, another group design by taking into account change and the participation of users. The author of this book thinks in parallel with philosopher A. Badiou (2002), and considers that being open for change is ethical. The adaption of buildings falls within the field of interior design. Although structural guidelines for making changes in frame systems represent a conservative approach, it can be stated that many structural restrictions within these guidelines can be avoided with the help of creative and innovative designs.

La Maison de Verre in Paris, France, which was rebuilt at the beginning of the twentieth century, is one of the best examples of creative and innovative designs (Neumann et al., n.d.). The old masonry building's ground floor was removed and the masonry walls were replaced by a steel frame, which carries the old masonry walls at the upper level. The old structure was supported during the construction of the steel frame.

However, it is much safer and more economical to plan change during the initial design of the building. Open building is an architectural approach that is based on planning change in architecture. Open building designers see the structure of buildings as a separate entity that does not change, whereas partition walls and other building systems are seen as changeable systems and parts. Thus, structure and infill are separate categories to be treated differently (Habraken, 1998; Leupen, 2005). However, many architects do not design for change and their buildings are later changed by interior architects.

In order to be able to list conservative structural guidelines for making changes in frame systems, the issues of change will be analysed as:

- Making subtractions from frame systems.
- Making additions to frame systems.

Most of these changes are only possible if the law of the country allows it.

MAKING SUBTRACTIONS FROM FRAME SYSTEMS

The elements that might take place in frame systems are columns, beams, slabs, stairs, rigid partition walls and lightweight partition walls. Subtraction of these elements can be analysed one by one (Hurol, 2013).

Subtraction of columns, beams and shear walls

Columns and beams form the frames. Frames should start from one end of the building and end at the other end in order to be able to resist horizontal loads. All frames should be connected to each other. It is not recommended to cut a column if there is another column at the top of it and if the beams adjacent to this column will not be supported by some other means. This is true whether the structural material is reinforced concrete or steel. Similar recommendations can be made for reinforced concrete or steel shear walls. It is not possible to cut shear walls or to partially remove them to have openings.

It is also not recommended to cut a beam even if the slabs on it are removed, because this might change the continuity of the frame and disturb the condition of the adjacent columns. Remaining elements will not be ready for the new condition. Furthermore, these actions are against the structural design project prepared for the initial design. Many professionals think that it is ridiculous to discuss cutting columns and beams, but it is known that this happens in real life.

Subtraction of slabs

Slabs work like diaphragms and transfer and distribute horizontal loads to columns. Thus, it is not recommended to have more than 1/3 of slab area as galleries. Thus, while taking into account the percentage of gallery area, some slabs can be removed from the structural system during the interior design of the existing building.

However, it is not possible to remove half of a reinforced concrete slab, because this removal disturbs the reinforcement in the remaining part. It might be possible to remove the whole slab or open small holes in a reinforced concrete slab. Steel slabs might be removed partially by removing the joists at the related part of the slab. It is better to get advice from a structural engineer before partial or full removal of slabs.

Subtraction of stairs

Interior designers see stairs as sculptural elements that enrich the space and they usually prefer to have lightweight stairs. Thus, it is very common for interior designers to remove the existing stair and replace it with another one. If the stair takes place within a gallery and if it sits on the beams, it can be removed. However, if the stair takes place in a shaft, which is formed by walls or frames, then it is more difficult to remove the stair, because without the structural connections, the hollow shaft might become weak. It is better to get advice from a structural engineer before removal of stairs.

Subtraction of rigid infill walls

Frames do not need any type of infill walls in order to realise their structural roles. Thus, it is possible not to have any infill walls. However, rigid infill walls affect the behaviour of columns and beams by eliminating their deflection. Lightweight infill walls do not affect the behaviour of frames. Thus, arbitrary placement of rigid infill walls can cause soft-storey and short-column problems (as explained in chapter 9). Similarly, partial removal of rigid infill walls during interior design can cause the same problems. Some of the recommendations to avoid soft-storey and short-column problems during the initial architectural design are not applied for interior design projects. For example, it is possible to improve the strength of reinforced concrete columns through retrofitting (Olivova & Bılcik, 2009), but it is usually not advisable for economic reasons. Thus, it is better to get advice from a structural engineer before the removal of rigid infill walls.

Subtraction of lightweight infill walls

Since the lightweight infill walls do not affect the structural behaviour of frame elements, they can be removed from the buildings without causing any problems. If there is any earthquake risk, it is more beneficial to adopt an open building design approach with regard to the use of lightweight partition walls.

MAKING ADDITIONS TO FRAME SYSTEMS

Once the partition walls are subtracted, it becomes possible to add other structures to the existing structures in order to articulate the space. These additions can be at various positions with respect to the existing building. They can be (Misirlisoy, 2011):

- Within the structure.
- Outside the structure.
- Starting from inside and extending outside the structure.

For example, there can be an addition to the atrium of an existing building and this addition will be within the structure. There can be an additional storey to the existing building and this addition will be outside the existing structure. There can also be an additional mass starting from outside and passing through the building. In all cases, it is better for these additional structures to have the following characteristics (Hurol, 2013):

- To make connection with the existing structure.
- To have an order.
- To use a lightweight structure.
- To request input from a structural engineer.

Connection with the existing structure

It is usually better to connect these additional structures to the joints of the existing structure. This means that the axis system of the additional structure will be related to the axis system of the existing building. This is necessary in order to avoid independent behaviour of the old and the new parts (for example, overturning). Achieving continuity usually helps both structures. It is possible to recommend that the axes system of the existing structure should be considered during the design of the additional structure. For example, if the additional structure is an additional storey, the structure of this storey should be connected to the joints of the structure of the existing building.

Order

The structural elements of the additional structure should form a system. This means that they have to be connected to each other. An understandable order makes continuity and formation of a system possible.

Lightweight structures

Since these additional structures are dependent on the existing structure, it is better to employ lightweight structures for these additional parts. Otherwise, the additional structure will give too much unexpected load to the existing structure. This recommendation suggests the use of steel as the structural material and trusses, geodesic domes and grid shells as the structural systems. Even if the interior designer wishes to have thick walls in her/his design, the effect of thickness can be achieved with the help of a steel skeleton covered by panels with a cavity between them.

Structural engineering input

All additional structures are required to have a structural engineering input. Their stability, strength, equilibrium and deformation/deflection should be checked, as well as their relationship with the existing structure. Is it safe for the existing structure to carry the load transfered by the additional structure?

STRUCTURAL GUIDELINES FOR SUBTRACTIONS AND ADDITIONS TO EXISTING FRAME SYSTEMS

Structural guidelines for making subtractions from and additions to existing frame systems are listed in Tables 10.1 and 10.2. These guidelines are based on the above characteristics.

Although most of the recommendations that take place in these guidelines are based on safety, there are many example projects that have not followed these recommendations. This is possible only if the problem created by not following the recommendation is compensated by some other means. The only recommendations that cannot be ignored are those instructing structural engineering input and advice. Other recommendations can be compensated with the help of creative and innovative approaches.

Table 10.1 Structural guidelines for making subtractions from existing frame systems and the associated value system

Structural guidelines	*Value system*
Subtraction of columns, beams and shear walls is not recommended.	Safety
Reinforced concrete slabs can be subtracted as a whole or small subtractions can be made. Steel slabs can be subtracted as a whole or partially. It is better to get advice from a structural engineer before deciding about the removal of slabs.	Safety
If the structure of the stair does not affect the building structure, the stair can be subtracted. It is better to get advice from a structural engineer before deciding about the removal of stairs.	Safety
Rigid infill walls can be subtracted without causing soft-storey and short-column problems in high-risk earthquake regions. It is better to get advice from a structural engineer before deciding about the removal of rigid infill walls if there is earthquake risk.	Safety
Lightweight infill walls can be subtracted.	–

Table 10.2 Structural guidelines for making additions to existing frame systems and the associated value system

Structural guidelines	*Value system*
The additional structure should be connected to the joints of the existing structure.	Safety
The additional structure should have an order.	Practicality
The additional structure should be a lightweight structure.	Safety
A structural engineering project should be provided for all additions.	Safety

CASE STUDY 16: PHOTOGRAPHER'S PENTHOUSE, USA

This case study was chosen to examine an addition to the top of an existing building (see Figures 10.1 and 10.2). Christoff Finio Architecture's Photographer's Penthouse in New York has a clearly different design in comparison to the existing building, which was built in 1820. Whereas the existing building is traditional, the Penthouse has a very modern feel. The fact that it is possible to

10.1 A sketch of case study 16: Photographer's Penthouse, New York, USA, 1992 (drawn with the help of Mournement, 2007)

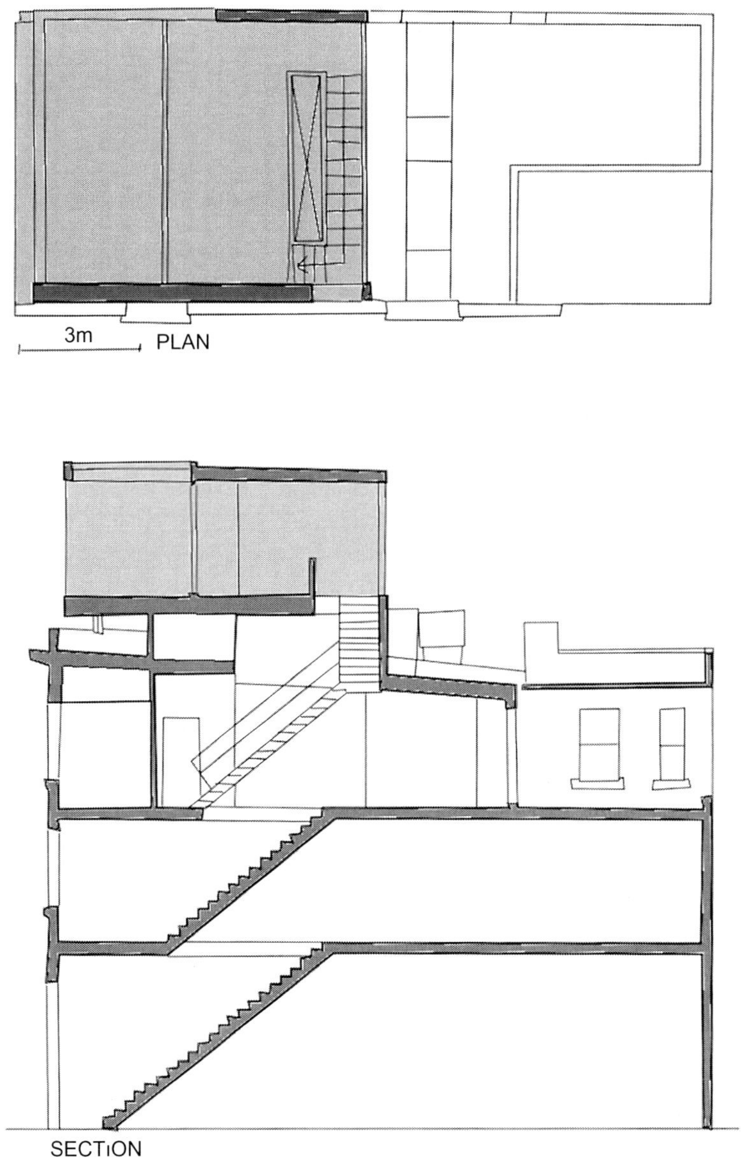

10.2 Plan and section of Photographer's Penthouse (drawn with the help of Mournement, 2007)

identify the additional part immediately is a requirement of historic conservation. The additional Penthouse was built to enjoy the view of the Hudson River: thus the major openings are towards the river. The dominant physical entity that determines the tectonic characteristics of the Penthouse is structure.

There are no subtractions made from the existing structure. An additional staircase and a room are added to the top of the existing building. As seen in Figure 10.2, the structure of the additional room is formed by the two side walls and the back wall, each corresponding to an axis of the existing structure. There is also a fourth axis, which is behind the front facade. A cantilever extends from this axis towards the front street. The Penthouse has a steel frame and the stair is metal. The walls of the space are insulated with high-density extruded polystyrene foam and filled with cast concrete. The exterior finish is synthetic stucco (Mournement, 2007). Thus, it can be stated that the Penthouse's relationship to structural guidelines is affirmative.

CASE STUDY 17: SUSPENDED BEDROOM, FRANCE

This case study was chosen to examine an addition inside an existing building (see Figures 10.3 and 10.4). Emmanuel Combarel Dominique Marrec Architectes' Suspended Bedroom was built in Paris in 2006. Since it was not possible to make additions to the top or sides of the existing building, the designers proposed this solution: a lightweight suspended structure that is carried by the slab system above. The main design concept is to have a suspended white cube, which divides the space into two. This addition makes this flat radically different from the other similar flats. The dominant physical entities in achieving this concept are structure and construction methods.

Since there might be columns close to the axes of the additional room, the vertical suspension elements may not

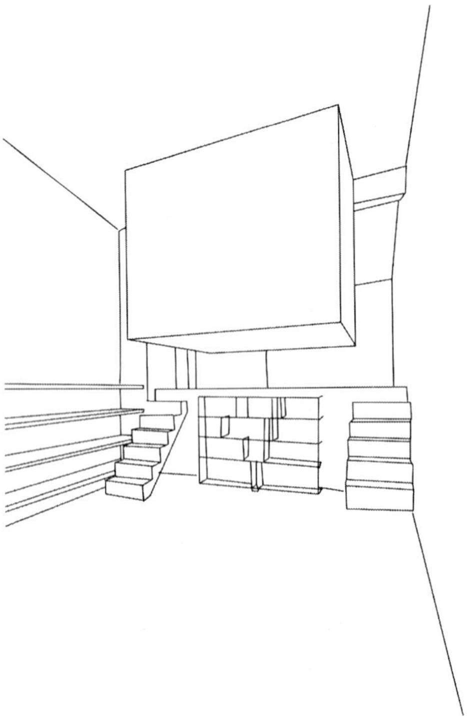

10.3 A sketch of case study 17: Suspended Bedroom, Paris, France, 2004 (drawn with the help of Mournement, 2007)

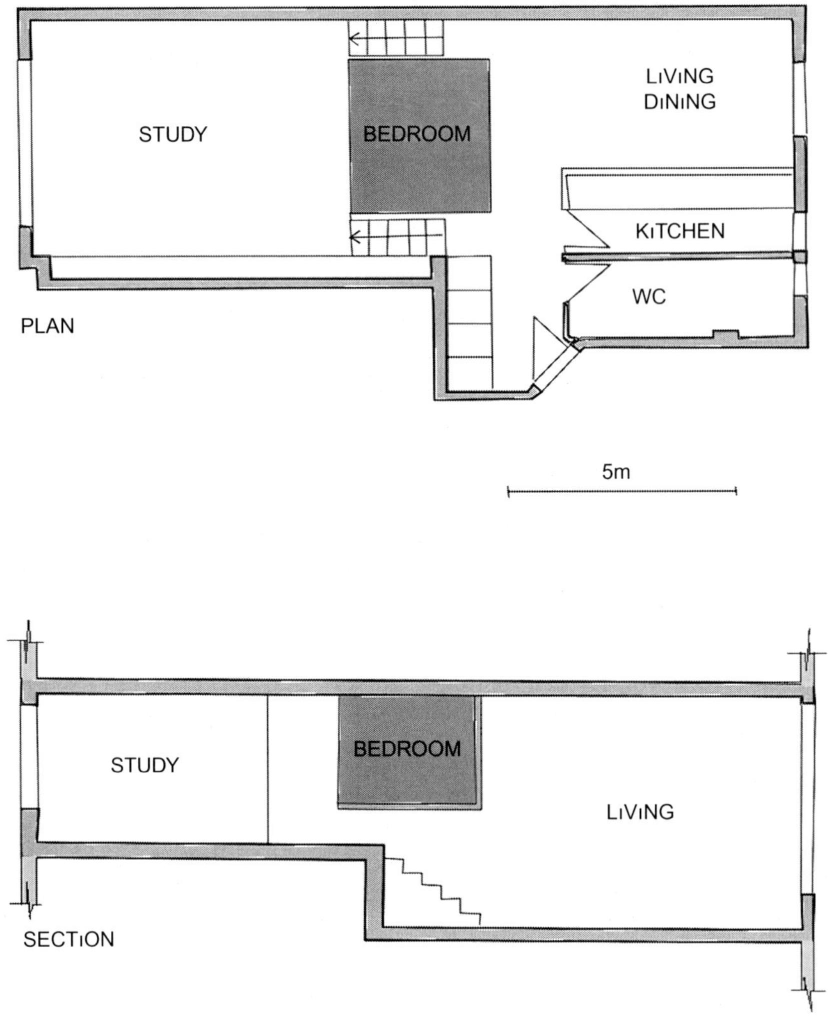

10.4 Plan and section of Suspended Bedroom (drawn with the help of Mournement, 2007)

give extra concentrated loading to the slab system above. The additional cube has a steel frame and it is cladded with timber panels. The finish is white polyurethane resin (Mournement, 2007). It can be stated that Suspended Bedroom has an affirmative relationship with structural guidelines.

CONCLUSIONS

The tectonics of interior architecture differs radically from the tectonics of architecture. There is a particular aesthetic involved in making extensions to existing buildings. Although the extensions coexist with the existing buildings, they usually differ from it with their scale, order and materials. Extensions often have a different order in comparison to the existing building, but the two orders are still related to each other. Extensions are usually formed with lightweight materials. The extensions usually take the attention of observers because of these differences with the existing buildings: extensions usually have high optical value. In this way, new values that are directly related to the users of the building are gained.

The existing building is the context of the extension. The extension has to be rooted in the existing building. This makes the issues of structure in interior architecture differ radically from the issues of structure in architecture. Structures in interior architecture are technologically challenging because each structural problem is a unique case.

REFERENCES

Badiou, A. (2002) *Ethics – An Essay On the Understanding of Evil*, trans. P. Hallward, Verso: London & New York.

Frampton, K. (2001) *Studies in Tectonic Culture: The Poetics of Construction in Nineteenth and Twentieth Century Architecture*, ed. J. Cava, The MIT Press: Cambridge, Massachusetts.

Habraken, J. (1998) *The Structure of the Ordinary – Form and*

Control in the Built Environment, ed. J. Teicher, The MIT Press: Cambridge, Massachusetts.

Hurol, Y. (2013) 'On Ethics and the Earthquake Resistant Interior Design of Buildings' *Science and Engineering Ethics*, 3 January, DOI 10.1007/s11948-012-9424-1.

Leupen, B. (2005) 'Towards time based architecture' in eds. B. Leupen, R. Heijne & J. van Zwol *Time Based Architecture*, 010 Publishers: Rotterdam.

Misirlisoy, D. (2011) *Analysis of the Structure and Design Relationship Between Contemporary Extensions and Remodelled Masonry Buildings*, unpublished Master thesis, Eastern Mediterranean University: North Cyprus.

Mournement, A. (2007) *Extensions*, Laurence King Publishing Ltd: London.

Neumann, S., Cowan, R., Compain, F. (n.d.) *Pierre Chareu – La Maison de Verre* (Pierre Chareu – The Glass House), film, (viewed 2015: www.youtube.com/watch?v=CFGluJ3fzGU)

Olivova, K., Bılcik, J. (2009) 'Strengthening of Concrete Columns with CFRP' *Slovak Journal of Civil Engineering*, Vol. 1: pp:1–9.

11
The Tectonics of High-Rise Building Structures

BY YONCA HUROL AND BAYDU CAN AL

The concept of high-rise building changes from place to place. Any buildings that are higher than the surrounding buildings are called high-rise buildings. The concept of skyscrapers is different from that of high-rise buildings, because skyscrapers are more slender: the height to width ratio (slendernes ratio) usually varies between 3 and 9. High-rise structures usually need special permission due to their height. They also need special design processes that are not covered by building codes. If the building height is over 80m, the International Building Code treats them differently in high-risk earthquake zones (ICC, 2000).

The story of modern high-rise building structures starts with the use of frame systems and elevators in high-rise buildings from the Chicago School of Architecture, USA. The increase in land value in Chicago led owners to demand higher buildings. William Le Baron Jenney's 10-storey Home Insurance Company Building was built in 1885. This building had a metal structure (cast iron columns and wrought iron beams). The metal frame was covered with brick cladding for fire safety purposes. Thus, this structure could have been seen as a contemporary masonry structure with cast iron vertical tie-beams. Steel frames started to be used in Jenney's Ludlington Building of 1891.

The economic boom in the 1920s and early 1930s saw a significant increase in skyscraper construction. Yet zoning restraints meant that these buildings were being designed with setbacks. In the early 1930s, there was big competition between the Empire State, Chrysler and General Motors buildings (all built in New York, USA). The Empire State building, which is 102 storeys and 443m high, won the competition and became the tallest building in the world until 1972. After the Empire State building, people stopped building such high skyscrapers due to the economic depression, and started to demand moderately high skyscrapers instead.

At the beginning of the 1960s, tubular structures were invented by Fazlur Khan and this led to another attempt to compete with the Empire State building. In 1972 the World Trade Center towers were built in New York and became the highest skyscrapers in the world. These framed tubes were 110 storeys and 417m high. In 1974, the Sears and Roebuck tower was built in Chicago as a bundled tube structure. This 110-storey and 442m high building became the highest building in the world until 1998. At the time of writing, the highest building in the world is Burj Khalifa in Dubai, which was completed in 2010. This building is 163 storeys and 828m high. The list of the highest buildings in the world can be found at URL1 (2015).

The economy of high-rise buildings differs from the economy of small-scale buildings. The cost of the structure is approximately

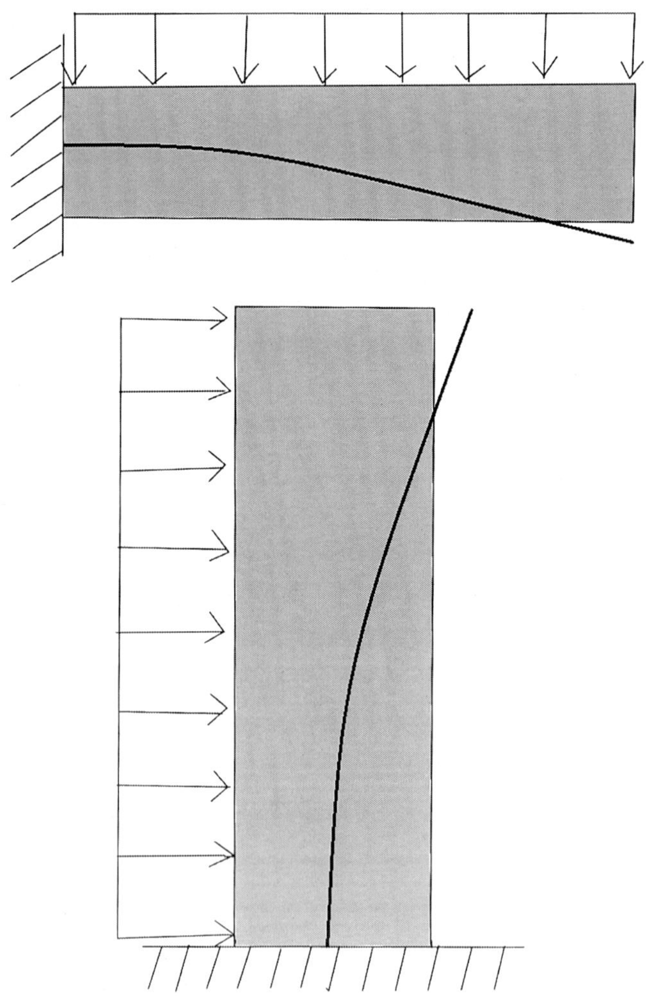

11.1 High-rise building structure and cantilevering beam

70% of the total cost for high-rise buildings. In comparison, the ratio for small-scale structures is around 30% (Khan, 1968). One of the reasons for this is that horizontal loads, such as wind load and earthquake load, determine the structural design of high-rise buildings.

High-rise building structures have similarities to cantilevering beams, as seen in Figure 11.1. Both the cantilevering beam and the high-rise building structure are affected by forces that are perpendicular to their axes and thus they develop bending stress.

THE CANTILEVER METHOD

Since high-rise building structures have similarities to cantilevers, their frames can be analysed by considering this similarity. The triangular bending stress distribution in beams is reflected in the axial force distribution in the columns in the Cantilever Method.

The Cantilever Method also has some similarities to the Portal Method. Both methods are approximate methods used to analyse the effects of horizontal loading. In both methods, the moment at the middle of columns and beams is assumed to be zero.

In the Cantilever Method, it is assumed that the axial force in each column is proportional to the column's distance from the centre of the frame. Unlike the Portal Method, the Cantilever Method is used for high-rise and slender-frame systems (Lin & Stotesbury, 1981: p.229).

The steps of the Cantilever Method are as follows:

1 Determine the centre of gravity for the frame by considering the area of the columns and the distance between the columns, as seen in Figure 11.2.
2 Find the axial force in the columns of each storey by considering that it will be proportional to the column's distance from the centre of gravity, as seen in Figure 11.3. Except for these initial steps, the method is very similar to the Portal Method, which is explained in chapter 9.
3 Find the shear in beams for all storeys by considering joint equilibrium, as seen in Figure 11.3.
4 Find the moment in beams and columns of all storeys as a factor of shear value, as seen in Figure 11.4.

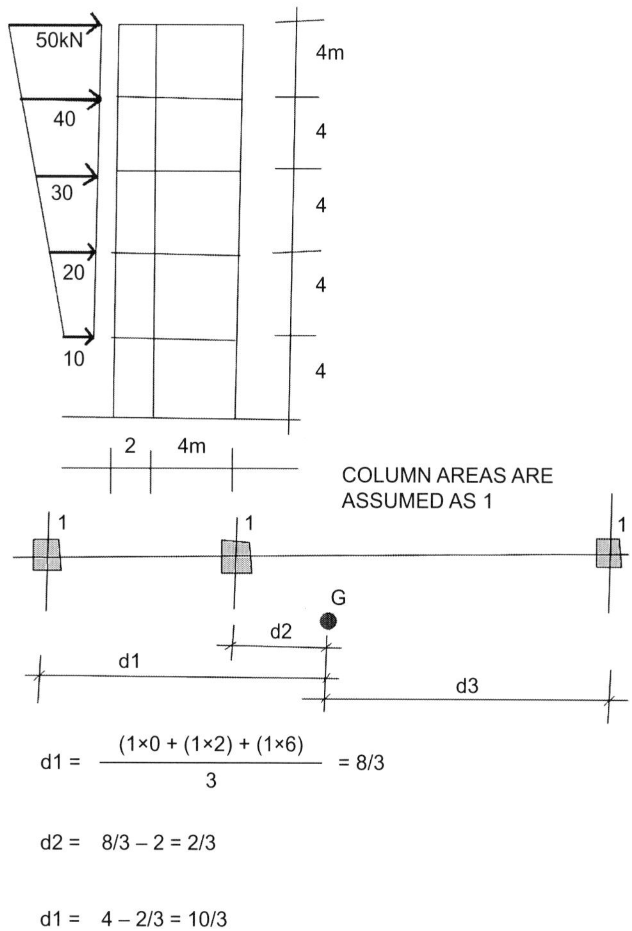

11.2 Finding the center of gravity for a frame

5 Find the shear in columns of the whole frame as a factor of the moment in columns, as seen in Figure 11.4.
6 Find the axial force in beams of the whole frame by considering joint equilibrium, as seen in Figure 11.4.
7 Draw N, V and M diagrams of the frame, as seen in Figure 11.5.

Consider that the moment at the top and the bottom of the columns of Figure 11.5 should be the same.

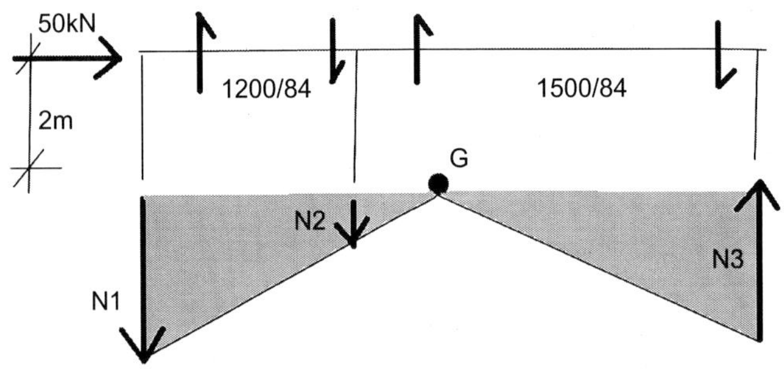

4th FLOOR M = 50 × 2 = 100 kNm

COUNTER MOMENT DUE TO AXıAL FORCE ıN COLUMNS = $(10/3)^2N + (2/3)^2N + (8/3)^2N = 168/9N$

168/N = 100kNm N = 900/168

N1 = 8/3 × 900/168 = 1200/84kN

N2 = 2/3 × 900/168 = 300/84kN

N3 = 10/3 × 900/168 = 1500/84kN

3rd FLOOR M = (50 × 6) + (40 × 2) = 380kNm

2nd FLOOR M = (50 × 10) + (40 × 6) + (30 × 2) = 800kNm

1st FLOOR M = (50 × 14) + (40 × 10) + (30 × 6) + (20 × 2) = 1320kNm

GROUND FLOOR M = (850 × 18) + (40 × 14) + (30 × 10) + (20 × 6) + (10 × 2) = 1900kNm

11.3 Finding the axial force in columns and the shear in beams for the top floor of a frame

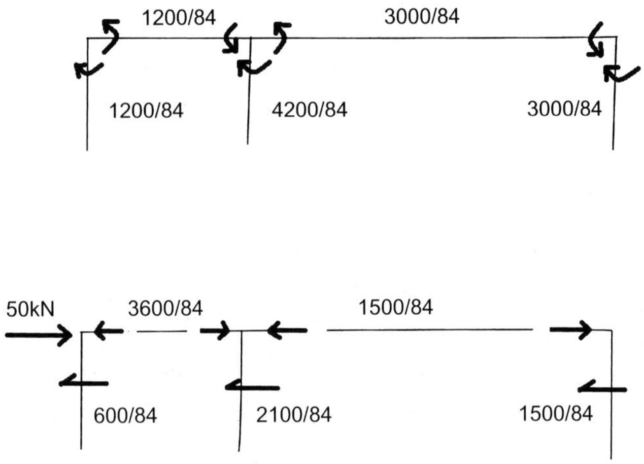

11.4 Finding the moment in beams and columns and finding shear in columns and axial force in beams for the top floor of a frame

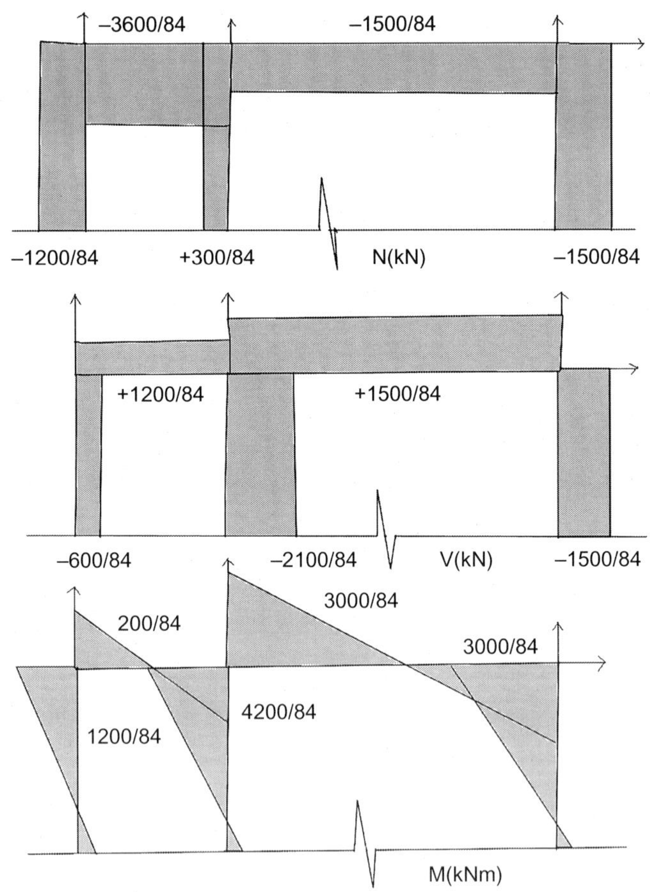

11.5 N, V and M diagrams of the top floor of the given system

STRATEGIES TO INCREASE HEIGHT

This book covers high-rise building structures by presenting different strategies to increase the height of buildings. These are:

- Use of aerodynamic form.
- Selecting the appropriate structural system.
- Use of damping systems.

Use of aerodynamic form

There is much available research that can help in visualising what happens if a high-rise building is subjected to wind. Some of these sources are: T. Lee et al. (2009); T. Tamura (2009); C.W. Park and S.J. Lee (2000); Y. Liu et al. (2005); V. Dousset and A. Potherat (2010); and T. Uffinger et al. (2010).

All buildings and structures create obstacles to wind flows, and air changes direction primarily towards the sides and the top in order to flow around the building. Flows created around a building depend on many factors, such as the height and the plan form of the building. As a result, the building experiences forces and moments. The forces and moments that affect the building structure are (Gunel & Ilgin, 2007):

- Along wind
- Across wind
- Torsional.

Figure 11.6 shows these three effects on buildings due to wind. It is standard practice to select one axis along (along wind) and another axis perpendicular to the direction of the flow (across wind). This is called the side force. Along wind creates pressure fluctuations on the windward and leeward surfaces of the building. This is called the drag of the building. Across-wind effect occurs due to transverse wind action on the two sides of the building and this is called the side force. These forces on the facade of the building could also cause a torsional moment around a vertical axis.

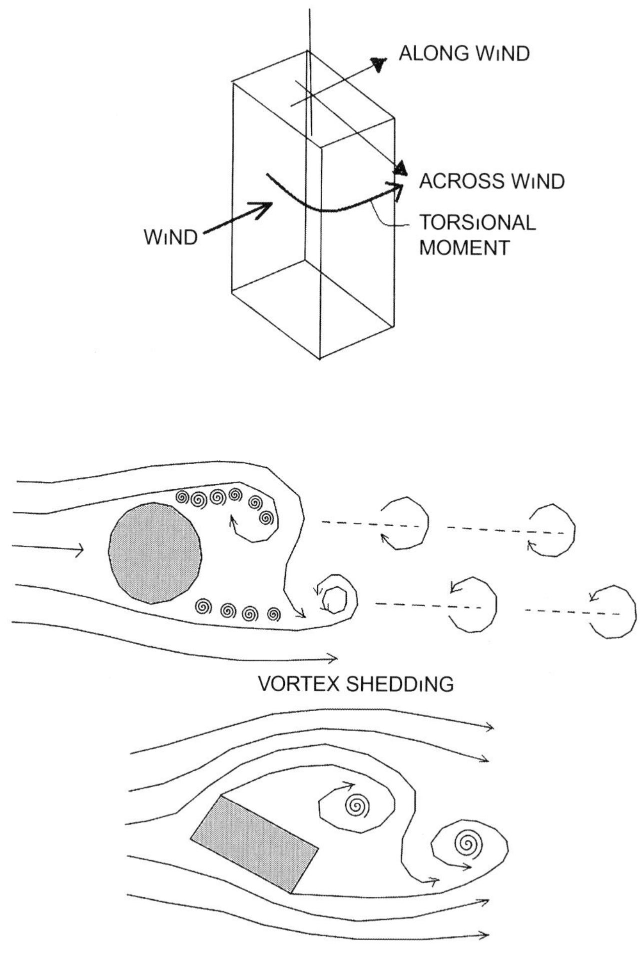

11.6 Wind effects on a building depending on the wind direction

There is also wind motion towards the top and the base (downwash) of the building. The motion towards the top causes flow recirculation at the top, which also affects building structure (see Figure 11.7). The motion towards the base (downwash) causes disturbance for the pedestrians at street level. Thus, it is a good idea to position an obstacle in front of this movement, as seen in Figure 11.8 (Kajarekar, 2009).

The magnitude of all these flows and their resulting forces is related to the wind speed, the plan form and the size of the

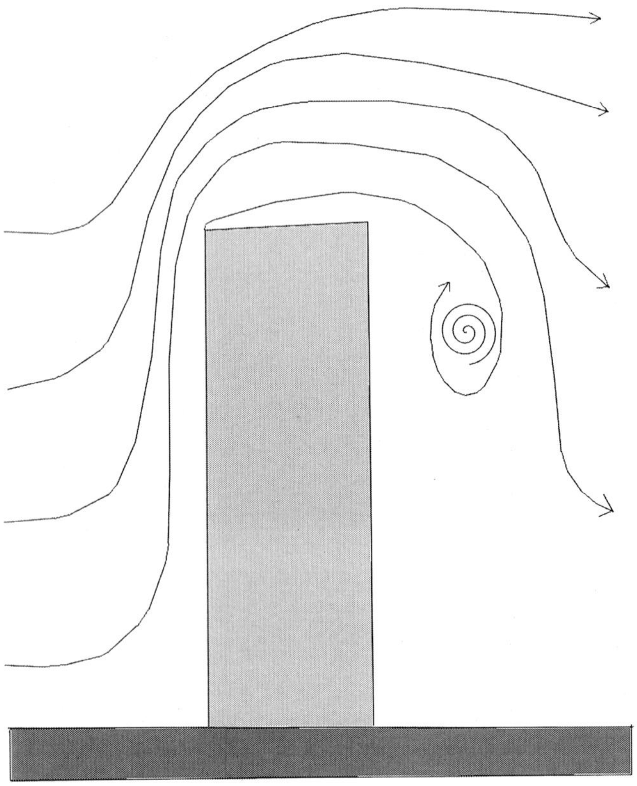

11.7 Wind movement towards the top of a building

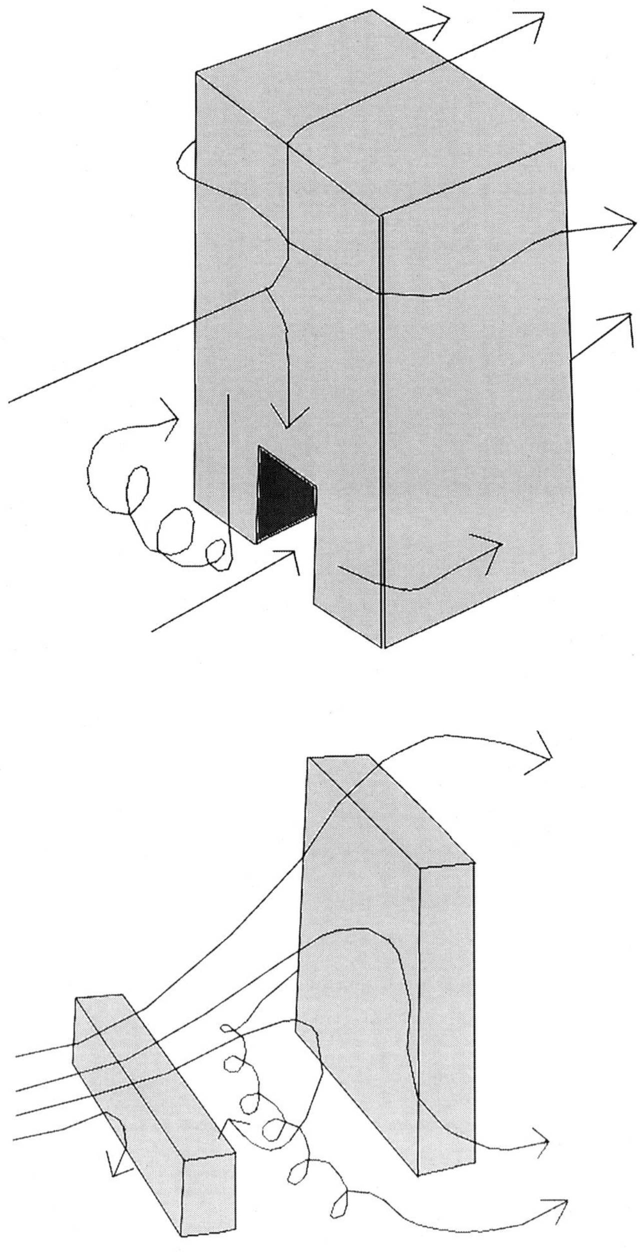

11.8 Wind movement around a building

building or structure. Depending on the plan form of the building, flow separation could occur. If the wind speed is low, these vortices caused by the flow separation are usually uniform. However, when wind speed increases, vortices are shed alternately on the two sides; first one side is affected from vortices, then the other side. This is called vortex shedding (see Figure 11.6) and the alternate occurrence of vortices may cause vibration of the building or structure.

There are many architectural strategies to decrease the wind effect on high-rise buildings:

- *Plan form:* It is better to have cylindrical, elliptical, crescent and triangular plan forms, which are less vulnerable to wind force

(Ali & Armstrong, 1995; Merrick & Bitsuamlak, 2009). These forms considerably decrease the possibility of the formation of vortices. Some building codes allow 20% or 40% reduction in wind load if these plan forms are used (Schueller, 1977). The Marina City Towers in Chicago (USA), Toronto City Hall in Toronto (Canada) and the US Steel Building in Pittsburgh (USA) can be given as examples of this approach.

- *Section form:* Having a tapered form, making setbacks and varying the plan shape decrease the wind load that increases with height (Baker, 2004). The John Hancock Building in Chicago (USA) has a tapered form. The Petronas Towers in Kuala Lumpur (Malaysia) and the Sears and Roebuck Tower in Chicago (USA) have setbacks. The Taipei 101 Tower in Taipei (Republic of China) has varying plan shapes along the height of the building (Ali & Armstrong, 1995; Schueller, 1977).
- *Surface roughness:* The use of smooth surfaces for the facades of high-rise buldings increases the wind effect on buildings. Use of rough surfaces decreases the wind effect, as in the case of golf balls. Rough surfaces delay flow separation, and as a result, drag is reduced (Lignarolo et al., 2011).
- *Treatments at the top:* To avoid any negative effects of wind at the top, high-rise buildings can be sculptured (Gunel & Ilgin, 2007). The Jin Mao Building in Shanghai, the Shanghai World Financial Center and the Wuhan Greenland Center (all situated in Republic of China), and the Petronas Towers in Kuala Lumpur (Malaysia), have various treatments at their top floors to decrease wind effects. The Shanghai World Financial Center is one of the case studies in this chapter.
- *Corner treatments:* Corners in plans of high-rise buildings can be recessed, cut, slotted or rounded in order to decrease wind effects (Kwok, 1995; Kawai, 1998; Tse et al., 2009). Since the sharpness of the corners are decreased, the effect of vortices also decreases. The Taipei 101 Tower in Taipei has such corner modifications.

It may be useful to search the above examples on the internet in order to see the relationship between their strategy against wind and their tectonic characteristics.

Facades of the buildings should be designed to resist aerodynamic forces and they should effectively transfer these forces to the main building structure.

Selecting the appropriate structural system

Selecting or designing the appropriate structural system is another strategy for increasing the height of skyscrapers. The most common structural systems that are used in high-rise building design are:

- Frames
- Shear walls
- Frame and shear wall systems
- Framed tubes
- Trussed tubes
- Tube in tubes
- Bundled tubes.

Framed tubes were invented by Fazlur Khan at the beginning of the 1960s. Later, Khan and Skidmore, Owings and Merrill developed the other types of tubes. The idea of tube systems can be understood by studying the moment of inertia of the structural plans of systems with frames, shear walls and tubes, as seen in Figure 11.9.

As explained in chapter 4, the moment of inertia (I) of any structural element increases by locating the structural material away from the centre of gravity. Structural material is evenly distributed over the plan of frame systems. Thus, the moment of inertia of frames is not high. Shear walls, which form shear cores at the middle of the plan, do not form plans with a high moment of inertia. In the case of tubular structures, however, structural material is concentrated on the elevations of the structure. Thus, tubular systems have the highest moment of inertia amongst these systems. Their resistance to moment is therefore higher than that of other structures. This characteristic makes tubular structures more economic for the highest structures.

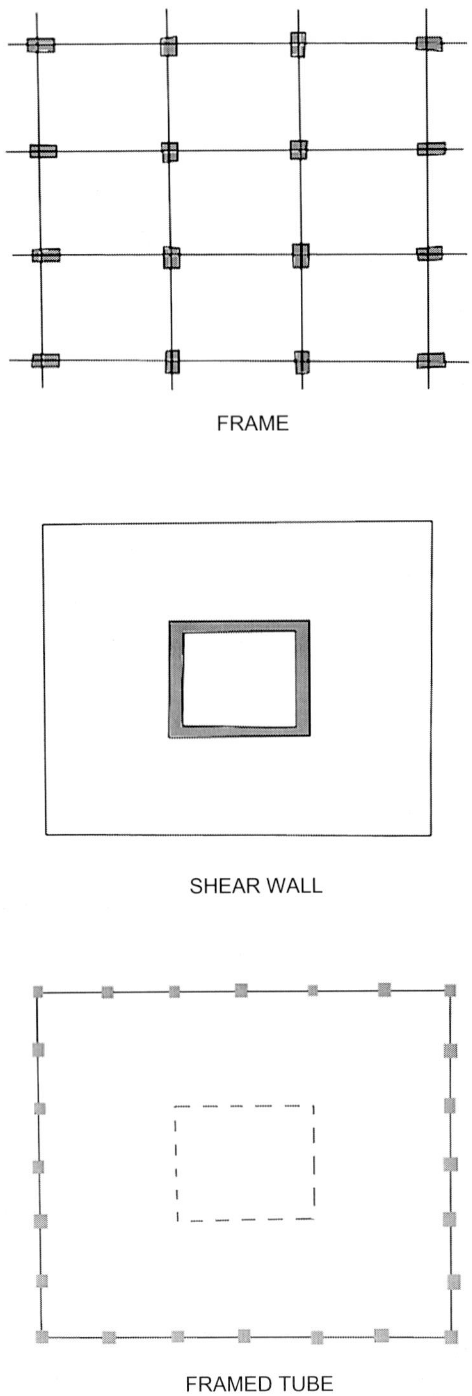

11.9 The moment of inertia for plans of frames, shear wall systems and tubes

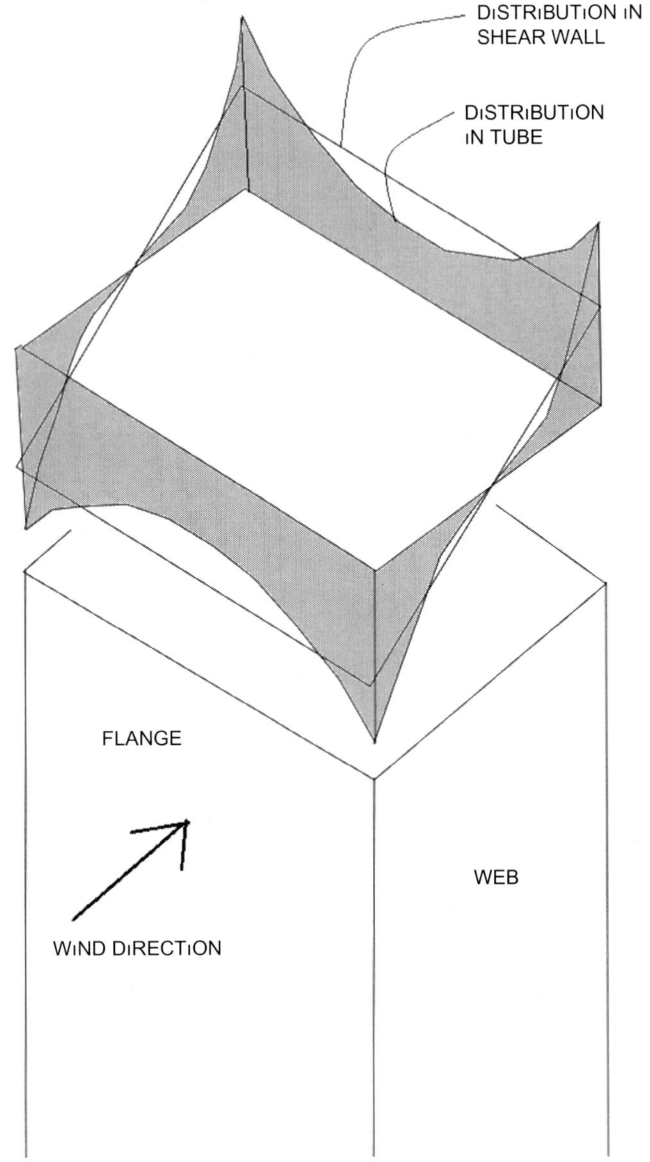

11.10 Shear lag in tubes

The main characteristics of tubes are:

- Placement of the structural material towards the outermost parts of the plan.
- Small distance between the columns.
- Use of rigid beams.

The distance between the columns of the 43-storey framed tube of the De-wit Apartments in Chicago, USA, is approximately 2m (Khan, 1968). This makes framed tubes very close to shear wall structures with small holes. However, since the openings exist, the distribution of axial forces in the columns cannot be in the same form as the distribution of axial forces in shear walls, as seen in Figure 11.10. This phenomenon of variation in the distribution of axial forces in

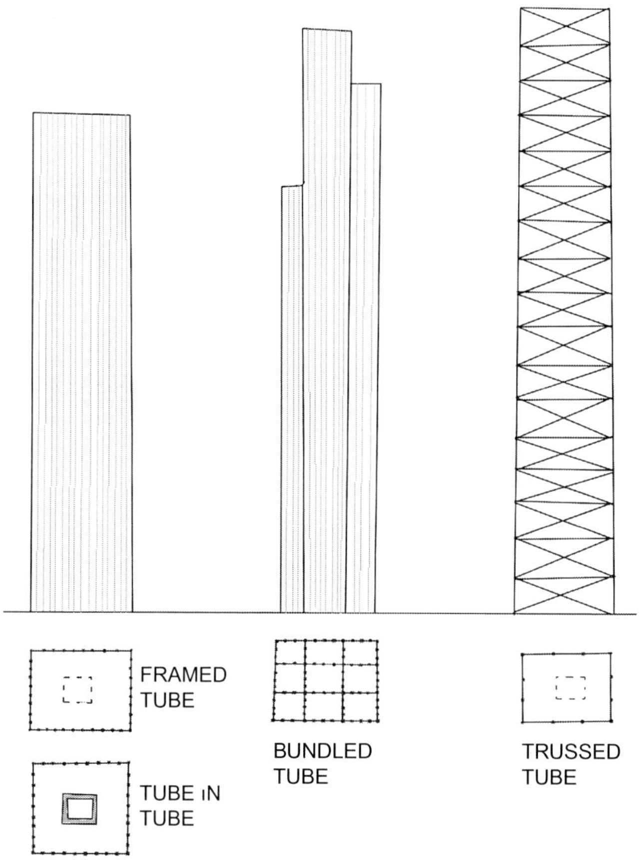

11.11 Structural configurations of various tubes

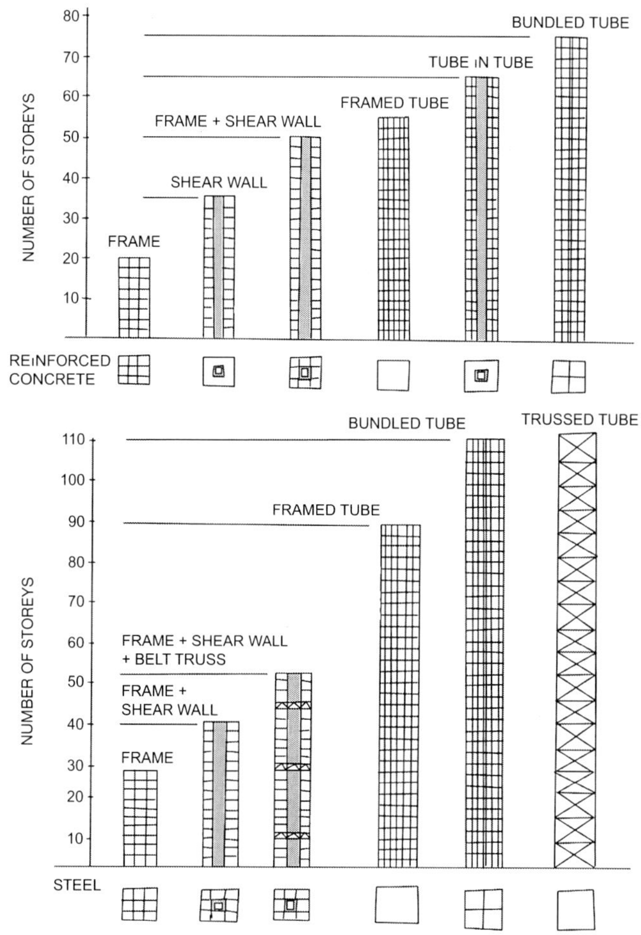

11.12 Height ranges of high-rise building structures (drawn with the help of Mir, 2001 and URL2, n.d.)

columns is known as shear lag (Khan, 1968; Soegiarso & Tjendera, 1997). Shear lag is the main weakness of tubular structures and the development of tubes has been based on the reduction of shear lag. The use of trussed facades and bundling the tubes decreases shear lag, and it becomes possible to build higher structures. Shear-lag problem also determines the distance between the columns of tubes. Distances between columns can increase up to 4.5m in bundled tubes and more in trussed tubes.

The structural configuration of various types of tubes can be seen in Figure 11.11. Bundled tubes can have various arrangements in height due to the presence of bundles of tubes. The Sears and Roebuck tower in Chicago, USA, is an example of this type of arrangement. Figure 11.12 shows the height ranges of high-rise building structures for steel and for reinforced concrete.

Examples of types of tubular structures are as follows:

- *Framed tube:* World Trade Center, New York, USA, 1972, 110 storeys. (This building is one of the case studies in this chapter.)
- *Trussed tube:* John Hancock Building, Chicago, USA, 1968, 100 storeys.

- *Bundled tube:* Sears and Roebuck Building, Chicago, USA, 1973, 108 storeys; and Zifeng Tower, Nanjing, Republic of China, 2010, 89 storeys.

The floor structure of skyscrapers is very important because it repeats many times within the structure. The floor structure also affects the height of the building. In many skyscrapers, a two-way trussed structure is used with a concrete top. This is a kind of composite floor system, which was used in the World Trade Center towers. The first case study of this chapter shows a section from the floor system of the World Trade Center. These floor structures work as diaphragms between the elements of the vertical structure. Unless a structure is used at the middle of the building, such as a shear core or inner frame, these floor structures have large spans. All tubular structures therefore contain inner structures in order to decrease span and depth of the floor structure. If these inner structures contribute to the resistance of the structure against horizontal loads, then the structure is called tube in tube. If the inner structure is there only to carry the vertical loads and not contribute to resistance against horizontal loads, then the structure is a hollow tube: a framed or trussed tube.

The design of the ground floor structure of skyscrapers is always a problem due to the contrast between the need for in and out interaction with the environment, and the frequently placed large columns. There are many strategies to solve this problem. Vierendeel trusses might be used to eliminate some of the columns, or the load of some columns can be transfered to other columns, as was the case in the ground floor of the World Trade Center towers.

If there are two different types of structures within the system, such as the outer tube and the inner shear core, these two structures are connected to each other by outrigger systems and belt trusses that repeat at certain intervals (Ali & Moon, 2007). These systems integrate the two structures and make them behave in harmony together, as seen in Figure 11.13.

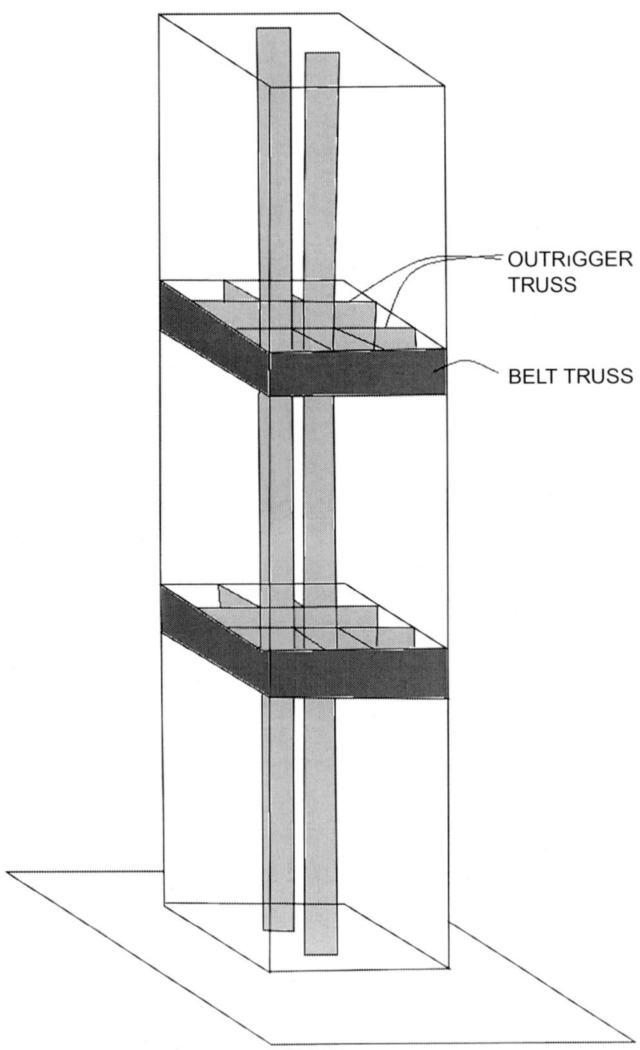

11.13 Use of outrigger systems and belt trusses to connect inner and outer structural systems

The highest skyscraper structures, which have been built since 2000, also contain structures that are different to simple tube applications. The Shanghai World Financial Center tower (118 storeys, built in 2008, Republic of China) developed the tube concept by distributing the perimeter columns to form a usable area. The Taipei 101 Tower (101 storeys, built in 2004, Republic of China) and the CTF Finance Center (111 storeys, to be completed in 2016, Guangzhou in Republic of China) have a core at the middle, a tube surrounding the facade, mega-columns strengthening the tube and outrigger trusses that connect the core and the outer structure. Burj Khalifa (163 storeys, 2010, Dubai) is different to all of these structures because it gains its strength from its form, in which three shear cores buttress each other. There are also attempts to develop structural materials, such as the use of fibre reinforced concrete.

Use of damping systems

Damping systems are added to skyscraper structures in order to control drift. There can be passive and active damping systems. If the system works naturally without any additional energy, it is called passive, and if there is a need for additional energy, it is called active. There are many types of damping systems, which are described by Kareem et al. (1999); and Ali and Moon (2007) in detail.

There can be damping details with viscoelastic materials that provide energy dissipation and restore force when deformed, and these can therefore be used to reverse the drift of the building. For example, there were such damping details between the columns and the floor trusses of the World Trade Center towers.

Secondary masses can be introduced to move according to the movement of the building. These can be concrete blocks, as in the case of the City Corp Center in New York, USA. These are placed at the 63rd floor of the building and weigh 410 tons. In the Hancock Tower in Boston, USA, the secondary masses are formed from steel boxes filled with lead weighing 300 tons (Kareem et al., 1999).

Other systems include water tanks with computer controlled hydraulic actuators, as in the case of Gold Tower in Kagawa, Japan. Nagasaki Airport Tower, also in Japan, contains small water tanks, which are distributed in a floor. Pendulums can also reverse drift, as in the case of Landmark Tower in Yokohoma in Japan (Kareem et al., 1999). These systems can sometimes affect the architecture of the building. The Taipei 101 Tower in Republic of China contains a steel pendulum that works as a tuned mass damper within the gallery between the 87th and 92nd floors.

STRUCTURAL GUIDELINES FOR HIGH-RISE BUILDING STRUCTURES

The structural guidelines listed in Table 11.1 are based on the above analysis.

Although the safety of high-rise building structures is very critical, it can be seen from Table 11.1 that most of the recommendations are based on economic value. This is because the

Table 11.1 Guidelines for structural design of high-rise buildings and the associated value system

Structural guidelines	*Value system*
Use an aerodynamic form to decrease wind load.	Economy
Use cylindrical, elliptical, crescent and triangular plan forms to reduce wind effect.	Economy
Have tapered forms, making setbacks and varying plan shape throughout height to reduce wind effect.	Economy
Consider the height ranges in Figure 11.12.	Economy
Use rough surfaces on the facades to reduce wind effect.	Economy
Design the top of the building to reduce wind effect.	Economy
Recess, cut, slot and round the corners to reduce wind effect.	Economy
Select the appropriate structural system according to the height of the building.	Economy
Have a plan shape with a high moment of inertia in order to increase resistance against bending.	Economy
Design the depth and strength of the floor structure to avoid extra height whilst providing diaphragm action.	Economy
Reduce the number of columns at entrance level without disturbing the structure.	Safety
Use outrigger systems and belt trusses to connect inner and outer systems to each other.	Safety
Use damping systems to control drift.	Economy

required strength can be achieved for every form – however, this might not be economic.

CASE STUDY 18: WORLD TRADE CENTER, USA

This case study was chosen because the building represents the importance of selecting a structural system. The towers of Minoru Yamasaki and Emery Roth and Sons' World Trade Center in New York were built in 1972 and they were 110 storeys and 411m high (see Figures 11.14 and 11.15). The plan dimensions of the buildings were 64m × 64m. The slenderness ratio of the buildings was 6.8. They were designed as steel framed tubes.

The framed tubes consisted of 244 exterior columns each 36cm × 36cm in plan. The distance between the axis of two adjacent columns was 100cm (Eagar & Musso, 2001). Beams were

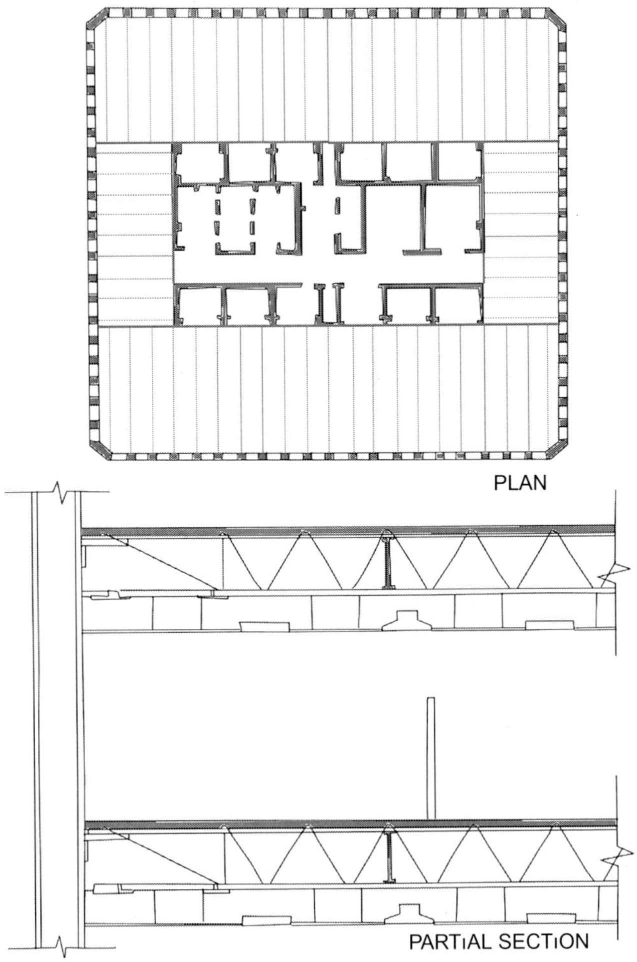

11.15 Plan and partial section of World Trade Center (drawn with the help of URL4, n.d. and URL5, 2003)

11.14 A sketch of case study 18: World Trade Center, New York, USA, 1972 (drawn with the help of URL3, 2011)

130cm deep: these were prefabricated as panels and the panels were joined on site. There was a 27m × 40m reinforced concrete shear core at the middle. This core was carrying only the vertical loads and not contributing to the resistance of the structure to horizontal loads. The walls of the core were thin, so that they were moving together with the movement of facade walls. W. Schueller (1977: p.188) says that: 'Although these buildings have interior cores, they act as hollow tubes because the cores are not designed to resist lateral loads . . .'.

Since the columns were very close to each other, the ground floor columns had to be arranged in a different manner in order to allow for an entrance to the building. So, the three columns were joined together like the branches of a tree. This decision gave a particular aesthetic character to the World Trade Center towers (see Figure 11.16).

As seen in Figure 11.15, 80cm-deep two-way trusses joined the exterior tube and the inner shear core. These trusses had concrete slabs at their top, forming a composite slab. The trusses were connected to the frame tube walls by details containing

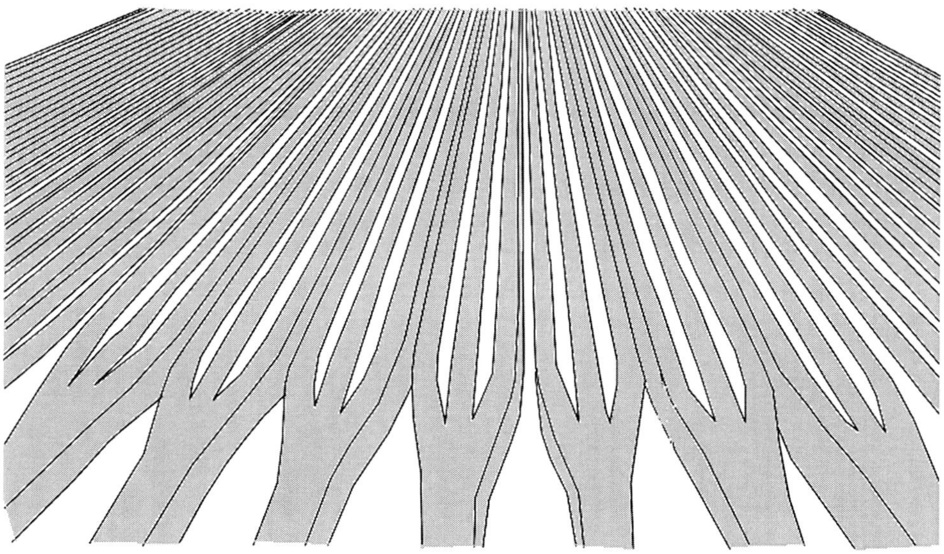

11.16 Facade of the World Trade Center towers (drawn with the help of URL6, 2014)

damping material. Thus, as the angle between the wall and the slabs changes, the details were creating forces to turn back to the original angle.

The World Trade Center towers were not designed to reduce wind. Instead, they were designed to have a plan shape that has a high moment of inertia in order to increase the resistance against bending. The dominant physical entity was the structural system.

The design concept of the World Trade Center towers could be described as having twin rectangular prisms and an emphasis on verticality. The closely placed columns, the hidden presence of the beams, and the tree arrangement of the columns at the bottom part of the skyscrapers, serve this concept. The arrangement of columns as tree branches gave a natural effect to the towers. Furthermore, the World Trade Center towers had a symbolic role representing New York City. By considering the building's height and its relationship with the wind, it can be said that the World Trade Center towers had a contravening relationship with the contemporary structural guidelines. However, when it was designed, it was designed to be an ideal structure.

CASE STUDY 19: SHANGHAI WORLD FINANCIAL CENTER, REPUBLIC OF CHINA

This case study was chosen because the building represents the importance of the use of aerodynamic form against wind load. The tower at Kohn Pedersen Fox and East China Architecture and Design Institute's Shanghai World Financial Center tower is a symbol of Shanghai, as well as being representative of contemporary skyscraper design (see Figures 11.17 and 11.18). It has an elegant

11.17 A sketch of case study 19: Shanghai World Financial Center, Shanghai, Republic of China, 2008 (drawn with the help of URL7, 2013)

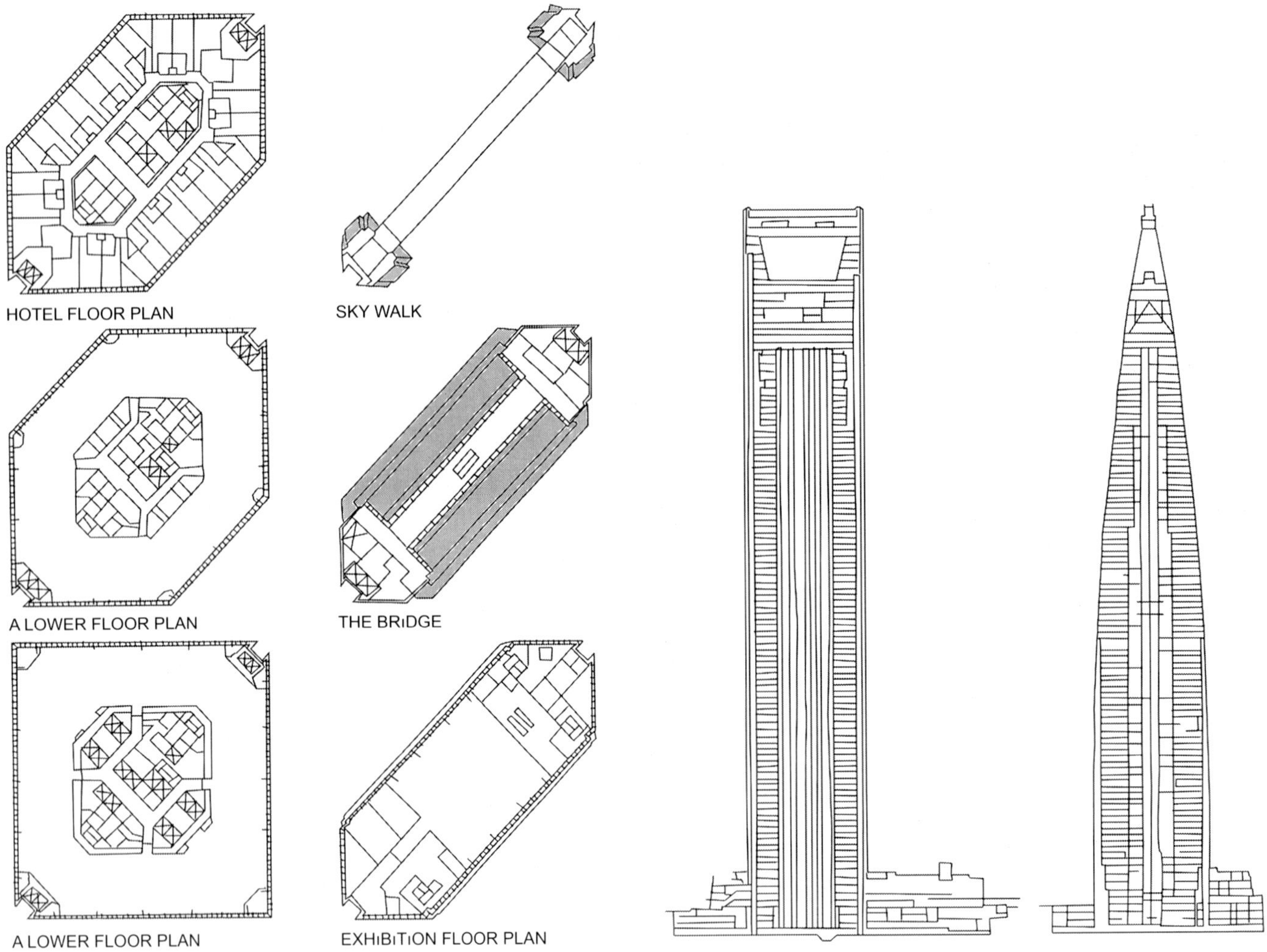

11.18 Plans and sections of Shanghai World Financial Center (drawn with the help of URL8, 2014)

smooth form finished with an aerodynamic arrangement at the top of the building. The aerodynamic form contributes to the aesthetics of the building. The architectural concept of the building is based on the design of the form of the building according to wind load. It is designed against wind, earthquake and terror. The dominant physical entities that determine design concept are the wind and the structural system.

It has a mega-structure (a diagonally braced frame) that embraces narrower columns and beams. The hierarchy of the structure is as follows (Paul et al., 2008):

- Major columns (steel and reinforced concrete), diagonals and belt trusses.
- A reinforced concrete shear wall at the middle.

- Outrigger trusses connecting the outer and inner structures.
- Three narrower columns between the major columns and narrower beams positioned between the belt trusses.

The mega-structure is perceivable from the outside and this contributes to the strong image of the building. The floors with outrigger trusses were used for skylobbies in order to express the structure within these floors.

The Shanghai World Financial Center tower was designed to reduce wind in an innovative way by:

- Having a tapered form.
- Having a varying plan shape through the height of the building.
- Designing the top of the building to decrease wind effects.

The building also has an appropriate structural system which is economical in relation to its height. Thus it can be said that the design of this building combines multiple strategies to increase economy. Shanghai World Financial Center has an affirmative relationship with structural guidelines.

CONCLUSIONS

High-rise buildings, in particular skyscrapers, express the technical development of the country and the economic power of the owner(s). They are landmark buildings and they represent the city in which they are located. Their aesthetics are based on their verticality and expression of structure. However, all high-rise buildings have three basic parts: base, column and capital.

The expression of the structure is altered at ground floor level in order to provide the entrance to the building and to eliminate any disturbing psychological effects of perceiving high-rise buildings from a close distance. The column part usually represents the verticality. The capital part always has different arrangements for various reasons, such as providing a helicopter platform, aerodynamic reorganisation of the top floors, or for compositional purposes. Skylobbies and mechanical floors are usually placed around outrigger trusses. All these architectural features are closely related to the design of the structural system.

During the twentieth century the design of high-rise buildings was mainly an engineering topic. However, twenty-first century design requires a teamwork approach that is involved with all dimensions of the building design.

PROBLEMS TO SOLVE

Draw the N, V and M diagrams of the following frames by using the Cantilever Method.

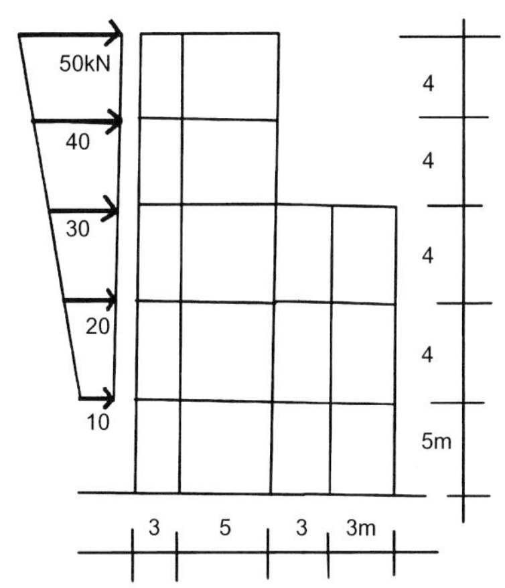

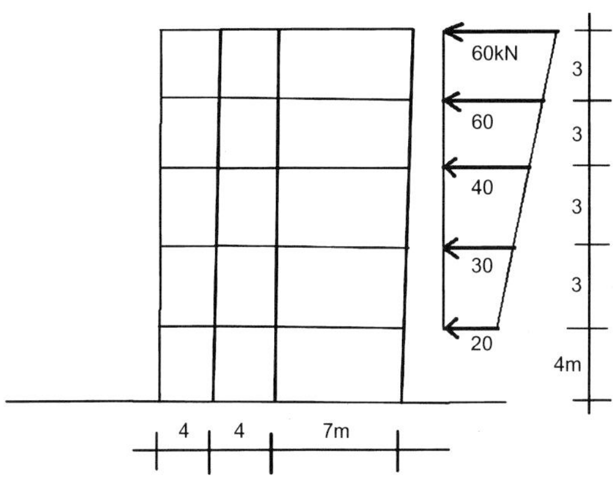

REFERENCES

Ali, M., Armstrong, P. (1995) *Architecture of Tall Buildings*, McGraw Hill: New York.

Ali, M.M., Moon, K.S. (2007) 'Structural Developments in Tall Buildings: Current Trends and Future Prospects' *Architectural Science Review*, Vol. 50, No. 3: pp.205–223.

Baker, W. (2004) *The World's Tallest Building – Burj Dubai*, Council on Tall Buildings and Urban Habitat: Seoul.

Dousset, V., Potherat, A. (2010) 'Formation Mechanism of Hairpin Vortices in the Wake of a Truncated Square Cylinder in a Duct' *Journal of Fluid Mechanics*, Vol. 653: pp.519–536.

Eagar, T.W., Musso, C. (2001) 'Why Did the World Trade Center Collapse? Science, Engineering and Speculation' *Journal of Metals*, Vol. 53, No. 12: pp.8–11.

Gunel, M.H., Ilgin, H.E. (2007) 'The Role of Aerodynamic Modifications in the Form of Tall Buildings Against Wind Excitation' *METU Journal of the Faculty of Architecture*, Vol. 24, No. 2: pp.17–25.

ICC (2000) *International Building Code*, International Code Council: Illinois.

Kajarekar, A.M. (2009) *Project Report on Study of Structural Shapes Depending on Wind Loading*, International Institute of Information Technology: Hyderabad.

Kareem, A., Kijewski, T., Tamura, Y. (1999) 'Mitigation of Motions of Tall Buildings with Specific Examples of Recent Applications' *Wind and Structures*, Vol. 2, No. 3: pp.201–251.

Kawai, H. (1998) 'Effect of Corner Modifications on Aeroelastic Instabilities of Tall Buildings' *Journal of Wind Engineering and Industrial Aerodynamics*, Vols 74-76: pp.719–729.

Khan, F. (1968) 'Column-free Box-type Framing With and Without Core' *IABSE Congress Report*, Vol. 8 (viewed 2 August 2014: http://retro.seals.ch/digbib/view?pid=bse-cr-001:1968:8::31)

Kwok, K.C.S. (1995) 'Aerodynamics of Tall Buildings' *Ninth International Conference on Wind Engineering*, New Delhi, India.

Lee, T., Lin, C.L., Friehe, C.A. (2009) 'Large-eddy Simulation of Air Flow Around a Wall-mounted Circular Cylinder and a Tripod Tower' *Journal of Turbulence*, Vol. 8 (viewed 30 July 2014: http://dx.doi.org/10.1080/14685240701383122)

Lignarolo, L., Lelieveld, C., Teuffel, P. (2011.) 'Shape Morphing Wind-responsive Facade Systems Realized With Smart Materials' *Proceedings of the Adaptive Architecture Conference*, 3–5 March 2011: London, UK.

Lin, T.Y., Stotesbury, S.D. (1981) *Structural Concepts and Systems for Architects and Engineers*, John Wiley and Sons: New York.

Liu, Y., So, R..M.C., Cui, Z.X. (2005) 'A Finite Cantilevered Cylinder in a Cross Flow' *Journal of Fluids and Structures*, Vol. 20: pp.589–609.

Merrick, R., Bitsuamlak, G. (2009) 'Shape Effects On the Wind Induced Response of High-rise Buildings' *Journal of Wind and Engineering*, Vol. 6, No. 2: pp.1–18.

Mir, M.A. (2001) 'Evolution of Concrete Skyscrapers: From Ingalls To Jin Mao' *Electronic Journal of Structural Engineering*, Vol. 1, No. 1: pp.2–14.

Park, C.W., Lee, S.J. (2000) 'Free End Effects On the Near Wake Flow Structure Behind a Finite Circular Cylinder' *Journal of Wind Engineering and Industrial Aerodynamics*, Vol. 88: pp.231–246.

Paul, K., Leslie, R., Sawteen, S. (2008) 'Shanghai World Financial Center: Without Compromise . . .' *CTBUH Eighth World Congress*, 3–5 March: Dubai, UAE.

Schueller, W. (1977) *High-rise Building Structures*, John Wiley and Sons: New York.

Soegiarso, R., Tjendera, E. (1997) 'Behaviour of High-rise Tubular Structures Due To Horizontal Loads' *Jurnal Teknik Spil F.T. Untar*, No. 2: pp.73–85.

Tamura, T. (2009) 'Large Eddy Simulation On Building Aerodynamics' *Seventh Asia-Pacific Conference on Wind Engineering*, 8–12 November: Taipei, Taiwan.

Tse, K.T., Hitchcock, P.A., Kwok, K.C.S., Thepmongkorn, S., Chan, C.M. (2009) 'Economic Perspectives of Aerodynamic Treatments of Square Tall Buildings' *Journal of Wind Engineering and Industrial Aerodynamics*, Vol. 97: pp.455–467.

Uffinger, T., Becker, S., Ali, İ. (2010) 'Vortex Dynamix in the Wake of Wall-mounted Cylinders: Experiment and Simulation' *15th International Symposium on Applications of Laser Techniques to Fluid Mechanics*, 5–8 July: Lisbon, Portugal.

URL1 (2015) *The Skyscraper Center* (viewed 13 May 2015: http://skyscrapercenter.com/buildings)

URL2 (n.d.) *Design and Construction Process* (viewed 28 October 2014: http://pmbook.ce.cmu.edu/03_The_Design_And_Construction_Process.html)

URL3 (2011) *Early Days of the World Trade Center – 1977* (viewed 28 September 2014: www.raischstudios.org/one-world-trade-center/2011/8/23/early-days-of-the-world-trade-center-1977.html)

URL4 (n.d.) *World Trade Center Disaster* (viewed 28 September 2014: www.ussartf.org/world_trade_center_disaster.htm)

URL5 (2003) *2 WTC* (viewed 28 September 2014: www.serendipity.li/wot/wtc_ch2.htm)

URL6 (2014) *World Trade Center Observatory* (viewed 28 September 2014: www.nyctourist.com/wtc_new1.htm)

URL7 (2013) *World Financial Center* (viewed 28 September 2014: http://puqyl.esfh.ir/castrol-oil-case/orld-financial-center.php)

URL8 (2014) *Shanghai World Financial Center* (viewed 28 September 2014: www.iam-architect.com/shanghai-world-financial-center)

PART 4
THE TECTONICS OF FORM-RESISTANT STRUCTURES

Different authors define the effect of form on structures in different ways. Form is important for all structures: even for a straight horizontal beam, the cross-sectional form affects the strength. For example, I beams are preferred for steel because I beams have a higher moment of inertia. However, for some structures such as shells and membranes, the resistance depends upon form.

As stated in chapter 5, tensile and compression structures fall within the group of form-resistant structures. The development of tension and/or compression within a structure is an effective way of responding to external loads. These structures usually have thin surfaces or light elements as a result of the strength gained through form. Engel (1997) calls these systems form-active structures and they consist of cables, membranes, pneumatic structures, arches, shells and grid shells. Salvadori (2002) also writes about strength acquired through form.

Part 4 of this book explains the structures for which form is the determinant, and contains the following chapters:

- Tectonics of tensile structures: containing cables, membranes, suspended glass systems, pneumatic structures and negatively curved shells.
- Tectonics of compression structures: containing steel and reinforced concrete vaults and domes, geodesic domes, grid shells and positively curved shells (all shells are described within this chapter in order not to separate positively and negatively curved shells from each other).
- Tectonics of tension and compression structures: containing 2D trusses, 3D trusses and space frames.
- Tectonics of folded plates.

Tension and compression structures cover all types of trussed structures (which are categorised together with bending structures in chapter 5), for which triangulation is a determinant. They are studied under form-resistant structures because they develop only tension or compression in them in order to overcome bending due to external forces. However, although folded plates work with bending stress, they are also considered in this section as their form determines their effectiveness.

REFERENCES

Engel, H. (1997) *Structure Systems*, 3rd edition, Gerd Hatje Publishers: Ostfildern, Germany.

Salvadori, M. (2002) *Why Buildings Stand Up? The Strength of Architecture*, 2nd edition, W.W. Norton & Company: New York.

12
The Tectonics of Tensile Structures

As mentioned in chapter 5, tensile structures usually have a negatively curved form and consist of the following systems:

- Cables and suspension structures.
- Cable trusses, bicycle-wheel structures and suspended glass systems.
- Membrane structures.
- Pneumatic structures.
- Negative curvature shells.

This chapter contains general characteristics and problems of tensile structures, together with descriptions and examples of each structure type.

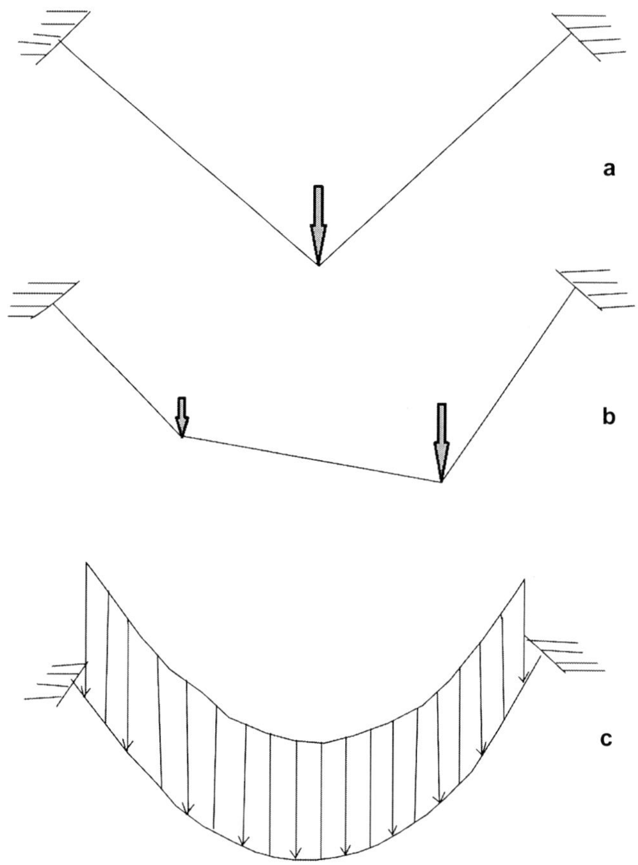

12.1 Change of form depending on load

GENERAL CHARACTERISTICS AND PROBLEMS OF TENSILE STRUCTURES

One of the main characteristics of tensile structures is that they alter their form depending on the changing load. These are called funicular structures. Figure 12.1 shows a cable changing its shape according to the load affecting it.

This change in form guarantees that the structure will respond to loads solely with tensile stress. It is also the reason for tensile structures being the most effective structure type for spanning long distances. When seen from this point of view, the change of form due to loading is an advantage. However, it also presents a disadvantage, because we do not usually want our buildings to change their form. This might be disturbing for the users and it might also cause some difficulties in selecting the covering material for the roof. There are load types such as wind load, which might change frequently. The covering surface of the roof will also be subjected to movement and extra loads due to the wind-load movement. This phenomenon is known as the wind instability problem of tensile structures. All tensile structures should be designed with a strategy against the wind instability problem. As mentioned in chapter 4, some stuctures might be allowed to move; but if movement is not wanted, the designer should find a way to stop it during the design process. The film of the collapse of Tacoma Narrows Bridge in Washington (USA) in 1940 due to wind instability can be seen in URL1 (2006). There are various strategies to avoid wind instability in cable structures. These are (Santoso, 2003):

- Use of a secondary cables series in the opposite direction (Figure 12.2a).
- Use of pre-tensioning involving pulling the cables towards each other by using shorter cables, or keeping them apart with short struts (Figure 12.2b).

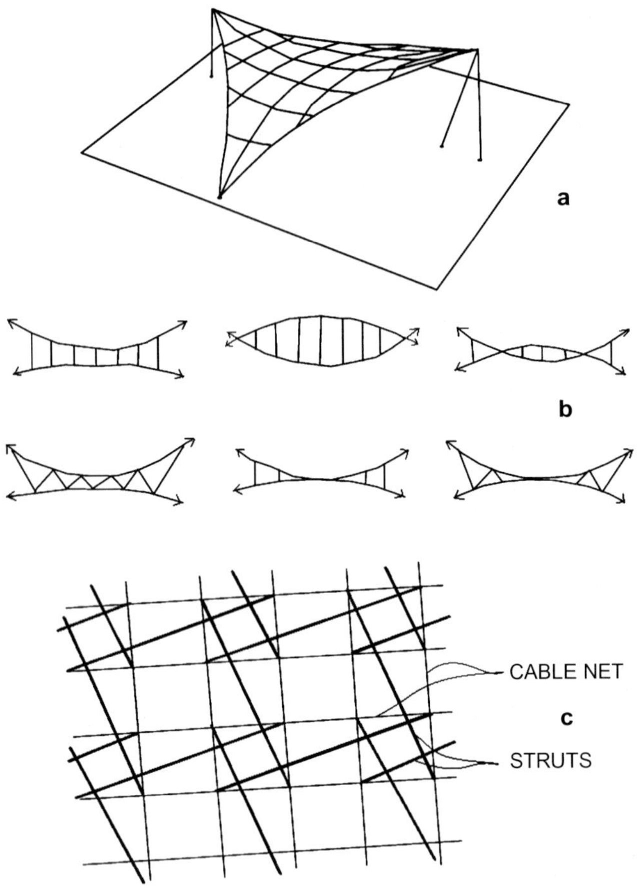

12.2 Examples of strategies to avoid wind instability

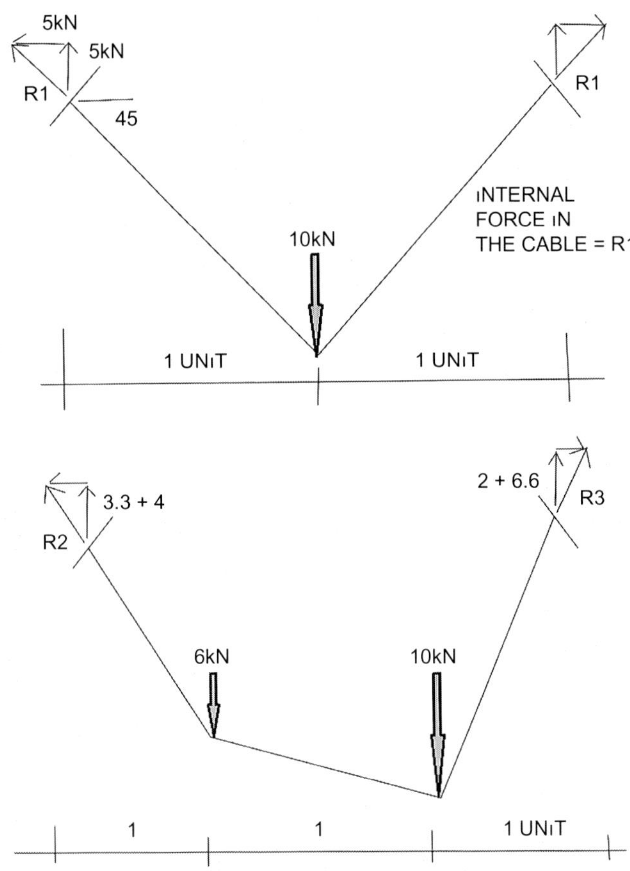

12.3 Reactions at the supports

- Use of tensegric shells by adding compression struts to the structure (Figure 12.2c).
- Spraying concrete (shotcrete) over the surface.

Tensegric shell struts work in compression but do not touch each other. There are many different applications of tensegric shells, but since the design of them is difficult, they are not as popular as other solutions (Vilnay, 1991).

The second common problem of tensile structures is known as the horizontal force problem. When a tensile structure is loaded, it develops reactions at its supports. Figure 12.3 shows some examples of the development of reactions. These reactions will always have the same angle to the body of the structure where it connects to the supports. This means that these reactions are composed of vertical and horizontal components. Horizontal force problem is caused by the horizontal components of these reactions. These forces have to be resisted by some elements within the structure. Every designer who designs tensile structures should have a strategy in order to resist these horizontal forces. Strategies against the wind instability problem and the horizontal force problem are discussed in detail, especially with regards to cable structures, in this chapter.

The best form for a cable, a suspension structure, a membrane or a negatively curved shell can be found in two ways:

- By loading a string according to the expected loading on the structure.
- By using the opposite of the form of moment diagram for the building form.

The first method was used by Spanish architect Antonio Gaudi. Gaudi designed some of his buildings, including the Sagrada Familia Church in Barcelona, with the help of this method. By hanging weights to strings in certain forms, he used the resultant form of these strings in his projects (see URL2, 2007). The second method can be applied by drawing the moment diagram, as explained in chapter 9. Then the best form for a tensile structure can be drawn as the opposite to this moment diagram, as shown in Figure 12.4. If the structure is in this form, no bending moment occurs in the structure and it works solely with tension. This means that knowing how to draw the moment diagram can help architects in determining the best form for different types of loadings.

All points of funicular structures are like hinges (the turnable joints at which moment is equal to zero). Thus, depending on the length of the structural element, the opposite form of the moment diagram can be achieved with different heights. If the element in Figure 12.4 is long, the final form will be a higher triangle. If it is short, then the final form will be a shallower triangle. Whatever the final form, there will be no moment in the funicular structure.

Another common point concerning all types of tensile structures is their architectural design process. One of the best architectural tools to design tensile structures is the making of physical models. If the system is not complete in itself, the model will highlight this problem. If you pull and push it to give the effects of various loadings, you can see if the structure is strong and stable against these types of loads.

The International Building Code (ICC, 2000) included guidelines for membrane structures; however, the 2012 version (ICC IBC, 2012) refers only to standards concerning cables and membranes.

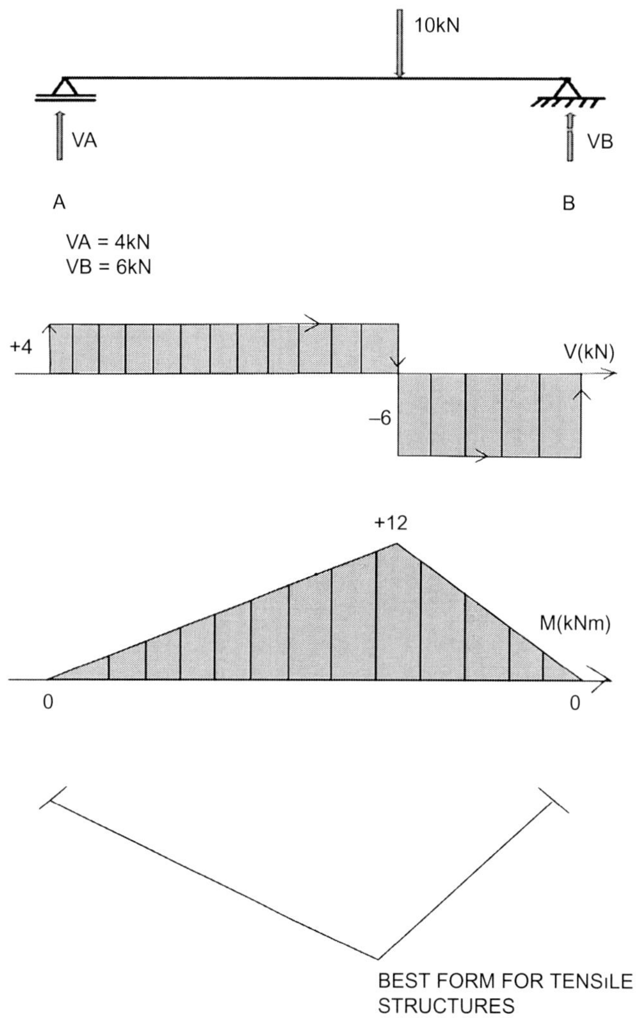

12.4 Drawing the best form for tensile structures

One of the issues covered by the International Building Code is the fire resistance of tensile structures: space dimensions and material types are limited in order to avoid fire problems.

CABLES AND SUSPENSION STRUCTURES

Cables and suspension structures usually consist of very different types of structural elements. For example, suspension bridges have supporting towers, deck beams, secondary cables that carry the deck beams, and the main cable. Some of these elements work with tension and some of them work with compression. Materials are also chosen according to the stress type in the elements. Since the longest spanning element is the main cable, these structures are called cable structures or tensile structures. The longest spanning element usually carries most of the other elements. Thus, its place in the hierarchy of structural elements is high.

It is not usually possible to understand the behaviour of cable and suspension structures before understanding their construction process. The design of the construction process is an important part of structural design. One can easily design impossible structures if one does not consider the construction process during the design stage.

To understand cable and suspension structures also requires understanding how these structures respond to wind instability problems and horizontal force problems. Most cable and suspension structures have strong solutions against these problems.

According to the above points, it is possible to analyse cables and suspension structures as follows:

- Analysis of the elements in the structure; stress type in each element and material used for each element.
- Hierarchy of structural elements, showing which elements carry which elements.
- Identification of the main spanning element.
- Analysis of the response of the structure to wind instability and horizontal force problems.
- Analysis of the construction process of the structure.

According to Standard ASCE/SEI 19-10, elements of cable structures should be replaceable. The same standard requires that the erection procedure of the structure is specified in the contract.

Since there are many approaches for the use of cable and suspension structures, this book analyses structures of four different examples in order to show the plurality of architectural solutions. These examples are:

- Akashi Kaikyo Bridge, Kobe, Japan, 1998, spanning 1991m: longest span in the world (URL3, 2009).
- Former Federal Reserve Bank Building, Minneapolis, USA, 1972, 12 storeys, spanning 95m (Lin & Stotesbury, 1981: p.386; URL4, 2013).
- Dorton Arena, North Carolina, USA, 1952, spanning 100m (Salvadori & Heller, 1975: p.115; URL5, 2014).
- Yale Hockey Rink, New Haven, USA, 1958, spanning 100m (Salvadori & Heller, 1975: p.115; URL6, 2014).

Analysis of Akashi Kaikyo Bridge

Figure 12.5 is a schematic sketch of Akashi Kaikyo Bridge. The structural elements in this bridge are: the deck, which is nearly horizontal (with slight positive curvature); the secondary cables, which carry the deck; and the main cable and the towers, to which the whole load is transferred. Since the deck is nearly horizontal, it is a bending structure. This is the reason for its design as a steel trussed element. Secondary cables and the main cable are under tension: thus, steel was chosen as the structural material of these elements. The towers are mainly in compression. It was possible to use steel or reinforced concrete for these towers, however the designer preferred steel in order to achieve a lighter outlook. The hierarchy of structural elements are in the order of: deck, secondary cables, main cable, towers; and this determines which elements carry which elements. The form of the main cable shows that there is distributed load on the cable. This is the load of the deck, which is transferred to the main cable through the secondary cables. Although the towers take the highest point in the hierarchy, the main cable is the main spanning element. The bridge spans 1991m and it is still the longest spanning structure in the world as at 2014.

Since a suspension structure is mainly designed for downward forces, the wind instability problem is critical for suspension bridges. The wind instability problem is solved by the thickness

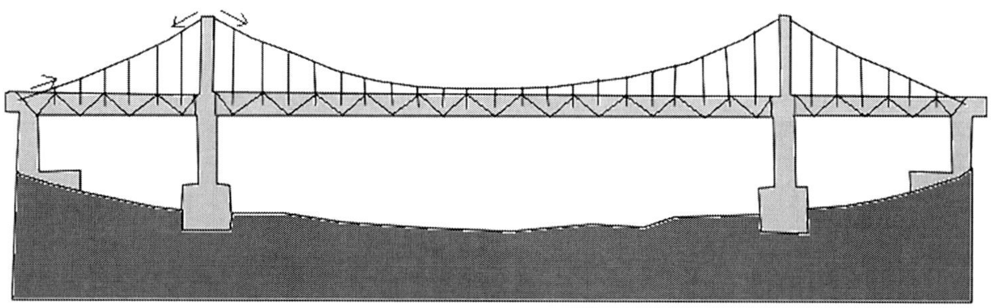

12.5 A schematic sketch of Akashi Kaikyo Bridge, Kobe, Japan, 1998

and slightly curved form of the deck trusses. Compared to the deck of Tacoma Narrows Bridge, the deck of this bridge is much thicker and stronger.

There are eight points in a suspension bridge that can have a horizontal force problem. Four of them are at the top of the towers, where the main span ends and the main cable turns down. Since the two sides of the cables are balancing each other, as shown in Figure 12.5, there is no horizontal force problem at these points. Thus, the towers do not have to resist any bending and they can be slender elements. Four other points are at the two ends of the structure, where the cables meet the foundations. Cables pull out the foundations at these points with an angle parallel to the angle of the cable. Thus, there are horizontal and vertical components to this force, which have to be carried by the foundations.

The construction process of suspension bridges needs the use of high technology. It is easy to guess that the foundations and the towers should be built first. This should be followed by the placement of the main cable. The last step consists of piece-by-piece placement of deck parts by hanging them to the secondary cables. The deck pieces are placed from both sides and they might be post-tensioned to ease the construction, as well as to increase the stability of the deck against wind.

Analysis of the former Federal Reserve Bank Building (now Marquette Plaza)

Figure 12.6 is a schematic sketch of the former Federal Reserve Bank Building. The elements that make up the structure of the former Federal Reserve Bank Building are: the floor systems, which are horizontal; the vertical elements, which transfer the load of the floors to the main cable; the main cable, which has a parabolic shape; the vertical towers at the two sides of the building; and the truss between the two towers. The floor systems have bending stress in them. The materials of the floor systems should be steel in order to decrease their weight. The bottom vertical elements, which take place under the main cable, are under tension. Thus, they should also be steel. The upper vertical elements, which are positioned over the main cable, are under compression and so should be steel struts. The steel main cable is under tension and it is in a parabolic form because it carries distributed loads in the vertical direction. The two towers are pulled inwards by the main cable. Thus, there is bending in the towers. This is why the towers are massive reinforced concrete elements. The truss between the towers connects the two towers and ensures unity of the structure. Since it is a horizontal element, it has bending stress in addition to the tensile and compressive forces, due to the forces applied by the towers. The hierarchy of these elements are in the order of: floor systems, vertical elements, main cable and towers. The truss can be included with the towers because the truss and towers form a frame together. The main cable is the main spanning element because it is the last horizontal element in the hierarchy.

It is not easy to guess how the wind instability problem is solved in this structure. The connection of the main cable to the two towers causes horizontal forces, which have to be resisted by the towers. These towers are different to the suspension bridge towers, which do not have any horizontal force problems. This is the reason for the presence of such strong towers.

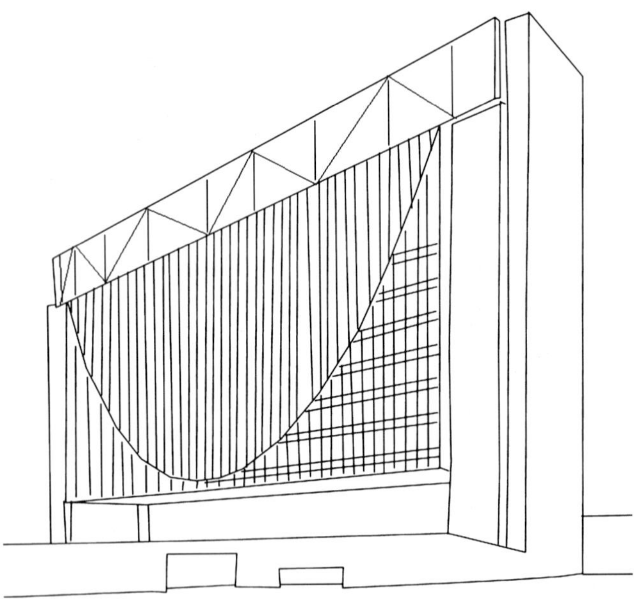

12.6 A schematic sketch of the former Federal Reserve Bank Building, Minneapolis, USA, 1972 (drawn with the help of URL17, 2014)

If one tries to guess about the construction process of this building, it can be stated that the towers and the truss at the top should have been built first. Then, the main cable should have been placed. The floors should have been placed together with the vertical elements in a special way at the end of the process.

Analysis of Dorton Arena

Figure 12.7 is a schematic sketch of Maciej Nowicki's Dorton Arena. Structural elements that take place within this structure are the cable-net, the two arches, and thin columns (URL7, 2014). The cable-net works in tension and it is steel. The main cables are between the two arches and the pre-tensioning cables are perpendicular to these. They are curved in opposite directions: the main cables are negatively curved and the pre-tensioning cables are positively curved. Together, they form a saddle shape. The two parabolic arches work in compression and they are reinforced concrete. The slender columns work with compression and they are made out of steel covered with concrete (Petroski, 2002; URL8, n.d.). It is difficult to discuss the hierarchy of elements in this structure, because the arches are kept in their positions by the cables and the arches reach the foundations. Thus, it can be stated that the vertical load due to the deformation of the structure is carried by the slender columns. It is also not easy to point to the main spanning element. Both the arches and the cable-net realise the spanning function together. The span of the structure is 100m.

The wind instability of the saddle-shaped roof is eliminated by the pre-tensioning cables. One of the most interesting characteristics of this structure is the solution it provides for the horizontal force problem. The main cables pull the arches inwards and downwards, while the arches pull the main cables outwards and downwards. Thus, the horizontal forces applied by the cables are balanced by the horizontal forces applied by the arches.

The construction process of this building can be followed with the photograph album given in URL9 (2010). According to these photographs, the construction process is as follows: first the foundations and the slender columns were built; second the arches were built, followed by the placement of the main cables between the arches; third the net was formed by adding the pre-tensioning cables; and finally the roof panels were placed and covered with roofing tar paper.

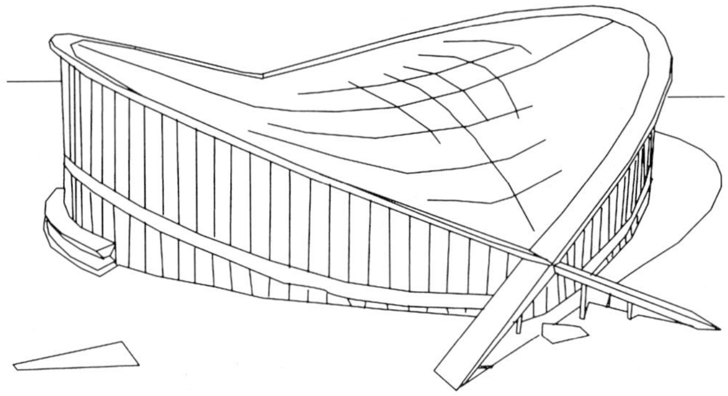

12.7 A schematic sketch of Dorton Arena, North Carolina, USA, 1952 (drawn with the help of NC State University, 2010)

Analysis of Yale Hockey Rink

Figure 12.8 is a schematic sketch of Eero Saarinen's Yale Hockey Rink. Structural elements in the Yale Hockey Rink (also known as Ingalls Rink) are: the arch in the middle; the walls on two sides; and the cable-net between the walls and the arch. The arch works with compression and its material is reinforced concrete. The walls are also reinforced concrete. The main cables of the cable-net are between the arch and the walls. Pre-tensioning cables, which are not shown in Figure 12.8, are perpendicular to the main cables. Both cables work with tension. The cable-net is covered with a timber surface and cover material at the top. The hierarchy of the structural elements is as follows: the cable-net is carried by the arch at the middle and the walls on two sides. Thus the main spanning element is the arch, which is 100m long (Huan, 2013).

The wind instability problem is solved by the pre-tensioning cables. Horizontal forces are applied to the middle arch from two sides, balancing each other. Horizontal forces applied to the side walls are balanced by the inclination of the side walls.

The construction process of Yale Hockey Rink started with the walls on two sides and the main arch. Next, the main cables were positioned and these were followed by the pre-tensioning cables.

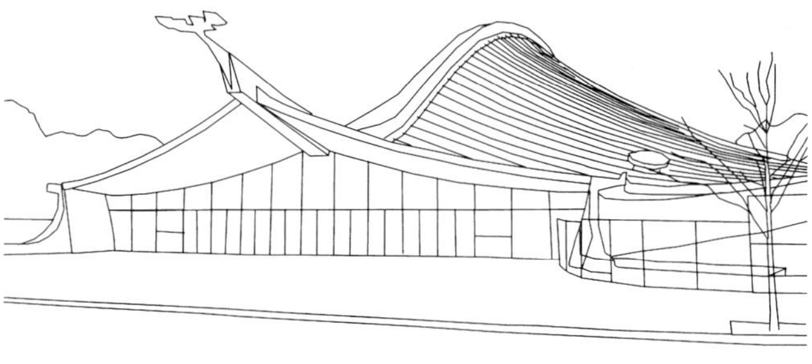

12.8 A schematic sketch of Yale Hockey Rink, New Haven, USA, 1958 (drawn with the help of URL18, 2011)

Structural guidelines for cables and suspension structures

Structural guidelines for cable and suspension structures are listed in Table 12.1. Since every structure is different, structural guidelines for cables and suspension structures have to be kept at a general level.

Table 12.1 Structural guidelines for cables and suspension structures and the associated value system

Structural guidelines	*Value system*
Wind instability problems should be solved by using a cable-net, pre-tensioning cables, tensegric shells, or by spraying shotcrete.	Safety
Horizontal force problems should be solved by balancing them or by providing necessary reaction to them.	Safety
Negatively curved forms are preferred in cable structures.	Economy
Cable and suspension structures should be used with long-span structures. Existing examples in building form are usually between 50 and 100m. The longest suspension bridge spans 1,991m.	Economy
Construction process of cable and suspension structures should be clearly defined during the design phase.	Practicality
Making structural models of cable and suspension structures improves their structural design.	Practicality

Case study 20: Zagreb Arena, Croatia

The case study selected for cables and suspension structures is UPI-2M Ltd's Zagreb Arena, which was built in Croatia in 2009. This building was chosen due to its tectonic quality and the use of pre-tensioning cables. A sketch of Zagreb Arena can be seen in Figure 12.9: plans and sections of the building can be seen in Figure 12.10.

The design concept of Zagreb Arena was to achieve a light building that is well integrated with the city. Continuity with the

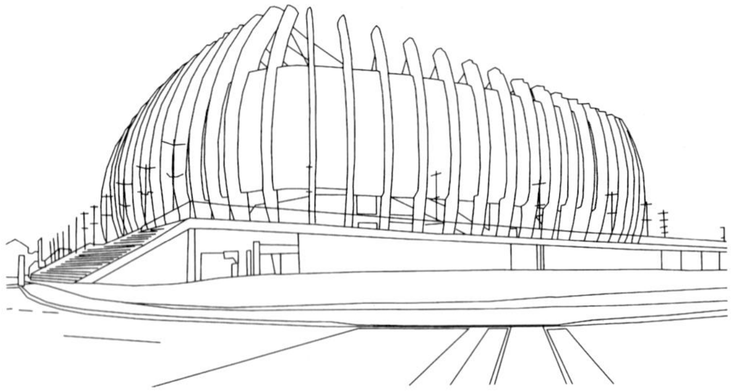

12.9 A sketch of case study 20: Zagreb Arena, Zagreb, Croatia, 2009 (drawn with the help of URL19, n.d.)

city was one of the designers' objectives. The slightly curved section, columns that look like ribs and the use of semi-transparent facade material sought to achieve this. The dominant physical entities in this building are the structural system and construction materials.

The structural elements used in Zagreb Arena are the curved columns and cables system, which carries the roof and the frame system, forming the inner spaces. The columns work with compression and bending due to the horizontal forces applied by the cables and the gravity effect. They are made out of prestressed and prefabricated reinforced concrete. These columns turn inwards at the top and thus decrease the span of the cables. They also take the horizontal and vertical loads applied by the cables. Thus, the horizontal force problem is solved by the columns (URL10, 2014; URL11, 2014).

The variation in the columns' height and their curved form give the main tectonic character to the building. The building looks like many hands holding a precious object. The columns are like fingers. These white columns carry semi-transparent facade elements, which also contribute to the tectonic character of the building. Cables are not perceived from outside or inside. Thus, the roof structure does not contribute much to the tectonics of the building.

When the form of the columns is evaluated from a structural point of view, it can be seen that this form of columns increases the magnitude of the reaction that they have to give against the pull of the cables. Thus, this form of columns cannot be structurally recommended. However, the columns are higher at the places where the pre-tensioning cables connect to them, thus allowing the pre-tensioning cables to have their required curvature.

The cable roof system contains two steel cables in tension against each other. The diagonal cables at the middle tense the two cables that are placed in the shorter direction of the plan. The span of these cables is 100m. Wind instability due to upward wind suction is avoided with the help of these pre-tensioned cables. The roof is not negatively curved due to the pre-tensioned

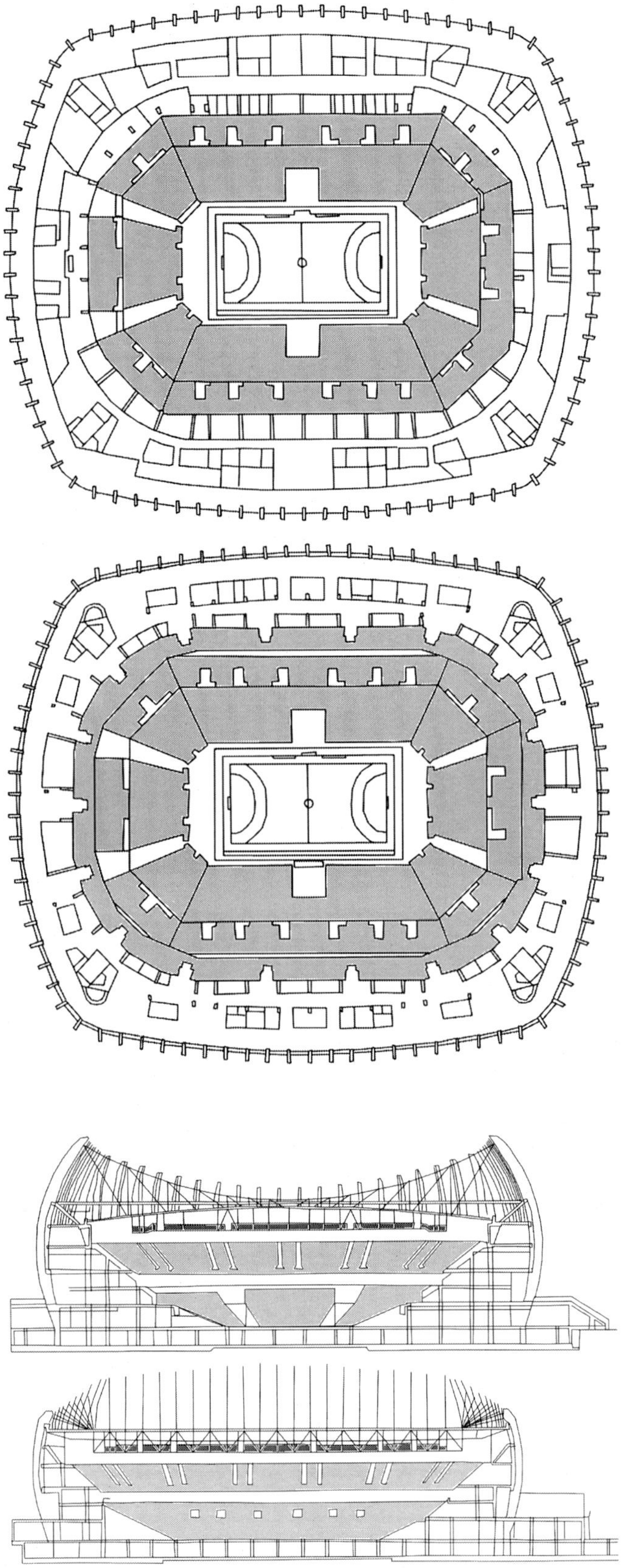

12.10 Plans and sections of Zagreb Arena (drawn with the help of URL20, 2011)

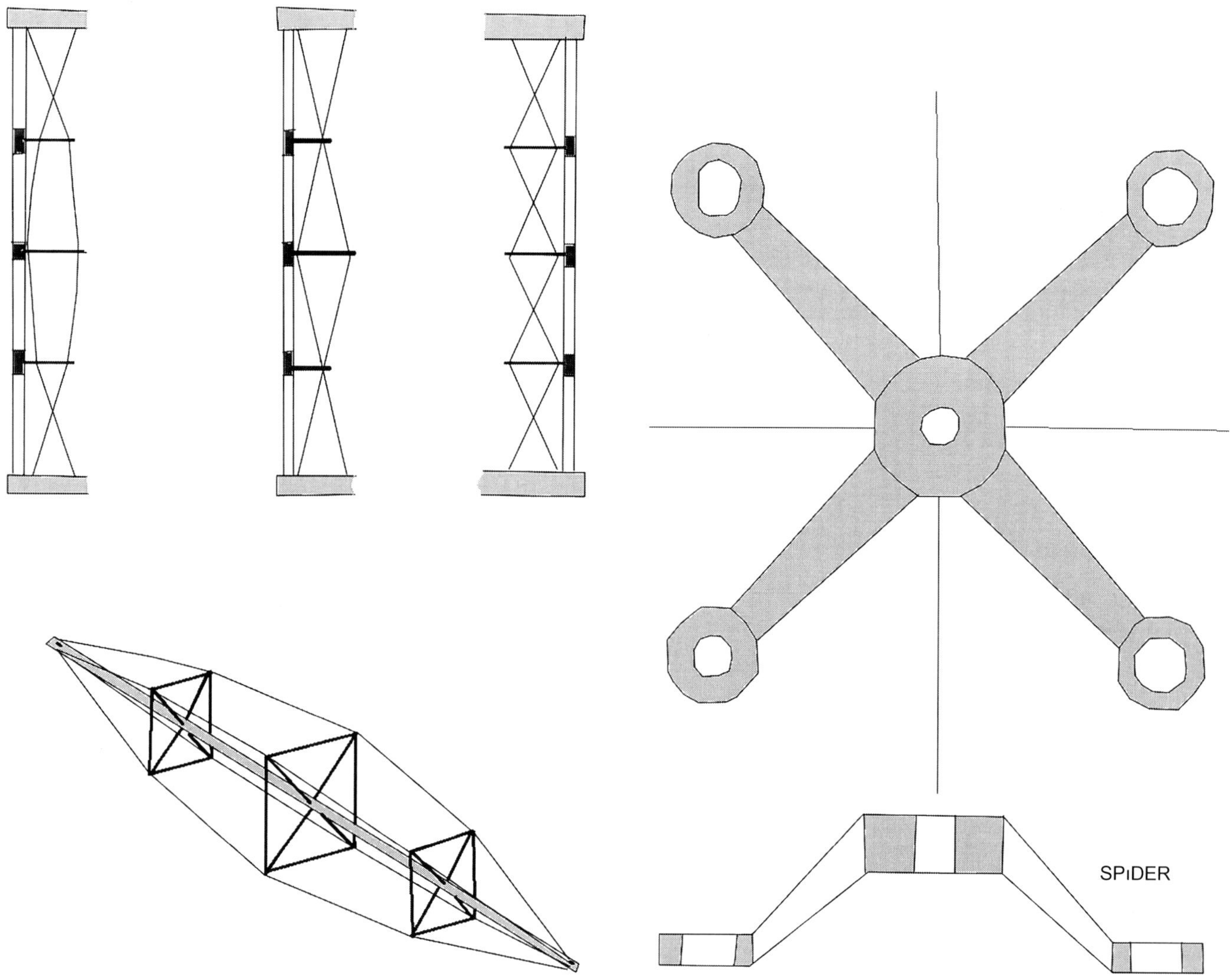

12.11 Types of cable truss

cables. The internal reinforced concrete frame also contributes towards carrying the forces applied by the roof cables (URL10, 2014; URL11, 2014). No special aesthetic consideration was given to the roof during its design because it is not seen from outside.

The construction process of Zagreb Arena started with the construction of the curved columns and the inner frame. Then the pre-tensioned cables and the roof surface were built. Detailed photographs of the construction of Zagreb Arena can be seen in URL12 (2008).

Since the form of the exterior columns is not parallel to recommendations given within structural guidelines for frame systems, the relationship of Zagreb Arena to structural guidelines is contravening.

CABLE TRUSSES, BICYCLE-WHEEL STRUCTURES AND SUSPENDED GLASS SYSTEMS

The cable truss is the main structural element that forms bicycle-wheel structures and cable-truss systems.

Structural behaviour of cable trusses

Cable trusses pre-tension two cables against each other with the help of some short compression elements. There can be many different applications of cable trusses, as seen in Figure 12.11. These cable trusses hold the glass surfaces from their corners with the help of spiders, as shown on the right side of Figure 12.11. The top three cable trusses in Figure 12.11 have various different arrangements of the two cables and the compression elements. However, the lower cable truss has a third type of structural

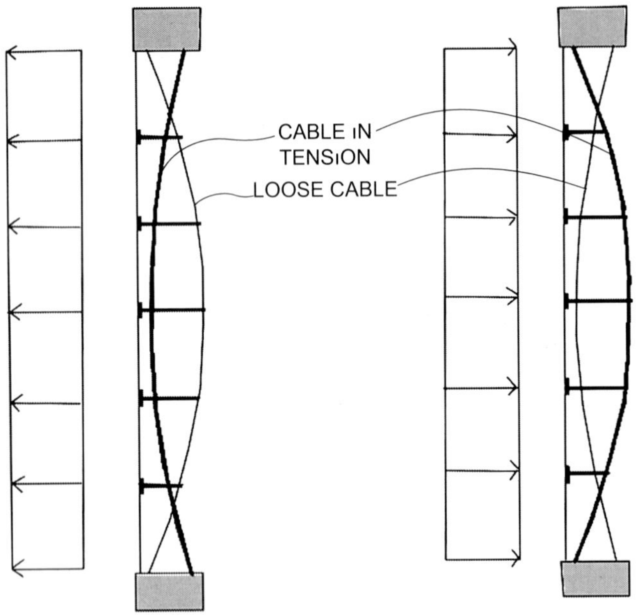

12.12 Structural behaviour of a cable truss

element besides the two cables and short compression elements. This element is a slender beam, which is positioned in the same direction as the two cables.

Figure 12.12 shows the structural behaviour of cable trusses. Depending on the direction of loading on the system, one of the cables takes more tension than the other (Atakara, 2002). The slender beam in the lower cable truss in Figure 12.11 works with bending when it is loaded perpendicular to its axis. This means that the slender beam and the cable truss support each other in resisting the moment due to the loading perpendicular to their axes. This also means that the beam dimensions can be reduced if they are supported with cable trusses.

Horizontal forces applied by the cables to the beams that support them balance each other. The wind instability problem is solved by the pre-tensioning cables.

Construction process of cable trusses

Josi et al. (2012) describe an experience concerning the construction of cable trusses as follows:

1 Pre-stretched cables are sent from the factory.
2 Trusses are assembled on the ground and the locations of clamps are marked on the cables.
3 Cables are disassembled.
4 Top cables are connected to the beam that will later connect the cable truss to the main building structure.
5 The beam is loosely connected to its place.
6 Lower cables are connected to their places.
7 The bottom beam is pulled down 5mm.
8 Cables are pre-tensioned by placing the compression elements.
9 The hoisting beam is locked in.

Depending on the type of cable truss and the type of structural element to which it is connected, the construction process of cable trusses might vary.

Bicycle-wheel structures

Cable trusses were being used as horizontal structural elements for roofs during the late 1950s. They started to be used for supporting vertical surfaces in the 1990s. Circular arrangement of cable trusses that carry horizontal surfaces are called bicycle-wheel structures (see Figure 12.13). In these structures, cable trusses are tensed between a tension ring and a compression ring. Bicycle-wheel structures are used in circular plans and they are very lightweight structures. Auditorium Utica, which spans 80m, was built in the USA in 1958 with the help of a bicycle-wheel structure. It is also possible to use cable trusses in many different arrangements for supporting horizontal surfaces.

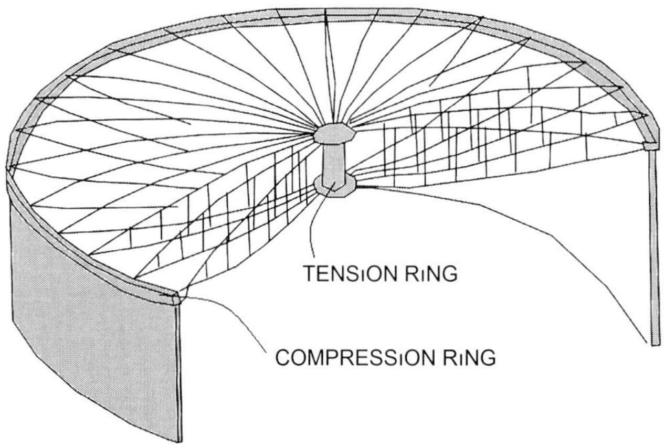

12.13 A bicycle-wheel structure

Suspended glass systems

The first example of suspended glass systems is Bernard Tschumi's Science and Technology Museum in Parc de la Villette, Paris (France), which was built between 1983 and 1986. He conceptualised the building with large glass surfaces. The idea was to have the most dematerialised glass surfaces. Based on Tschumi's concept, Peter Rice designed the suspended glass systems. Rice's design minimised the amount of structural material used in the structure of the glass surface. If we use the same terminology as used in chapter 3, Rice designed this structure with the approach of evolutionary structural optimisation. Rice says that he did not try to make the structure aesthetic: rather he used mathematics to achieve the natural forms with minimum structural material (Rice, 1994; Rice & Dutton, 1995). This building is studied as a case study in this chapter.

After the first use of suspended glass systems, many different applications developed. Atakara (2002) classifies suspended glass systems into three categories according to the location of the suspended glass systems in the building:

- Suspended glass systems that take place between the floors of the building (see Figure 12.14).
- Suspended glass systems that are connected to the main building structure with the help of a secondary lightweight structure (see Figure 12.15).
- Suspended glass systems that are integrated with the building structure.

If cable trusses are used between floors, there is no need for a secondary structure. Early examples include: the entrance to the Cnit-La Defense in Paris, France, built 1991; the Maritime Museum in Nagasaki, Japan, built 1994; and the Greenhouses of Parc Citroen in Paris, France, built 1992. As at 2014, there are many examples of these structures in many cities around the world.

If the cable trusses are not on one of the structural axes of the main building (they can be outside or inside, away from the structural axes), it becomes necessary to have a secondary structure to connect the cable trusses to the main building structure. This secondary structure is usually a steel lightweight frame. Bernard Tschumi's Science and Technology Museum in Parc de la Villette and Banque Populaire de l' Ouest et de l'Armorique in

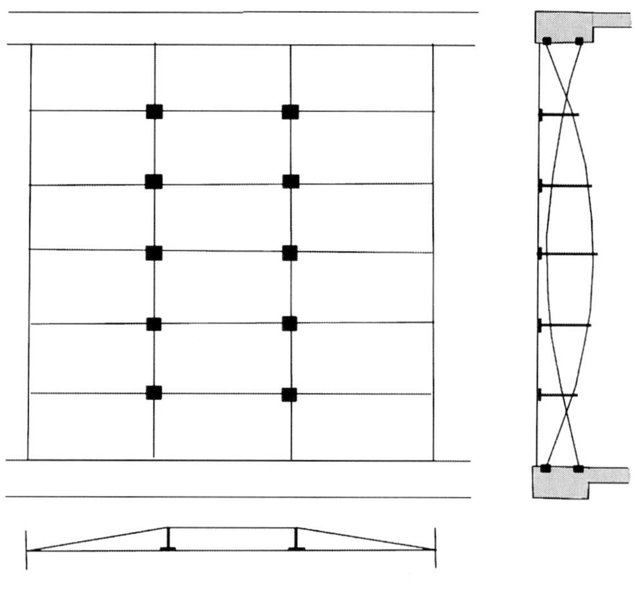

12.14 Use of cable trusses between floors

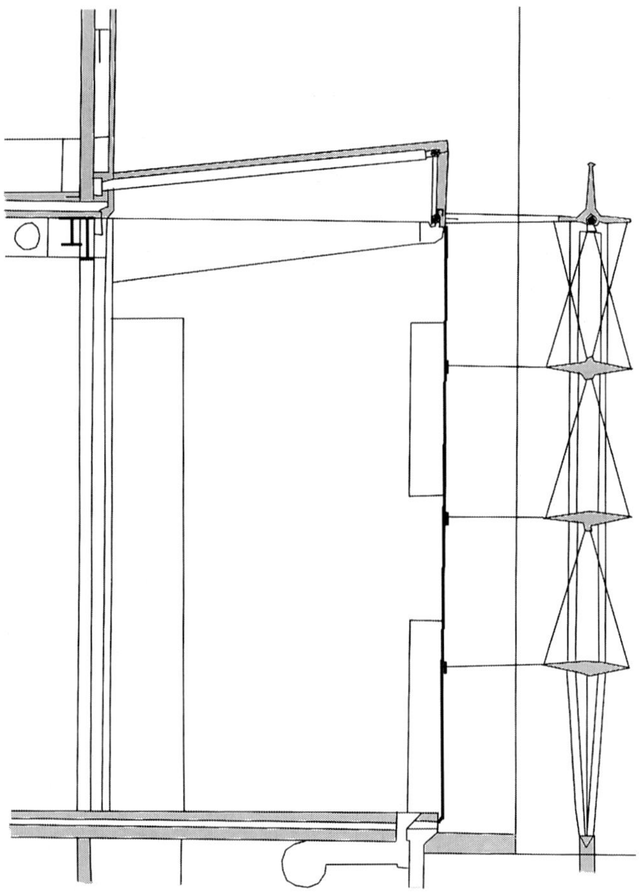

12.15 Use of cable trusses in Banque Populaire de l'Ouest et de l'Armorique, Montgermont, France, 1990 (drawn with the help of Rice & Dutton, 1995)

Montgermont (France), which was built in 1990, are early examples. In the first building, the distance of cable trusses from the main building is around 8m and is dependent upon the capacity of the beams in the secondary structure. In the second building, which is shown in Figure 12.15, the distance between the building and the cable trusses is limited by the length of the compression elements in the cable trusses. Suspended glass systems are dependent secondary structures.

The third application of cable trusses uses the secondary lightweight structure of the suspended system in an integrated manner with the main building structure. Early examples incude: 50 Avenue Montaigne in Paris, France, built 1993; Channel 4 Headquarters in London, UK, built 1994; and Club in Tokyo, Japan, built 1992. Since these are complicated structures that do not have a common configuration, it is not possible to produce schematic sketches.

Structural guidelines for bicycle-wheel structures and suspended glass systems

Structural guidelines for bicycle-wheel structures are listed in Table 12.2 and structural guidelines for suspended glass systems are listed in Table 12.3.

Table 12.2 Structural guidelines for bicycle-wheel structures and the associated value system

Structural guidelines	*Value system*
Circular plan shape is preferred.	Practicality
Structure should be lightly loaded.	Economy
Structure can span around 90m.	Safety
A tension and a compression ring are needed to take the horizontal forces exerted by cable trusses.	Safety
Making structural models of bicycle-wheel structures improves their structural design.	Practicality

Table 12.3 Structural guidelines for suspended glass systems and the associated value system

Structural guidelines	*Value system*
As dependent structures, suspended glass systems should either be in direct contact with the main building structure, or a secondary structure should connect them to the main building structure.	Safety
The distance from the main building structure is limited by the length of compression elements in the secondary structure.	Safety
Making structural models of suspended glass systems improves their structural design.	Practicality

Case study 21: Science and Technology Museum, France

Bernard Tschumi and Adrien Fainsilber (architects) and Peter Rice's (structural engineer) Science and Technology Museum was chosen as a case study due to the importance of its tectonic characteristics. The building, which is located in Parc de la Villette in Paris, can be seen in Figure 12.16, and the suspended glass system part, which repeats, can be seen in Figure 12.17. Parc de la Villette is an important urban design project. Thus, the Science and Technology Museum is sited within an architecturally well-known urban project.

Since the building is a science and technology museum, the architects wanted to have minimum barriers between the inside and outside, but did not rate existing glass technologies of that time. Tschumi's idea of a dematerialised glass surface led to the invention of a suspended glass system by Rice. Thus, the building can be seen as the product of an innovative collaboration between its architect and structural engineer.

The dominant physical entities in this building are the structural system of dematerialised glass surfaces and its details. The glazed parts are 24m in height and there are steel horizontal elements at every 8m. The glass surface is 8m away from the main building structure. When it is looked at from the inside, there is a natural continuity between inside and outside. When it is looked at from the outside, transparency, cables and spiders attract attention. The details in the glazed surface give the impression of an artistically

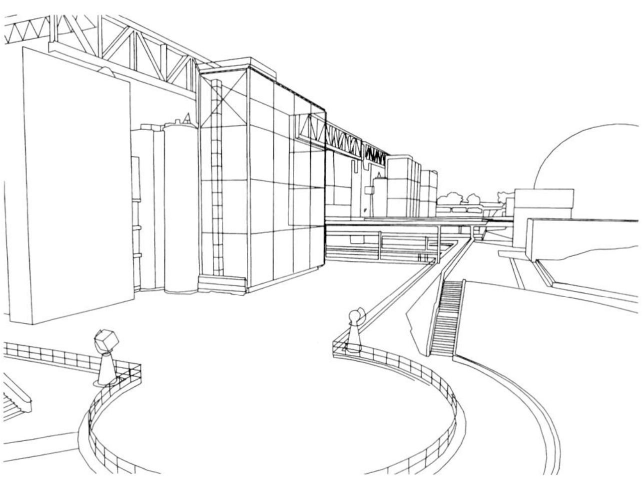

12.16 A sketch of case study 21: Science and Technology Museum, Paris, France, 1983–1998 (drawn with the help of URL21, 2013)

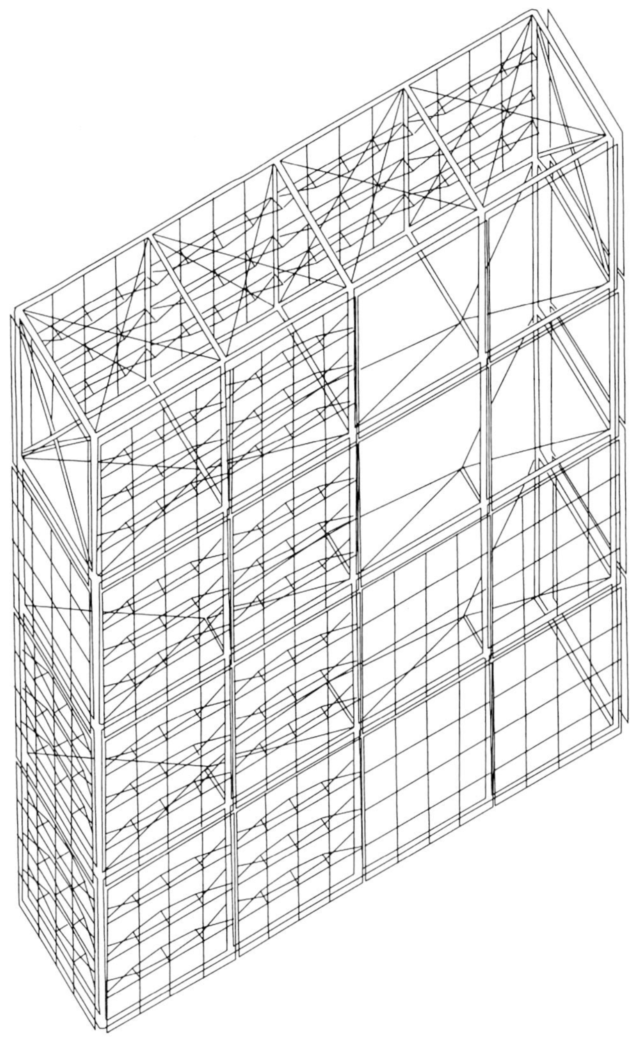

12.17 A schematic sketch of the suspended glass system in the Science and Technology Museum (drawn with the help of Rice & Dutton, 1995)

ornamented facade. Thus, the details designed by Rice are tectonic details, which create structural as well as aesthetic effects. The relationship between the Science and Technology Museum and structural expectations of that time is contravening.

MEMBRANE STRUCTURES

Membrane structures have similarities with cable structures. Especially when they are used for large spans, they start to have similar elements to cable structures. However, the architectural image of membranes is different from that of cables. Membranes look similar to the sails of boats and sails represent freedom. Thus, the use of membranes always gives a festive character to architectural space.

Membranes are used in many different types of buildings, including stadiums, university campus buildings, churches, etc. The general prejudice concerning limitations for their use only as temporary structures is no longer correct.

Structural behaviour and construction of membranes

The simplest example of a membrane is a tent. A small membrane consists of three structural elements as seen in Figure 12.18:

- Membrane fabric, which is in tension.
- Strut in compression.
- Supports at sides, elements of which are either in tension or compression.

Standard ASCE/SEI 55-10 defines the materials used for membrane surfaces as follows:

> The membrane material is usually fabric made of woven or laid yarns, but it can also be a film or foil. The yarns are most commonly made of nylon, polyester, glass, aramid, polyolefin, or PTFE fibers, which may be parallel or twisted together. Films are commonly made of any of the materials used in fabric coatings. In woven fabrics, the yarns pass alternatively over and under each other. In laid fabrics, yarns are placed on top of each other. Fabrics are usually coated with polyvinylchloride (PVC), polytetrafluoroethylene (PTFE), polyolefins, or silicon.

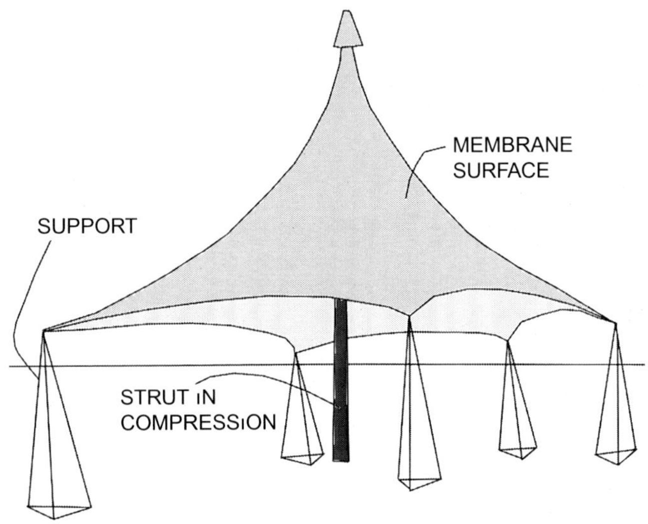

12.18 Structural elements of membranes

The same standard also requires that elements used in membrane structures are made out of non-combustable materials.

However, when the membrane becomes larger, the surface has to be supported against wind instability. In this case, the membrane fabric is used together with cables or pre-tensioned cables (as seen in Figure 12.2). These cables divide the membrane surface into pieces according to its geometry. These types of stable membranes have started to be used in buildings. According to Standard ASCE/SEI 55-10, membrane structures can be erected on roofs and can be attached to existing buildings.

Horizontal forces at the top of the membrane balance each other. However, horizontal forces created at the sides are taken by side supports.

The strut of a membrane can also be designed in various different forms. The following applications, which are shown in Figure 12.19, are frequently seen:

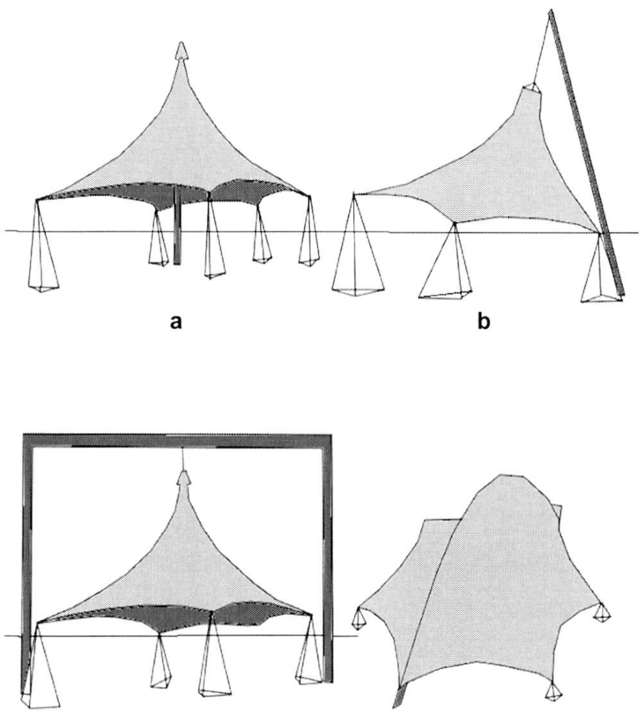

12.19 Different applications of membranes

a. If the strut is at the middle, it works with compression.
b. If there is an inclined strut outside, it works with compression and bending.
c. If there is a frame to support the membrane, it works with bending. (This can also be an arch that works with compression.)
d. If the membrane covers an arch or a frame, these structures have compression and bending respectively.

Depending on the size and location of the strut, there are very different applications creating different tectonic effects. The struts that have compression should be designed against buckling. Since membranes are light and elegant structures, there is a tendency to design all of their elements to be thin and slender.

The construction process of small membranes can be as follows:

1 The strut, frame or arch that will support the frame is placed.
2 The supports at the sides are built.
3 The membrane is tensed between these supports.

According to Standard ASCE/SEI 55-10, fabrication drawings should be made by the designer, and detailed installation procedures should be supplied by the erector.

Organisation of membrane units

There are two different types of membrane structures. These are:

- A singular but large membrane structure.
- A structure achieved by repetition of membrane units.

The following buildings are examples of the first type of membranes (Tian, 2011): Park Dome in Kumamoto, Japan, built 1997; Millenium Dome in Greenwich, UK, built 1999; Yao-Yuan County Arena in Taiwan, Republic of China, built 1993; Good Shepherd Lutheran Church in Fresno, USA, built 1982; and Hyogo Prefectural Tajima Dome in Hyogo, Japan, built 1998.

The following buildings are examples of the second type of membranes (Tian, 2011): San Nicola Stadium in Bari, Italy, built 1990; Don Valley Stadium in Sheffield, UK, built 1990; San Diego Convention Center in San Diego, USA, built 1989; 'Nuage Leger' La Grande Arche in Paris, France, built 1989; Columbus'92 'Bigo' in Genoa, Italy, built 1992; and Campus Center, University of La Verne in La Verne, USA, built 1973.

Searching for these examples on the internet can be useful to understand the tectonic characteristics of membrane structures.

The first type of membranes can be understood with the help of cable structures, but the organisation of the second type of membranes needs further information. These structures are compositions of units. These units can be like the ones seen in Figure 12.19, but there can be many different types of units depending on the creativity of the designer, as seen in Figure 12.20. Columbus'92 'Bigo' combines the inclined strut type with steel arches to hang the repeating membrane units. The struts are in a

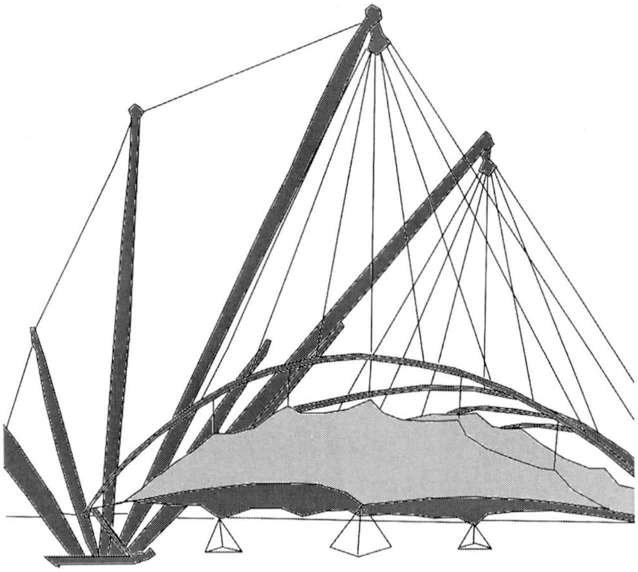

12.20 Columbus'92 'Bigo', Genoa, Italy, 1992 (drawn with the help of URL22, n.d.)

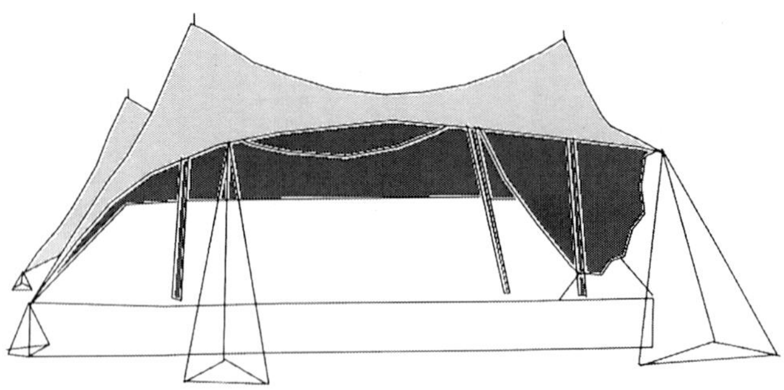

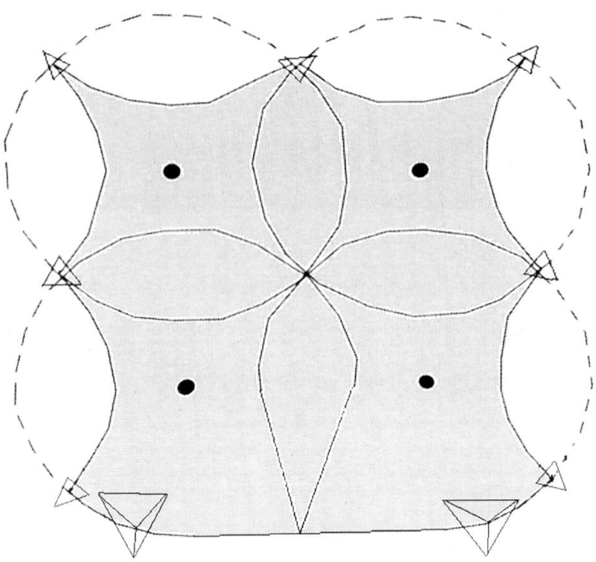

12.21 An example of order of form in membranes

cigar shape and the membrane canopy spans 60m × 40m without any vertical support. The membrane surface is formed by Teflon PTFE coated glass fibre.

The same units can be repeated in the same or in different sizes, or different units can be used in similar or in different sizes. Figure 12.21 shows an example of the use of the same type of units in the same size. Whatever the type of membrane, these structures can span between 10m and 80m (Engel, 1997).

Structural guidelines for membrane structures

Structural guidelines for membrane structures are listed in Table 12.4.

Table 12.4 Structural guidelines for membrane structures and the associated value system

Structural guidelines	*Value system*
Maximum span for membrane structures is around 80m.	Economy
Depending on the span, wind instability problems can be solved by the membrane only, by additional cables, and by additional pre-tensioned cables.	Comfort and safety
Strut form, material and cross-sectional characteristics change depending on the structure's location and inclination.	Safety
Compression elements in membranes should not be slender in order to avoid buckling.	Safety
Making structural models of membrane structures improves their structural design.	Practicality

Case study 22: Olympic Stadium in Munich, Germany

Gunther Behnish, Frei Otto, Herman Peltz and Carlo Weber's Olympic Stadium in Munich was chosen as a case study due to the interesting organisation of repeating units. A sketch of the stadium can be seen in Figure 12.22; a schematic plan and three-dimensional form of the repeating units can be seen in Figure 12.23.

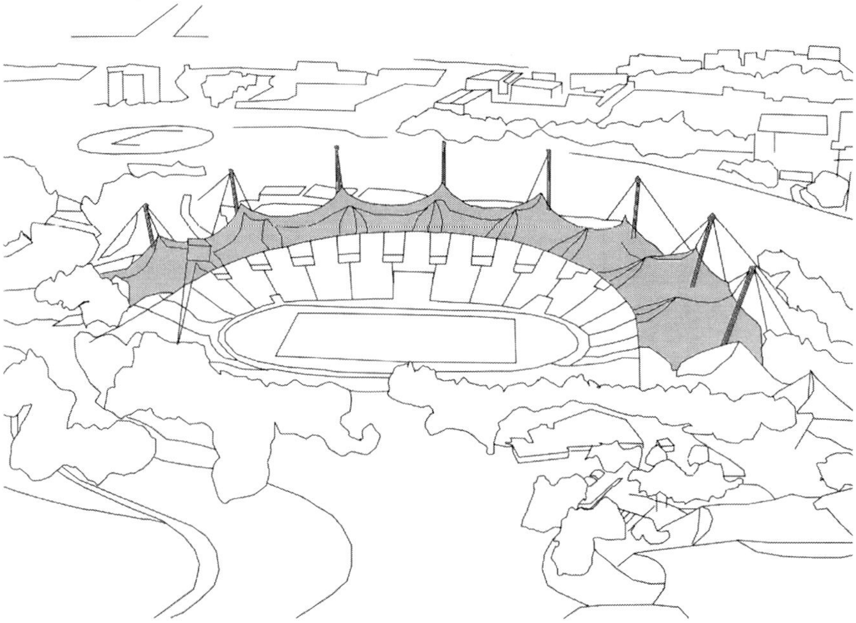

12.22 A sketch of case study 22: Olympic Stadium in Munich, Germany, 1972 (drawn with the help of URL23, n.d.)

The three important architectural decisions made during the design of this stadium were:

- To give a festive atmosphere to the stadium by using membranes.
- To eliminate any visual obstable in front of the audience.
- To solve the wind instability problem.

Since the form of the stadium was given beforehand, a very clever idea was needed to eliminate wind instability problems without creating any visual obstacles in front of the audience. The curved cable at the front and the inclined cables at the back give a pre-tensioning effect to the membrane surfaces. The dominant physical entity in Olympic Stadium in Munich is the structural system.

Membranes of this structure are formed by cables and they are covered with acrylic glass panels, which is a possible roof material for the climate in Germany. The details on the membrane surfaces give an ornamental effect to the elegantly curved surfaces. The relationship of the Olympic Stadium in Munich to structural guidelines is affirmative.

PNEUMATIC STRUCTURES

Pneumatic structures are air-inflated structures. Their tensed surfaces have similarities to the tensed surfaces of membranes. Thus, the standards defined by Standard ASCE/SEI 55-10 concerning materials to be used for membrane surfaces are also valid for many pneumatic structures.

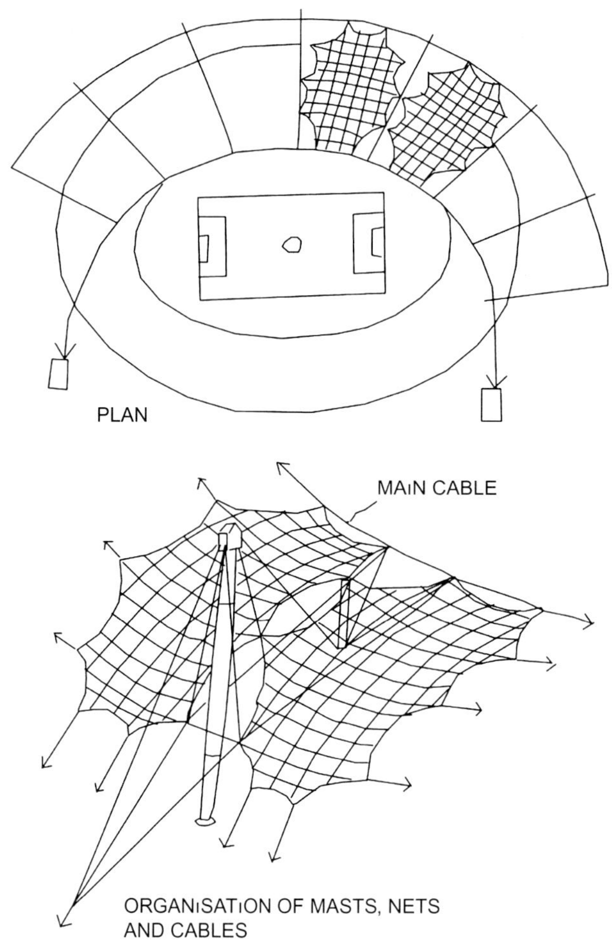

12.23 Drawings of Olympic Stadium in Munich (drawn with the help of URL24, n.d.)

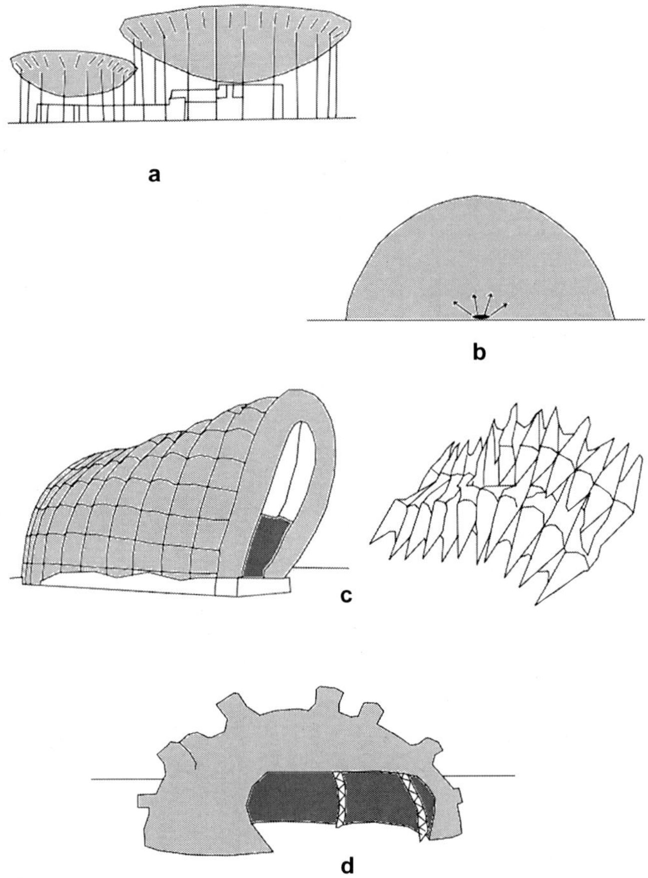

12.24 Types of pneumatic structures

Types of pneumatic structures

Pneumatic structures can be used for various different applications in architecture, as shown in Figure 12.24:

a. A pneumatic surface as a roof.
b. A continously air-filled pneumatic dome.
c. A pneumatic structure composed of linear pneumatic units.
d. A pneumatic structure with a supplementary structure.

The first type of pneumatic structure is usually used to give an identity to the building under it. An example is the Expo 2002 in Neuchatel, Switzerland (URL13, 2001).

The second type of pneumatic structure has been used frequently for circuses. These are deployable structures. There should be a double entrance to the dome in order to avoid air escaping. First the base of the dome is prepared, then the fabric is connected to this base through the circumference of its circle. Air is then filled into the closed space under the fabric. This type of pneumatic structure is sometimes used as formwork in the construction of reinforced concrete domes.

The third type of pneumatic structure eliminates the air escape problem of the second type. An example is the Finmeccanica Pavilion in Farnborough, UK, which was built in 2006 for the first time. Here, pneumatic cushions are used on its facades. A film of the construction process for this building can be found in URL14 (n.d.). Sprachpavilion in Austria (URL15, 2010) also has pneumatic cushions.

The fourth type of pneumatic structure contains a supplementary structure within it. Most of the contemporary pneumatic structures have a steel or aluminium structure in them in order to avoid wind instability problems and to span longer distances. Depending on the type of structure that supports the lightweight pneumatic surface, these structures can span up to 220m. The Finmeccanica Pavilion in Farnborough has a steel structure inside it and the facades are supported by steel trusses. Cables were also used in this structure against wind suction (URL14, n.d.). The Sprachpavilion in Austria is a smaller structure, but it still contains a secondary trussed structure together with pneumatic cushions (URL15, 2010).

Structural guidelines for pneumatic structures

Structural guidelines for pneumatic structures are listed in Table 12.5.

Table 12.5 Structural guidelines for pneumatic structures and the associated value system

Structural guidelines	*Value system*
A secondary structure solves wind instability problems.	Safety
Maximum span for pneumatic structures is 220m.	Economy

Case study 23: Swarovski Pavilion, Switzerland

Veech Media Architecture's Swarovski Pavilion in Basel was chosen as a case study due to the determinant nature of its architecture. Figure 12.25 is a sketch of the pavilion, and Figure 12.26 shows the plan and sections.

Swarovski Pavilion takes place in an ordinary steel structure. The inner pneumatic structure is supported by this outer structure.

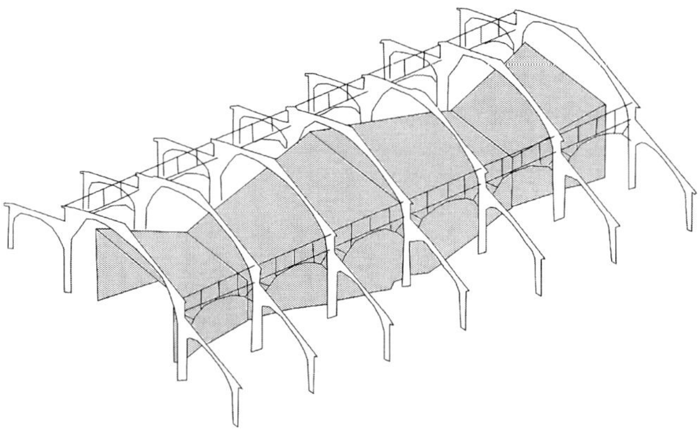

12.25 A sketch of case study 23: Swarovski Pavilion, Basel, Switzerland, 2008 (drawn with the help of URL16, 2009)

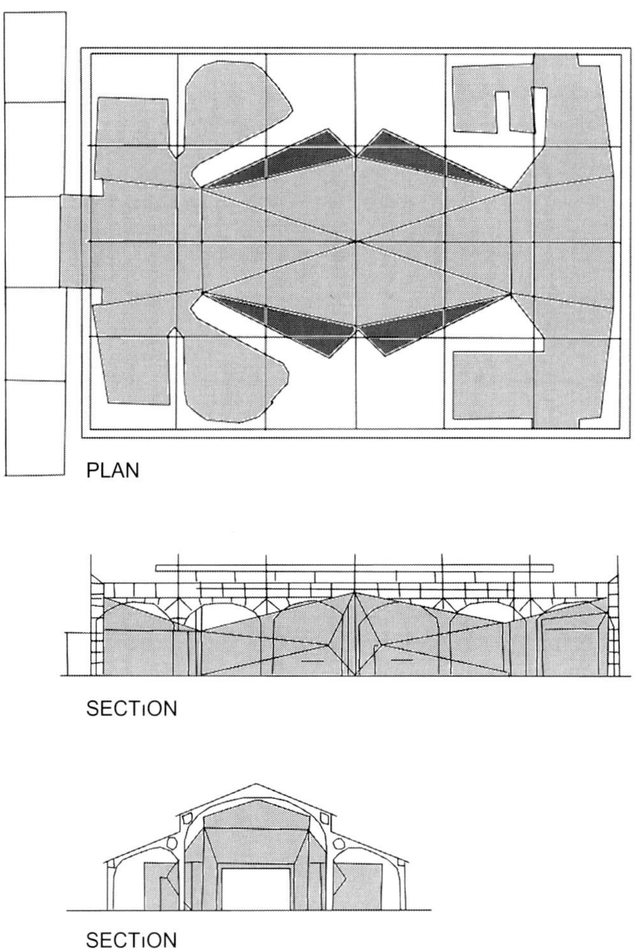

12.26 Plan and sections of Swarovski Pavilion (drawn with the help of URL16, 2009)

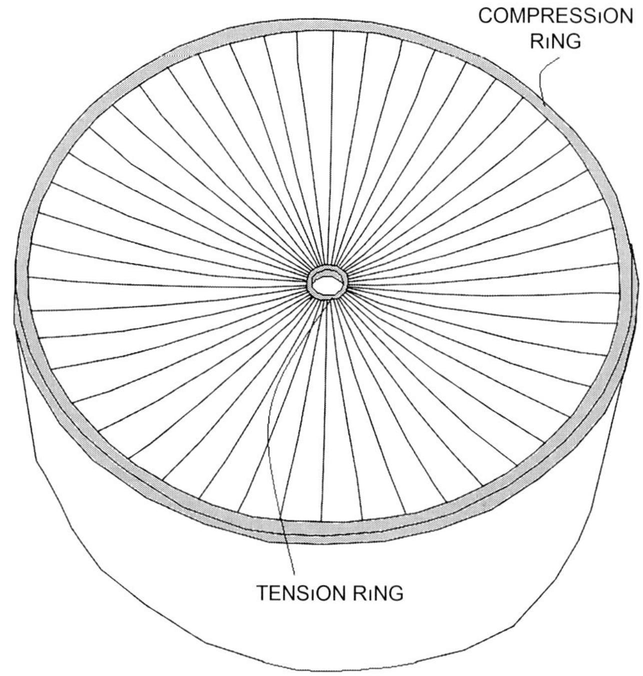

12.27 Negative curvature shell with cables

However, the form of the inner structure is different to its exterior structure. Although the outer structure is in an ordinary form, the inner pneumatic structure is in the form of a gemstone and thus it presents a valuable spatial experience. The idea of a gemstone provides the architectural concept for this project. The dominant physical entity in Swarovski Pavillion is the structural system.

The inner structure is a pneumatic membrane with inflated facets. These facets are hung to the outer structure (URL16, 2009). The outer structure guarantees the stability against horizontal forces. Since Swarowski Pavilion follows structural guidelines for pneumatic structures, it has an affirmative relationship with structural guidelines.

NEGATIVE CURVATURE SHELLS

The form of negative curvature shells is apparent from its name. They are usually covered with concrete and they have either cables or a special mesh reinforcement in order to resist the tension in the structure. Figure 12.27 shows an example with cables before the concrete has been poured. This type of structure is studied in more detail in chapter 13.

CONCLUSIONS

An analysis of the case studies and other examples in this chapter shows that there are three different approaches to the tectonics of tensile structures:

- Architecturally creative solutions are produced to solve the wind instability and horizontal force problems in tensile structures. This is usually the case with cable and membrane structures. Dorton Arena, Yale Hockey Rink and the Olympic Stadium in Munich can be given as examples for this approach.
- Not using the ideal solution in order to achieve aesthetics is another approach. An example for this approach is the Zagreb Arena's curved columns. The Swarovski Pavilion was also designed in order to achieve spatial quality only: the forms used in this building are not the ideal forms for pneumatic structures.
- Innovative collaboration between architects and structural engineers, combining creativity with innovation, produces designs for tensile structures. Examples for this approach include: the invention of suspended glass systems, as well as their development and new applications; and innovative and creative uses of recent pneumatic structures.

The other interesting issue concerning the tectonics of tensile structures is the ornamental effect of their details. These details are at a human scale and attract people's attention.

PROBLEMS TO SOLVE

Find the best form for the following loadings for tensile structures.

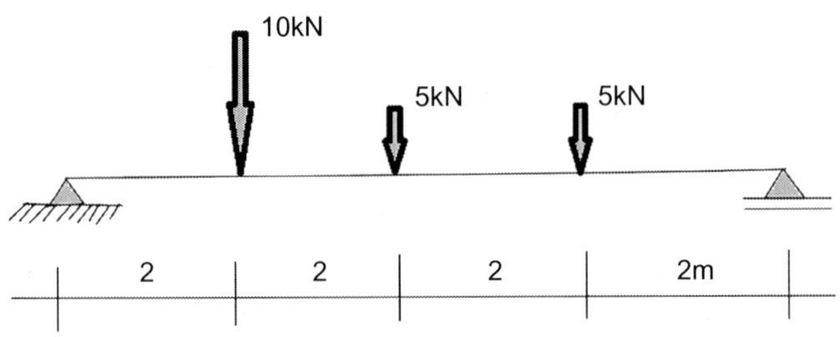

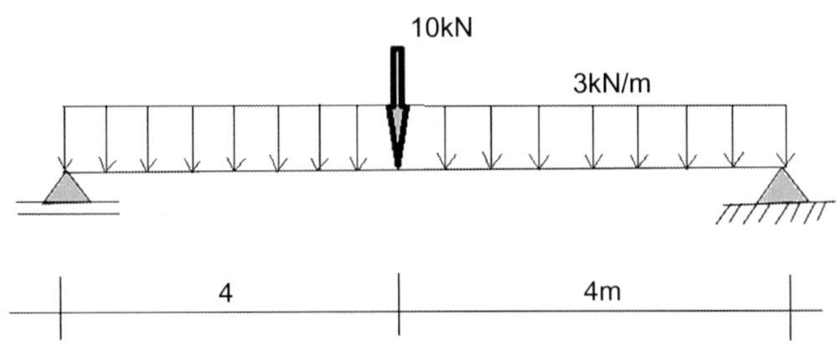

REFERENCES

Atakara, C. (2002) Spatial Characteristics of Suspended Glass Systems With Prestressed Cable Truss, unpublished Masters thesis, Eastern Mediterranean University: North Cyprus.

Engel, H. (1997) *Structure Systems*, 3rd edition, Gerd Hatje Publishers: Ostfildern, Germany.

Huan, E.C. (2013) *David S. Ingalls Rink* (viewed 18 August 2014: www.engr.psu.edu/ae/thesis/portfolios/2014/cih5144/Building%20Stat/Building%20Statistics%20Combined.pdf)

ICC (2000) *International Building Code*, International Code Council: Illinois.

ICC IBC (2012) *International Building Code*, 2nd edition, International Code Council: Illinois.

Josi, G., Montgomery, J., DiBattista, J., Anderson, G., Erdevicki, D. (2012) 'Design and Construction of a 30 Meter High Glazing

Wall Supported By Cable Trusses' *Third International Structural Specialty Conference*, 6–9 June: Edmonton, Canada.

Lin, T.Y., Stotesbury, S.D. (1981) *Structural Concepts and Systems for Architects and Engineers*, John Wiley and Sons: New York.

NC State University (2010) *Mathew Novicki Drawings and Other Material* (viewed 28 September 2014: http://news.lib.ncsu.edu/changinglandscape/2010/07/30/matthew-nowicki-papers/)

Petroski, H. (2002) 'Dorton Arena' *American Scientist* (viewed 18 August 2014: www.americanscientist.org/libraries/documents/20051220153327_306.pdf)

Rice, P. (1994) *An Engineer Imagines*, Artemis: London.

Rice, P., Dutton, H. (1995) *Structural Glass*, E&FN Spon: London.

Salvadori, M., Heller, R.A. (1975) *Structure in Architecture*, Prentice Hall: New Jersey.

Santoso, K. (2003) *Wide Span Cable Structures*, Master thesis, Massachusetts Institute of Technology: USA.

Standard ASCE/SEI 19-10 *Structural Applications of Steel Cables for Buildings*, American Society of Civil Engineers: Reston, VA.

Standard ASCE/SEI 55-10 *Tensile Membrane Structures*, American Society of Civil Engineers: Reston, VA.

Tian, D. (2011) *Membrane Material and Membrane Structures in Architecture*, Master thesis, The University of Sheffield: UK.

Vilnay, O. (1991) 'Design of Tensegric Shells' *Journal of Structural Engineering*, Vol. 117, No. 7: pp.1885–1896.

URL1 (2006) *Tacoma Narrows Bridge Collapse 'Gallopin' Gertie'* (viewed 17 August 2014: www.youtube.com/watch?v=j-zcz JXSxnw)

URL2 (2007) *A Different Type of String Theory: Antonio Gaudi* (viewed 17 August 2014: http://memetician.livejournal.com/201202.html)

URL3 (2009) *The Akashi Kaikyo Suspension Bridge – Japan* (viewed 17 August 2014: www.technologystudent.com/culture1/akashi1.htm)

URL4 (2013) *Federal Reserve Bank of Minneapolis* (viewed 17 August 2014: www.rrj.com/projects/by-rrj-service/79-projects/159-federal-reserve-bank-of-minneapolis)

URL5 (2014) *J.S. Dorton Arena* (viewed 17 August 2014: www.ncstatefair.org/facilities/dorton.htm)

URL6 (2014) *Stadium and Arena Pictures* (viewed 17 August 2014: http://architecture.about.com/od/greatbuildings/ig/Stadium-and-Arena-Pictures/Ingalls-Rink.htm)

URL7 (2014) *Dorton Arena* (viewed 18 August 2014: www.asce.org/People-and-Projects/Projects/Landmarks/Dorton-Arena/)

URL8 (n.d.) *J.S. Dorton Arena* (viewed 18 August 2014: www.arcaro.org/tension/album/dorton.htm)

URL9 (2010) *Dorton Arena – Images from the Lewis P. Watson Collection* (viewed 18 August 2014: www.flickr.com/photos/wakegov/sets/72157624828355984/detail/?page=4)

URL10 (2014) *Arena Zagreb* (viewed 19 August 2014: www.worldbuildingsdirectory.com/project.cfm?id=1494)

URL11 (2014) *Arena Zagreb UPI-2M* (viewed 19 August 2014: www.archdaily.com/80556/arena-zagreb-upi-2m/)

URL12 (2008) *Zagreb Arena* (viewed 19 August 2014: www.ingra.hr/files/ARENAZAGREB_preview.pdf)

URL13 (2001) *Expo 02* (viewed 24 August 2014: www.tensinet.com/database/viewProject/4183.html)

URL14 (n.d.) *Farnborough 2014 Finmeccanica Pavilion Time Lapse* (viewed 24 August 2014: www.youtube.com/watch?v=0IiXt_kOpYM)

URL15 (2010) *Sprachpavilion Inflatable Mobile Structure* (viewed 24 August 2014: www.veech-vma.com/2008/exhibition-event/sprachpavillon-inflatable-mobile-structure/)

URL16 (2009) *Veech Media Architecture Wins With 'Ambient Gem' Swarovski Pavilion* (viewed 24 August 2014: www.bustler.net/index.php/article/veech_media_architecture_wins_with_ambient_gem_swarovski_pavilion/)

URL17 (2014) *Federal Reserve Bank* (viewed 28 September 2014: http://structurae.net/structures/federal-reserve-bank)

URL18 (2011) *Ingalls Hockey Rink, Yale University* (viewed 28 September 2014: http://everydaythingsetc.com/2011/08/05/ingalls-hockey-rink-yale-university/)

URL19 (n.d.) *Sport Stadiums* (viewed 28 September 2014: www.pinterest.com/gailbaumoehl/sport-stadiums/)

URL20 (2011) *Arena Zagreb* (viewed 28 September 2014: http://housevariety.blogspot.com.tr/2011/01/arena-zagreb-by-upi-2m.html#.VCgOmE1xljo)

URL21 (2013) *Cit des Sciences et de l'Indrustrie* (viewed 28 September 2014: www.travelmagazine.org/?p=570)

URL22 (n.d.) *Grand Bigo* (viewed 28 September 2014: www.hotel-4venti.it/public/wp-content/uploads/Genova-portoantico.jpg)

URL23 (n.d.) *Munich Olympic Stadium* (viewed 28 September 2014: www.pinterest.com/pin/560838959816993008/)

URL24 (n.d.) *Olympic Games 1972 Munich* (viewed 28 September 2014: www.tensinet.com/database/viewProject/3779.html)

13 The Tectonics of Compression Structures

As mentioned in chapter 5, compression structures usually have a positively curved form, and contemporary compression structures consist of the following systems:

- Steel vaults and domes
- Geodesic domes
- Shell structures
- Grid shells.

This chapter contains general characteristics and problems of compression structures, together with descriptions and examples of each structure type.

GENERAL CHARACTERISTICS AND PROBLEMS OF COMPRESSION STRUCTURES

Since the strength of compression structures depends on their form, this is the most important characteristic of these structures. Any positively curved form is possible for modern compression structures. However, the best form for a specific loading combines the highest strength with the least amount of structural materials. The best form for compression structures has similarities with finding the best form for tensile structures, and can be found in two ways:

- By loading a string according to the expected loading and taking the reflection of this form as the best form (see URL19, 2007). Chapter 12 contains more information about this method.
- By using the form of the moment diagram as the building form.

The second method can be applied by drawing the moment diagram, as shown in chapter 9. The best form for a compression structure can be drawn in the same form as the moment diagram shown in Figure 13.1. If the structure is in this form, there can be only negligible bending moment in the structure and it can be stated that it works solely with compression. This implies that knowing how to draw a moment diagram can help architects to determine the best form for different types of loadings.

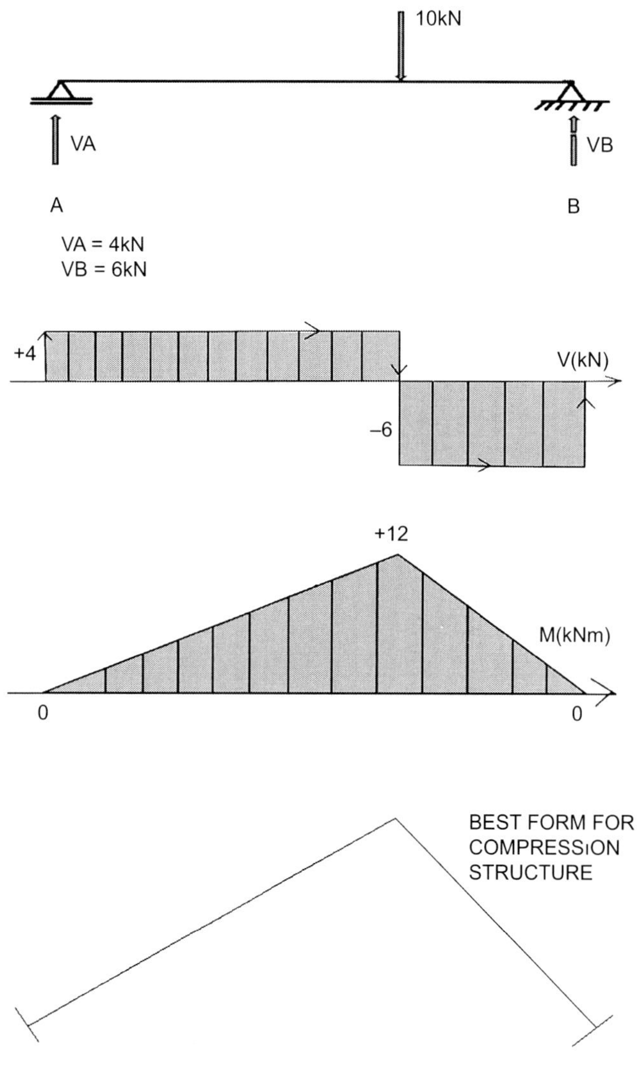

13.1 Drawing best form for compression structures using a moment diagram

In contrast to tensile structures, all points on compression structures do not work as hinges. This means that they can develop bending moment. Thus, finding the best height for the form of the moment diagram is needed. There are also other methods for finding the best form for compression structures (see Adriaenssens et al., 2014). If the structure is in the best form, then the structure works solely with compression (except for edges). If not, bending will also occur in the structure.

Another characteristic of form in compression structures is their curvature. A structure that is not sufficiently curved (or sufficiently high) cannot perform as a compression structure. For instance, shallow forms work with bending rather than compression. Figure 13.2 shows the difference between slightly curved and sufficiently curved forms. A slightly curved structure can only perform as a bending structure.

Since compression structures do not change their forms like tensile structures, they do not have a wind instability problem. However, it is still necessary to consider wind load, as it is for all structures. The general problems of compression structures are:

- Horizontal force problem
- Buckling.

The horizontal force problem in compression structures is similar to the horizontal force problem in tensile structures. The load on the structure is transferred downwards, parallel to the form of the structural element, as seen in Figure 13.3. Thus, reactions at supports are parallel to the structural element at that end. The horizontal components of these reactions are the cause of horizontal force problems. As mentioned previously for the horizontal force problem in masonry arches in chapter 7, there are many

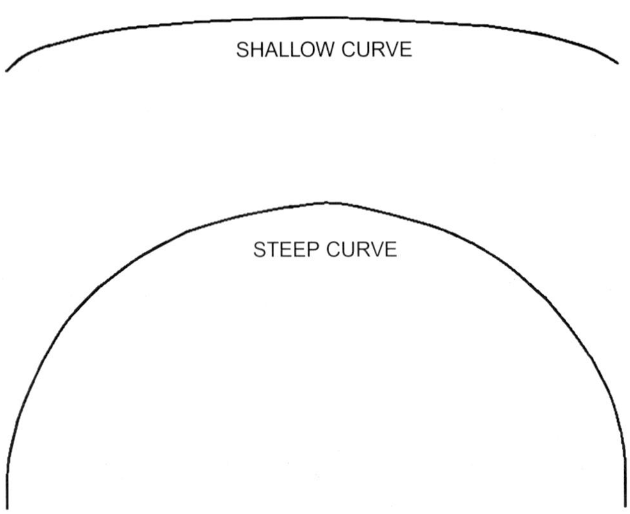

13.2 Curvature

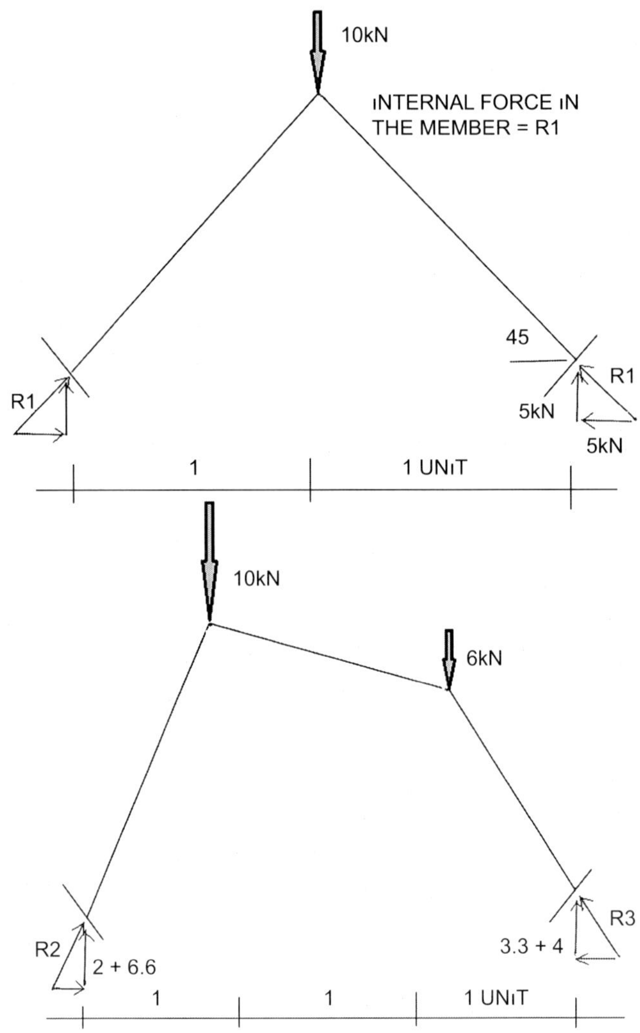

13.3 Horizontal forces in compression structures

solutions (Figure 7.6 shows some of these). Solutions to horizontal force problems are usually seen as opportunities to create tectonic effects, as seen in the case studies in chapter 7.

Buckling is the second general problem of compression structures. As explained in chapter 4, a buckling problem can be solved by avoiding slender compression elements. Compression structures are less effective than tensile structures due to the buckling problem. Compression structures therefore need to be thicker than tensile structures.

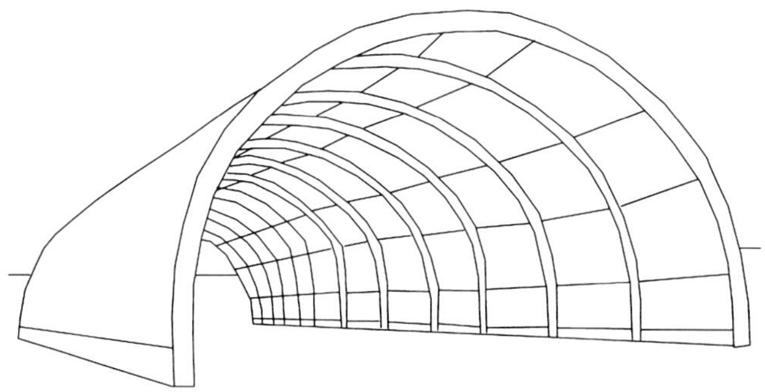

13.4 Structural elements in steel vaults

STEEL VAULTS AND DOMES

Since reinforced concrete vaults and domes are studied in connection with shell structures later in this chapter, this section will only cover steel vaults and domes.

Steel vaults

A steel vault usually consists of steel arch ribs, and beams connecting these ribs. Steel arches, as used in bridges, can span hundreds of metres. For example, Zdakov Bridge, which was built in the Czech Republic in 1967, spans 362m.

Arch ribs work in compression. The beams that connect the arch ribs in vaults work against horizontal forces that are perpendicular to the arches, while the main axes of the structure are formed by the arches. Figure 13.4 shows typical structural elements in simple steel vaults. Horizontal forces applied by the arches in the vault can be taken in many different ways. Foundations might be designed specifically to take these loads. There may be steel rods between the legs of the vault under ground level.

There are many different and creative applications of steel vaults. Some of the best examples can be seen on the atriums of shopping malls. The glass roof in the West Edmonton Mall in Alberta, Canada, is supported by a steel vault. Sometimes the streets between historical buildings are covered with glass vaults supported by steel elements. However, such glass roof systems are not applicable in some climatic conditions.

Steel domes

Steel domes can also be formed by half arches (ribs), as seen in Figure 13.5. A compression ring connects the arches to each other at the top of the dome. This compression ring takes the horizontal forces applied by the half arches and solves the detail problem at the point of junction of many elements at one point. The horizontal forces at the bottom of the steel dome are usually taken by a tension ring. The arches can be connected to each other by elements in the form of horizontal rings, in order to divide the height of the half arches into pieces.

An example of this type of steel dome is the Coliseum in Charlotte, USA, which spans 110m. The height of this dome is 40m (Crane, 1956). The Indonesia National Convention Center has trussed ribs supported by a compression and a tension ring and it spans 90m and is 10m high (Lin & Stotesbury, 1981: p.407).

The Louisiana Superdome (also known as the Mercedes-Benz Superdome), USA, is a steel trussed dome built in 1975 and spans 222m. The lamellar steel trusses are supported by a compression ring at the middle and a tension ring at the bottom (American Institute of Steel Construction, 1976). It is also possible to have elliptical steel domes. The elliptic dome of Guangzhou Gymnasium, Republic of China, built in 2001, spans 160m. There is a central truss girder dividing this dome into two segments. This truss girder supports trussed ribs. The trussed ribs are connected

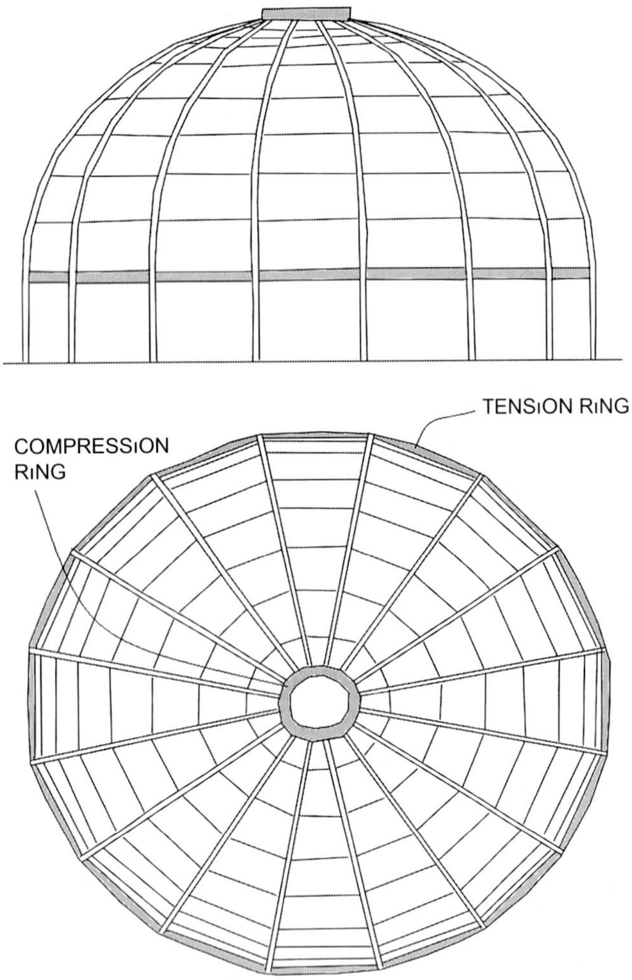

13.5 Structural elements in simple steel domes

to each other with circular trusses forming rings around the structure. The construction photographs of Guangzhou Gymnasium can be seen at URL1 (n.d.).

Structural guidelines for steel vaults and domes

Structural guidelines for steel vaults are listed in Table 13.1, and structural guidelines for steel domes are listed in Table 13.2.

Table 13.1 Structural guidelines for steel vaults and the associated value system

Structural guidelines	*Value system*
Best form for steel vaults is in the form of the moment diagram of the expected loading.	Economy
Sufficient height and curvature should be given to the vaults.	Economy
Horizontal forces applied by the arches in the vaults have to be taken by additional structural elements.	Safety
Prevent buckling of the arches in the vaults by avoiding slenderness of the arches.	Safety
Horizontal forces perpendicular to the arches should be taken by beams connecting these arches.	Safety

Table 13.2 Structural guidelines for steel domes and the associated value system

Structural guidelines	*Value system*
Best form for steel domes is in the form of the moment diagram of the expected loading.	Economy
Sufficient height and curvature should be given to the domes.	Economy
Horizontal forces applied by the arches in the domes can be taken by a compression ring at the top and a tension ring at the bottom.	Safety
Prevent buckling of the arches in the domes by avoiding slenderness of the arches.	Safety
Steel domes can span around 100m. Steel domes with trusses can span around 220m.	Economy

Case study 24: The dome at Parliament Building in Berlin, Germany

Norman Foster + Partners' dome at Parliament Building in Berlin (Reichstag) was chosen as a case study due to the multidimensional architectural importance of this dome. A sketch of the dome is shown in Figure 13.6, and a plan showing how the

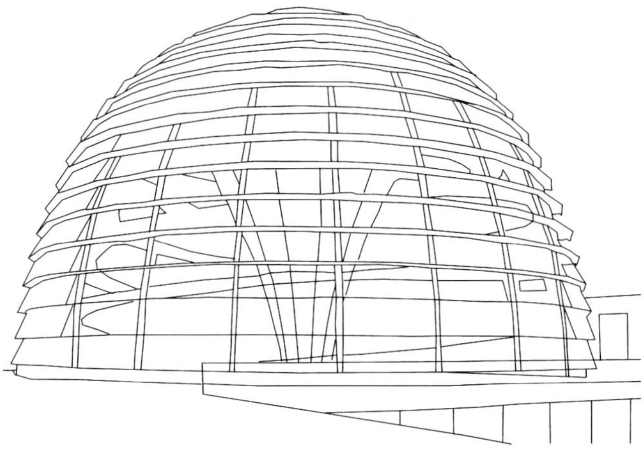

13.6 A sketch of case study 24: the dome at Parliament Building in Berlin, Germany, 1992–1999 (drawn with the help of *Encyclopedia Britannica*, 2014)

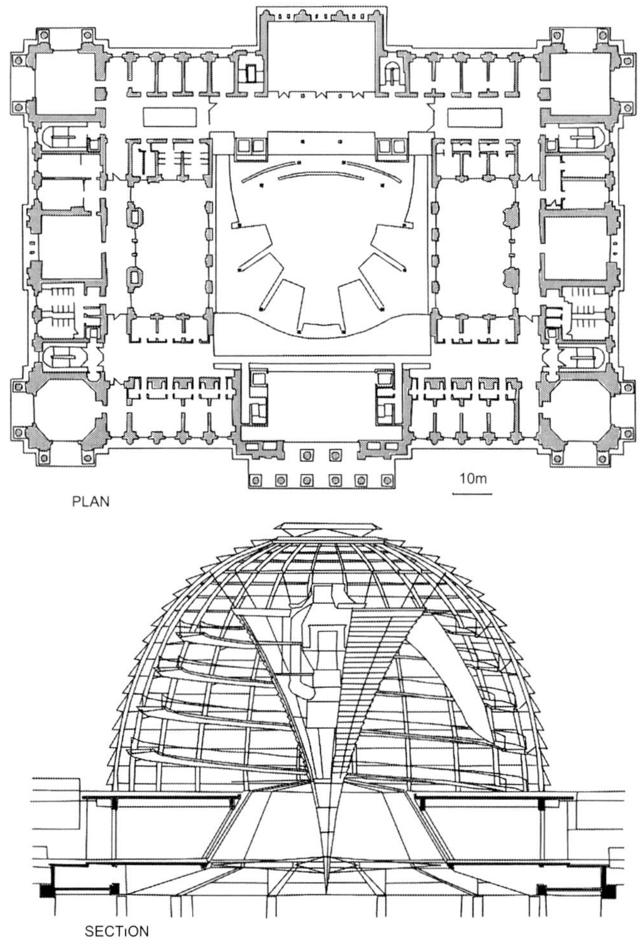

13.7 Plan and section of the dome at Parliament Building in Berlin (drawn with the help of Foster + Partners, 2014)

dome sits on the historical building and a section of the dome is shown in Figure 13.7.

The glass dome at Parliament Building in Berlin is built on a historical building that was ruined during the Second World War. The dome symbolises:

- Reunification of Germany.
- People's position above the parliament, by putting a visitors' platform inside the dome, which is situated above the main hall of the parliament.
- Transparency of government in comparison to the past political regimes in Germany.

The steel ribs are connected with a compression ring at the top and a tension ring at the bottom. The twin helical ramp inside the dome binds the ribs and forms the spiral walkway up to the visitors' platform at the top. The dome spans 40m and its height is 23.5m (Altin, 2001).

Another important issue regarding this dome is its environmentally-friendly design. Passive lighting and ventilation was incorporated into the design and the designers won the Pritzker Prize in 1999 for their work on this building. The transparent dome acts like a sunlight collector. There are hundreds of mirrors and movable shading devices on and in this dome. Daylight is transferred to the main hall and natural ventilation is provided. The electrical needs of the building are provided by photovoltaic panels and a generator that is powered by biofuel from vegetable oil (Bainbridge & Haggard, 2011). All these mirrors, movable shading devices, and other features, contribute to the tectonics of the

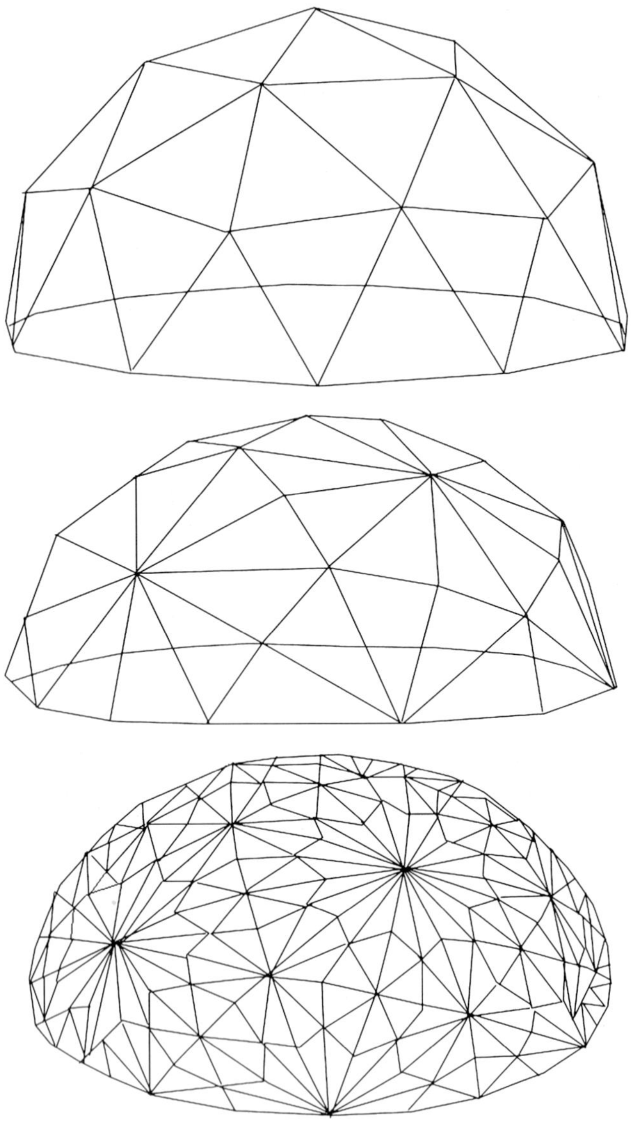

13.8 Different possibilities for geodesic domes

dome. The dominant physical entities of the dome in Parliament Building are the environmental control systems.

The structure enables the realisation of all these symbolic expectations, as well as being environmentally friendly, while following all structural guidelines for steel domes. Thus, the dome at Parliament Building in Berlin has an affirmative relationship with structural guidelines.

GEODESIC DOMES

Geodesic domes can be studied by looking at:

- Geometry and the structural behaviour of geodesic domes.
- Examples of geodesic domes.
- Construction methods of geodesic domes.

Geometry and the structural behaviour of geodesic domes

The geometric form used in geodesic domes is a polyhedron and usually an icosahedron. Other polyhedrons, which are compositions of hexagons and pentagons (an organisation of different triangles), can also be used to achieve similar structures that can be elliptical. However, the geometry of geodesic domes should allow equal distribution of the loads through the surface. Figure 13.8 shows different types of geodesic domes. The last illustration in Figure 13.8 can be constructed with the use of new technologies, but might not be economical.

There are different geometric applications, such as Grimshaw's Eden Project, which was built in Cornwall, UK, in 2001. The Eden Project contains overlapping spherical forms. There is also a deconstructed application of geodesic domes, which has a much smaller span in comparison to the circular forms (see URL10, n.d.).

The structure of a geodesic dome has the rigidity of a triangle. They contain triangles with different dimensions. The struts, which take place in the structure, form a strong network to transmit forces throughout the surface. They work in compression. Geodesic domes are light but strong structures in relation to their weight.

Examples of geodesic domes

Geodesic domes can be used for small structures as well as for large ones. Materials include aluminium, timber and steel. They can be formed with struts or panels. The largest geodesic dome in the world is Fukuoka Dome, which was built in Japan in 1993. It is used as a baseball field and it spans 216m.

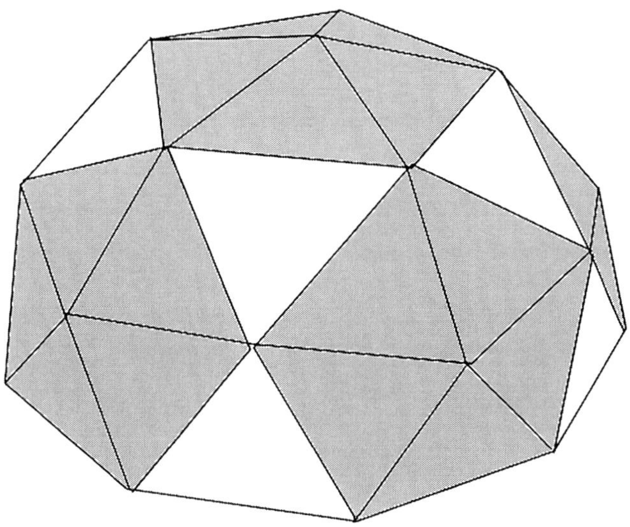

Other examples of geodesic domes include: Walt Disney World's Spaceship Earth, Epcot, USA, built 1982, spanning 48m; Climatron Conservatory, Missouri, USA, built 1960, spanning 58m; EcoCamp Patagonia Hotel, Chilean Patagonia, built 2000; Poliedro de Caracas Sports Arena, Venezuela, built 1974, spanning 143m; and Nagoya Dome, Japan, built 1997, spanning 187m.

Construction methods of geodesic domes

There are different methods for the construction of strut and panel geodesic domes. Small panel geodesic domes can be constructed piece by piece, as seen in Figure 13.9. Larger geodesic domes with struts can be built with reference to their geometric characteristics. The patent US 6295785 B1 concerns a method for construction of geodesic domes. Figure 13.10 shows that this method is based on the geometry of the strut organisation. First the top unit and four legs are built, and then the missing parts are completed (URL2, 2001). There are many films on the internet that show various ways to construct geodesic domes.

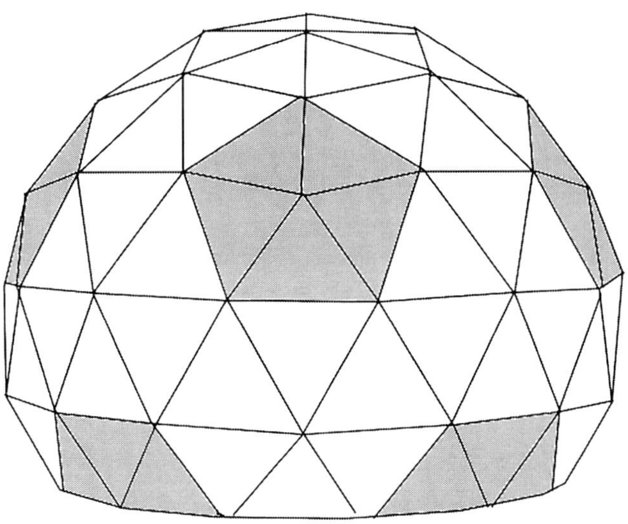

13.9 Construction of small geodesic domes made out of panels

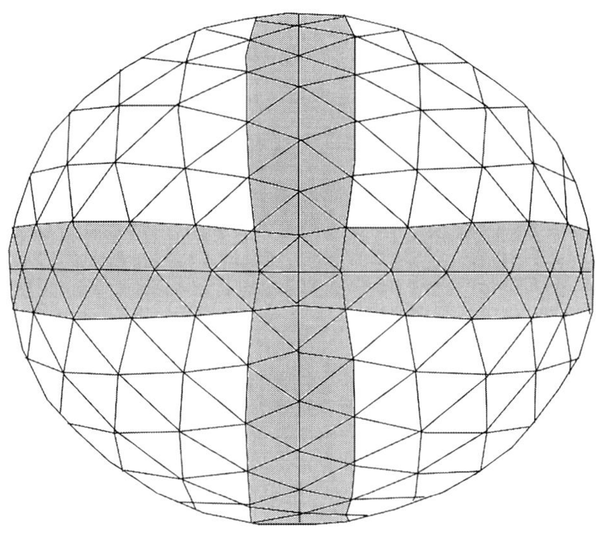

13.10 Construction of large geodesic domes using organisation of the elements

13.11 A sketch of case study 25: US Pavilion at Expo '67, Montreal, Canada, 1967 (drawn with the help of URL12, 2012)

Structural guidelines for steel geodesic domes

Structural guidelines for steel geodesic domes are listed in Table 13.3.

Table 13.3 Structural guidelines for steel geodesic domes and the associated value system

Structural guidelines	*Value system*
One of the geodesic dome geometries should be used.	Impossibility
Dimensions of elements should be calculated according to the type of geometry, span and height of the geodesic dome.	Practicality
Circular or elliptical plans can be used.	Impossibility
Geodesic domes should not be heavily loaded.	Safety
If there are buildings under geodesic domes, they should not touch the geodesic dome.	Safety
Maximum span of geodesic domes is 216m in 2014.	Economy
Prevent buckling of elements due to compression by avoiding slenderness of the elements.	Safety

Case study 25: US Pavilion at Expo '67, Canada

Buckminster Fuller and Shoji Sadao's US Pavilion at Expo'67 in Montreal was chosen as a case study because it presents the tectonic idea of a geodesic dome that houses another building within it. Figure 13.11 is a sketch of US Pavilion at Expo'67; the plan and section of the dome is shown in Figure 13.12.

US Pavilion at Expo '67 spans 76m and the dome's height is 62m. The dominant physical entity is the structural system. It is particularly eye-catching within its environment. Circular objects attract attention because they are similar to cosmic objects in the sky. Putting an object into a circle attracts attention to that object. The US Pavilion at Expo '67 attracts attention with its circular shape and it also demands attention for the building within it. This is the main architectural concept. Although all structural guidelines are followed, the symbolic meanings of the form makes the building tectonically valuable. Thus, the US Pavilion at Expo '67 has an affirmative relationship with structural guidelines.

SHELL STRUCTURES

Shell structures are usually constructed from reinforced concrete. They can be categorised according to:

- *Curvature type*: Shells can be positively curved, negatively curved or they can combine positive and negative curvatures, as seen in Figure 13.3.
- *Thickness*: Shells can be thin or thick.
- *Form*: Shells can be in simple (vault or dome), or more complicated, forms.

Positively curved parts of shells work with compression and negatively curved parts work with tension. There is reinforcement at all parts of the shell and this reinforcement is usually in the form of a mesh.

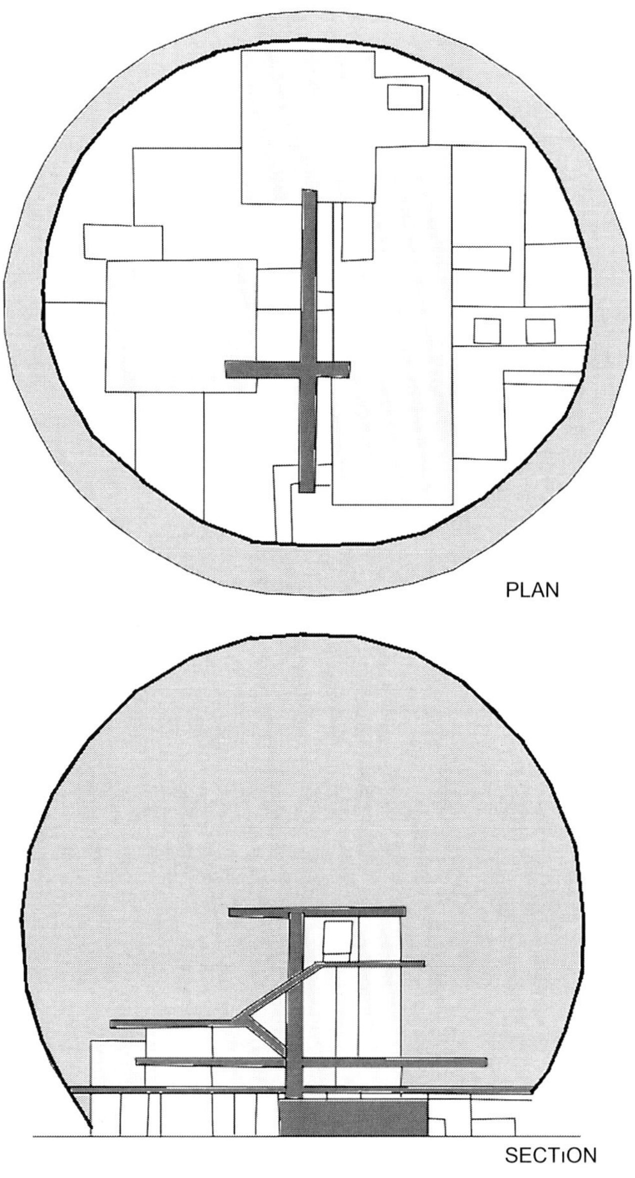

13.12 Plan and section of US Pavilion at Expo '67 (drawn with the help of URL11, n.d.)

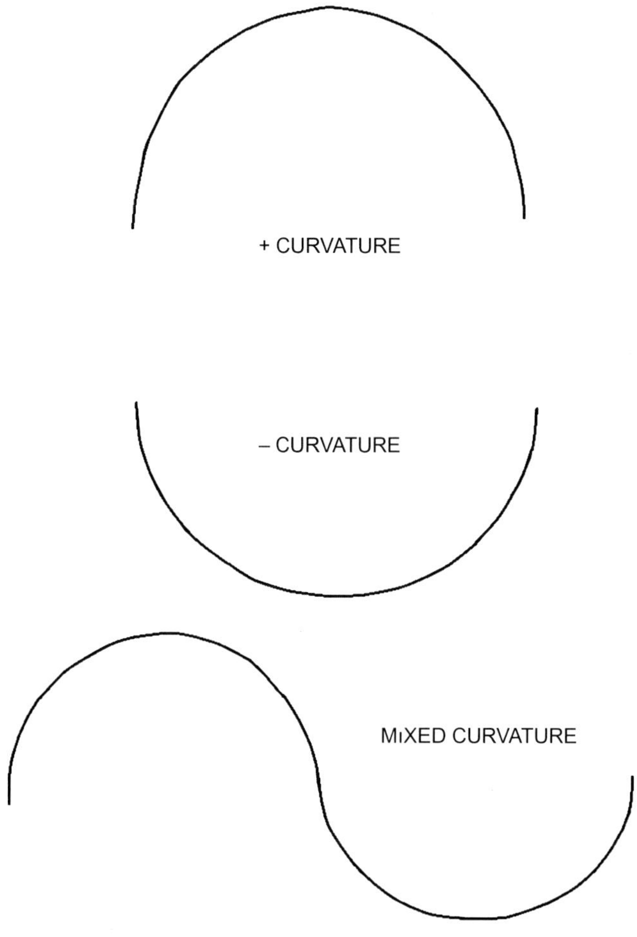

13.13 Curvature type of shell structures

Thin and thick shells

Thin shells are a few centimetres thick and they can span more than 200m. These are shells in the best form. Thus, it becomes unnecessary to have a thick surface. From the 1990s, these structures were designed by engineer–architects such as Heinz Isler and Felix Candela. These designers used mathematical analysis to find the most natural form. After the 1990s, developments in computer technology made form-finding a digital issue, and so architects started to design thin shell structures.

The Sydney Opera House in Australia, which was designed by Jorn Utzon in 1957, is an example of a shell structure designed by an architect. However, the construction of the Sydney Opera House was not completed until 1973 due to the complications created by its structure. The thickness of the ribbed surfaces of thick shells can be around 50cm. These structures are usually not in the best structural form.

Form of shell structures

Shell structures can be in various different forms. A shell can be in the form of a simple vault, a conical or spherical dome, or an inverted dome, as seen in Figure 13.14. The thickness of all shell structures increases at the edges of the shell and at the parts where it reaches its supports, because these parts of the shell support the other parts. There can be edge beams at the edges. The shell dome of the University of Illinois Assembly Hall, USA, which was built in 1963, spans 120m and the height of the dome is 38m. The Seattle Kingdome, USA, is also a shell dome and it was built in 1972. The Oklahoma State Fair Arena, USA, which spans 130m, has a roof in the form of an inverted dish. This concrete dish contains cables. Thus, it is a negative curvature shell, as shown in Figure 12.27 (see chapter 12).

A shell can also be in the form of a piece of one of the simple forms, as shown in Figure 13.15. Kresge Auditorium in Cambridge, USA, was designed in 1953 by Eero Saarinen and it spans 38m. CNIT Exhibition Hall in Paris, France, was built in 1958, and it spans 216m.

The pieces cut out of simple forms can be added to each other to form more complicated shell forms, as seen in Figure 13.16. There can be edge beams at the edges to connect different forms. St Louis Airport, USA, which spans 40m, combines many vaults to form a shell structure. Eero Saarinen's TWA Airport Building, which was built in New York, USA, between 1956 and 1962, is another example of this type of shell. The TWA Building combines four shells. This building is studied in more detail as a case study later in this chapter.

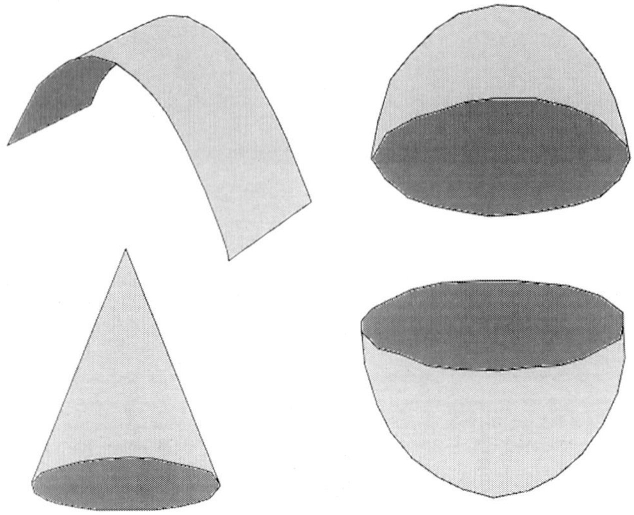

13.14 Simple forms for shell structures

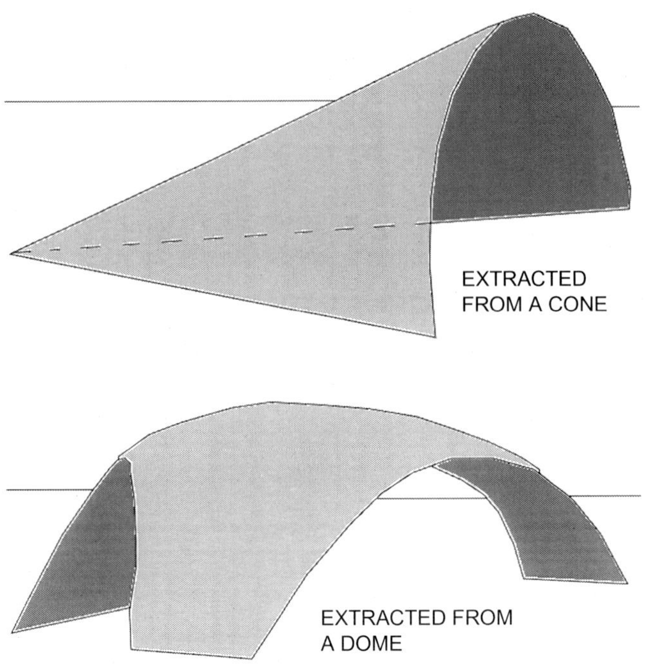

13.15 Pieces of simple forms

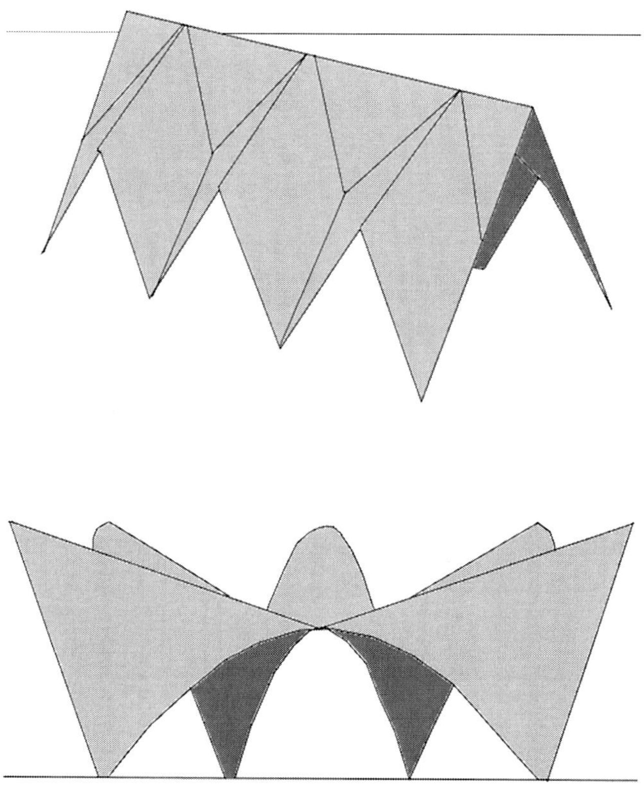

13.16 Addition of pieces of simple forms

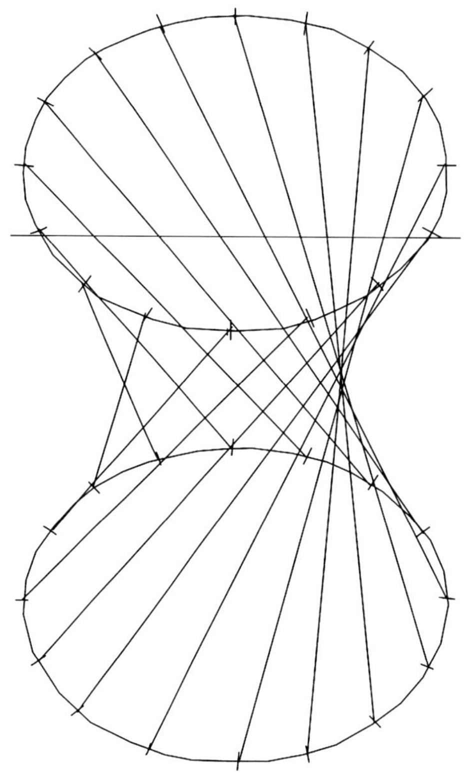

13.17 Hyperbolic paraboloid form

Hyperbolic paraboloid form, which is seen in Figure 13.17, can also be used to achieve a shell structure or a tower. Since it is not practical to have reinforced concrete hyperbolic paraboloid towers, they are made from steel instead. Thus, they are not shells.

Pieces of hyperbolic paraboloid forms, as shown in Figure 13.18, are frequently used in shell structures. London Velopark, UK, which was built in 2011, spans 250m. Other examples of the use of saddle forms include: Scotiabank Calgary Saddledome, Canada, 1983; Warszawa Ochota Railway Station, Poland, early 1960s; Church Army Chapel, London, UK, 1965; Scandinavium, Sweden, 1971; George Watson's Music School Auditorium, Edinburgh, UK, mid-1960s; and Candela's Chapel Lomas de Cuernavaca, Mexico, 1958.

The pieces of hyperbolic paraboloid forms can be added to each other to form shell structures, as seen in Figure 13.19. Felix Candela's Los Manantiales Restaurant, which was built in Xochimilco, Mexico City, in 1958, is an example of this form. The span of this building is 42m and the depth of the shell is 4cm (Burger & Billington, 2006; URL3, 2008).

Shell structures can also be formed from more complex additions of various types of forms, as seen in Figure 13.20. Heinz Isler's Sich Company Building, which was built in Geneva, Switzerland, in 1961, has a shell with a complex form. Many buildings by Santiago Calatrava and Zaha Hadid can be included in this category. It became easier to achieve these more complex forms with the help of computer technology and parametric architecture.

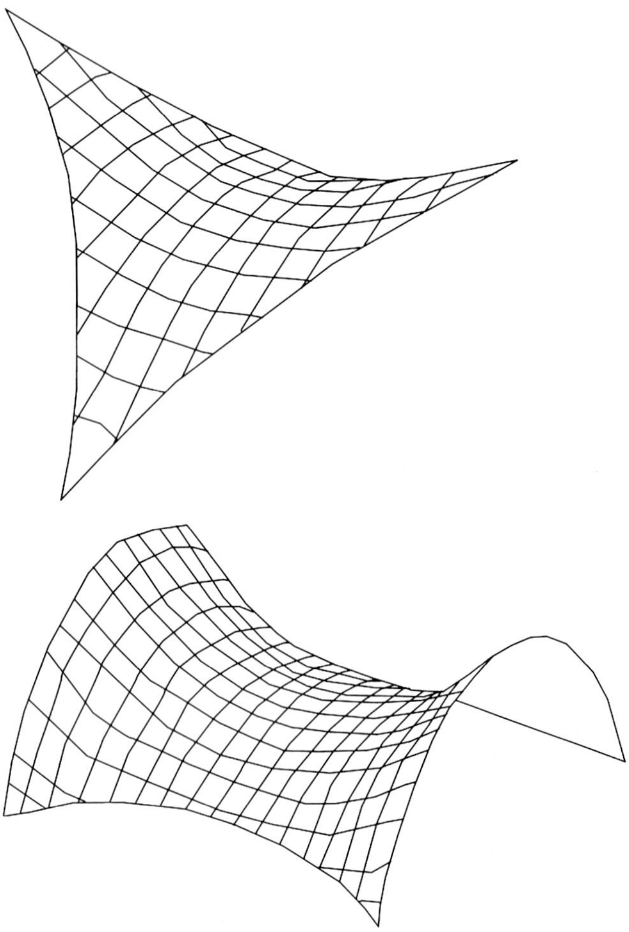

13.18 Pieces of hyperbolic paraboloid forms

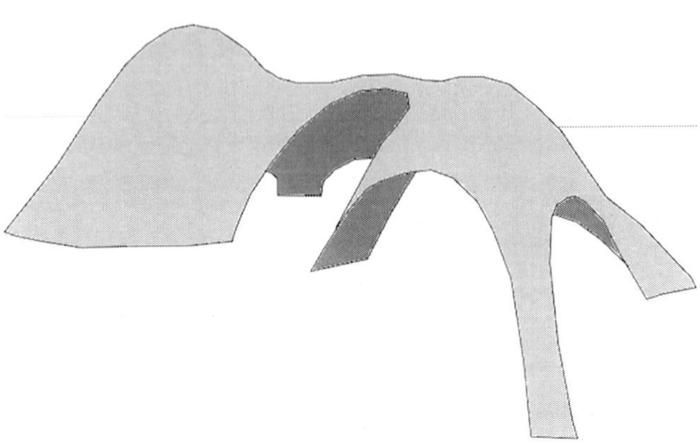

13.20 An example of a complex form for a shell structure

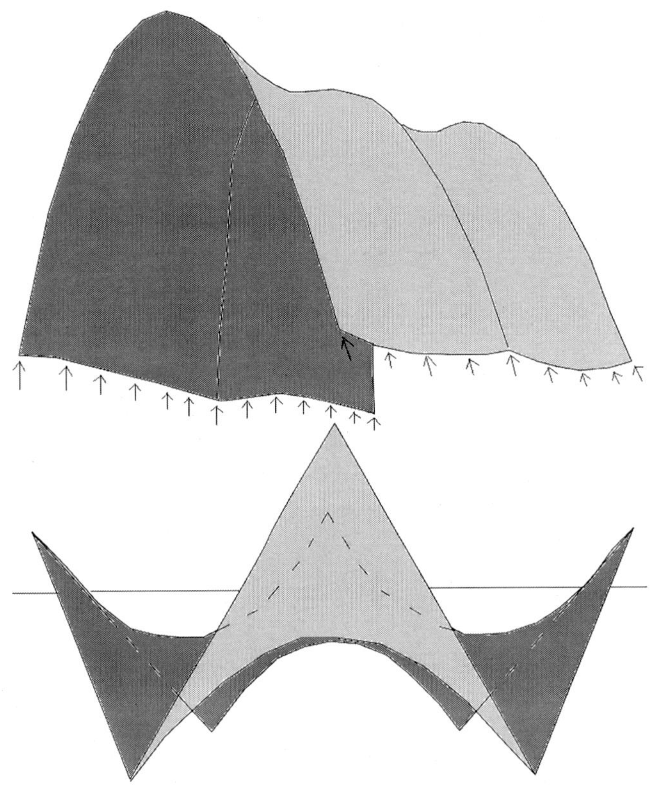

13.19 Addition of pieces of hyperbolic paraboloid forms

Ways to increase the span of shell structures

The strength of reinforced concrete shell structures can be increased by: using the best form; corrugating the surface of the shell with small folds, as seen in Figure 13.21; using ribbed shells (ACI318-95, 1995) as in the case of CNIT Hall in Paris, France; and folding the edges of the form, as seen in Figure 13.22. The Assembly Hall of the University of Illinois, USA, which spans 120m, and the CNIT Hall, which spans 216m, have corrugated shells. Many buildings by Heinz Isler, such as Wyss Garden Center, which was built in Solothurn, Switzerland, in 1961, have folded edges (URL4, 2009).

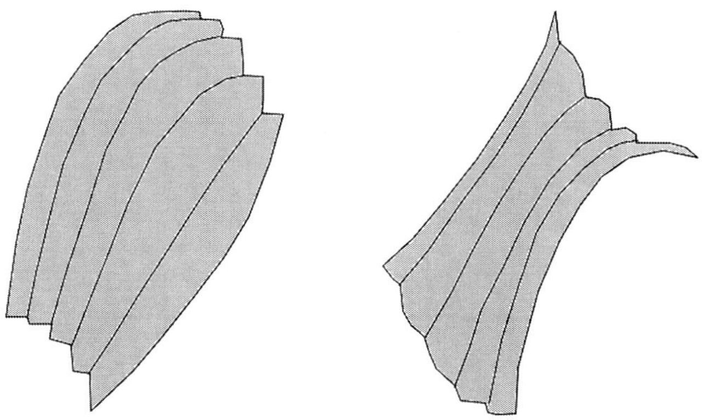

13.21 Corrugation of the surface of shell structures

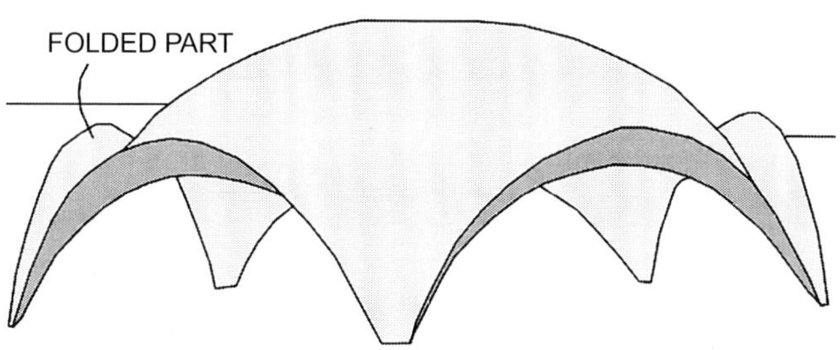

13.22 Folding the edges of a shell form

Construction methods of shell structures

The method of construction of shells depends on their form and span. Small semi-spherical domes can be built by using pneumatic structures as formwork. First the pneumatic structure is placed, then the reinforcement mesh is placed and shotcrete (concrete that is applied with a spraying technique) is applied. Finally, the pneumatic formwork is removed.

Inverted dishes can be built with the help of cables, compression and tension rings. First a compression ring is placed over supports; and a tension ring at the middle is placed with the help of a temporary support. Cables are put into their places and then the temporary support is removed. Finally, a permanent formwork is placed between the cables and shotcrete is applied. The resultant structure works in tension under gravity loads, but it works in compression (like a dome) if an upwards vertical load is applied.

It is easier to build shell forms that can be achieved with the help of straight lines. If this is the case, formwork is constructed with straight elements, the reinforcement is placed and the shotcrete is applied. Finally, the formwork is removed.

Structural guidelines for shell structures

Structural guidelines for shell structures are listed in Table 13.4.

Table 13.4 Structural guidelines for shell structures and the associated value system

Structural guidelines	*Value system*
Form of shell structures should be defined geometrically or digitally for the purpose of making their construction possible.	Practical
Curvature and height of the shell structure should be sufficient to work as a compression or tension structure.	Economy and aesthetics
If the best form (or a similar form to the best form) is used, the thickness of the structure can be a few centimetres.	Economy and aesthetics
Corrugation and folding can be used to increase the strength of the structure. Adding ribs can also increase strength.	Economical for long-span shells
Span can be up to 250m.	Economy
Horizontal forces at the edges can be taken by increasing the thickness of the shell or by providing edge beams.	Safety
Buckling of the thin surface should be avoided.	Safety

Case study 26: Kimbell Art Museum, USA

This building was chosen as a case study because it is an example of a shell structure in the form of a reinforced concrete vault.

13.23 A sketch of case study 26: Kimbell Art Museum, Fort Worth, USA, 1967–1972 (drawn with the help of Andrew Andersons' photograph in Paroissien, 2007)

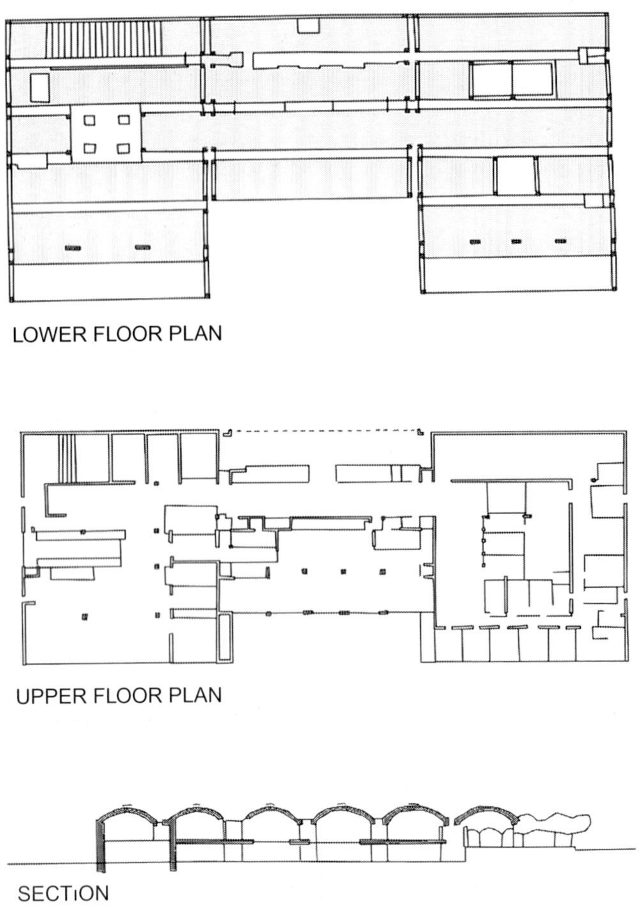

13.24 Plans and section of Kimbell Art Museum (drawn with the help of URL13, 2013)

Figure 13.23 is a sketch of Louis Kahn's Kimbell Art Museum in Texas; floor plans and a section of the building are shown in Figure 13.24.

Kimbell Art Museum is a very interesting case when the tectonics of its structure are considered. Kahn believed that a vault would be a relevant room for a museum. However, he did not choose the best structural form for his vault. He chose a cycloid barrel vault form in order to avoid domination of the form over the visitors of the building (URL5, 2014). However, the curvature and the height of the vault are structurally acceptable.

The colour of the concrete surfaces of the vault was selected after making many tests in order to satisfy both the technical and aesthetic requirements (URL5, 2014). Thus, the concrete surfaces were not painted to give colour; instead, the ingredients of the concrete were changed to achieve the required colour effect.

Use of natural light was also very important for Kahn. Thus, he provided narrow plexiglass skylights with reflectors at the top of his vault. The two sides of the vault were connected to each other by concrete elements at every 3m (URL5, 2014). This connection guaranteed the barrel vault behaviour. The dominant physical entity in Kimbell Art Museum is natural light.

Since Kahn was not trying to achieve a very long-span vault, the only structural guidelines relevant to him as regards reinforced concrete vaults were the principle of unity of the vault and the principle of curvature of the vault. Although Kahn did not follow these principles literally, he compensated for the problems he created. Kimbell Art Museum therefore has a contravening relationship with structural guidelines.

Case study 27: Small Sports Palace, Italy

This building was chosen as a case study because it is an example of a shell structure in the form of a reinforced concrete dome. Figure 13.25 is a sketch of Annibale Vitellozzi (architect) and Pier Luigi Nervi's (structural engineer) Small Sports Palace in Rome; a plan and partial section of the building are shown in Figure 13.26.

The perfect geometry and geometric arrangement of its structure is the main characteristic and architectural concept of this

13.25 A sketch of case study 27: Small Sports Palace, Rome, Italy, 1958 (drawn with the help of URL14, 2014)

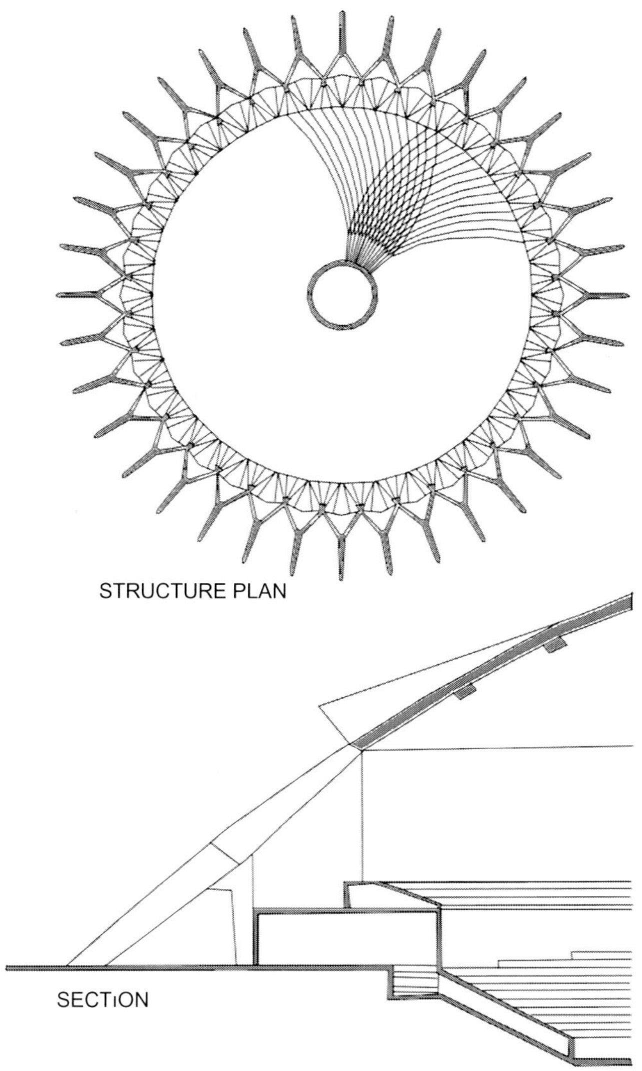

13.26 Plan and partial section of Small Sports Palace (drawn with the help of URL15, n.d. and URL16, 2013)

building. The span of the structure is 60m. The height of the dome is 21m and its thickness is 12cm (URL6, 2014). The surface of the dome has curved ribs. The horizontal forces applied by the shallow dome are taken by the Y-shape diagonal buttresses. The edges of the dome are folded to increase its strength. Natural light comes in from the compression ring at the top of the dome. All of these decisions made for the Small Sports Palace can be understood to aim towards achieving a perfect structure. Thus, the dominant physical entity is the structural system. Aesthetic considerations are rather abstract and mathematical. The relationship of the Small Sports Palace to structural guidelines is affirmative.

Case study 28: TWA Airport Building, USA

This building was chosen as a case study because it is an example of a shell structure in the form of a reinforced concrete shell. Figure 13.27 is a sketch of Eero Saarinen and Associates' TWA Airport Building in New York; plans and elevation of the building are shown in Figure 13.28.

The design of the TWA Airport Building does not depend on creating a perfect structure. Rather, the building symbolises a bird in flight and the excitement of travel: this is the design concept of the building. Four sections of the thin shell roof cover a plan of 67m × 96m. The height of the building is 15m (URL7, 2014). Two of the roof shells are in the form of cantilevers, representing the two wings of a bird. The Y-shaped columns take the horizontal forces applied by these shell pieces. The edge beam turning around the cantilever is much thicker around the support, and thinner at the end of the cantilever. The dominant physical entity in the TWA Airport Building is the structural system. The building represents the architecture of the 1960s, which was based on optimisation of design and domination of architectural decisions over structural engineering decisions. The TWA Airport Building has a contravening relationship with structural guidelines.

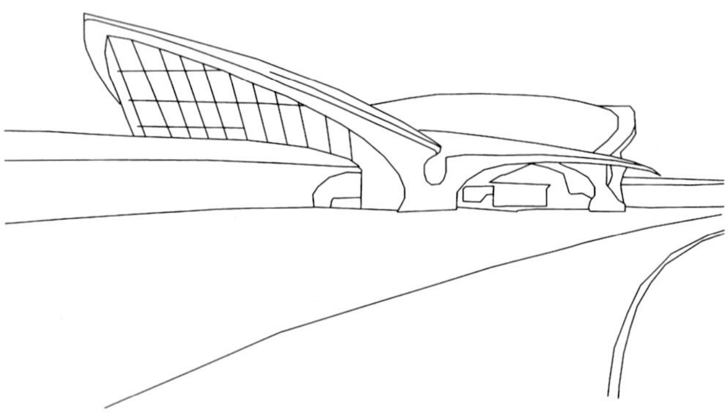

13.27 A sketch of case study 28: TWA Airport Building, New York, USA, 1956–1962 (drawn with the help of Lofgren, 2013)

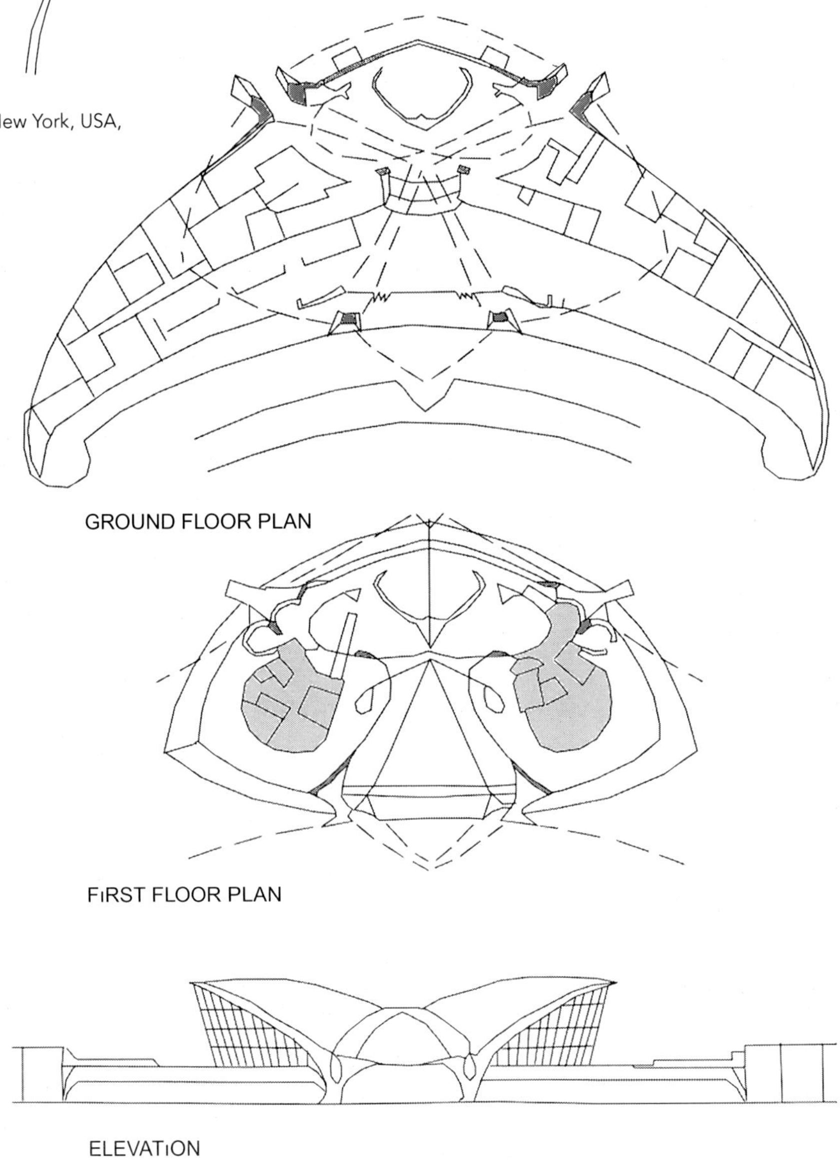

13.28 Plans and elevation of TWA Airport Building (drawn with the help of URL17, n.d. and Savela, 2011)

GRID SHELLS

Grid shells are usually made out of steel, aluminum or timber. They are lightweight structures with thin undulated surfaces. Grid shells can be analysed by studying their form and construction details.

Form and organisation of grid shells

The difference between grid shells and shell structures is the removal of unnecessary material away from the surface. Grid shells have to have sufficient curvature to resist their own weight. They usually form a double layer of undulated grid surfaces. Figure 13.29 shows an example of grid shells. Since deep elements cause difficulty in bending during the achievement of the expected form, double thin elements are used instead of a single thick element (Raynon, n.d.). Examples of timber grid shells are (Paoli, 2007):

- The Savill Building in Windsor Great Park, UK: a three-domed sinusoidal gridshell, which is 25m wide and 90m long.
- Mannheim Multihalle, Germany: built in 1975, it is 35m wide, 72m long and 15.5m high.
- Pavilion in Naples School of Architecture courtyard, Italy.
- Norwich Union Headquarters, UK.
- Japan Pavilion, Expo 2000, Hannover, Germany: 73.8m long, 25m wide and 15.9m high.

Examples of steel grid shells are:

- The cover of the Great Court at the British Museum in London, UK, by Foster + Partners.
- Yas Hotel, Abu Dhabi: built in 2009, it is 217m long.
- The Admirant, Eindhoven, Germany, by M. Fuksas.
- Zlote Taracy, Warsaw, Poland.

These structures can span up to 90m economically (Engel, 1997).

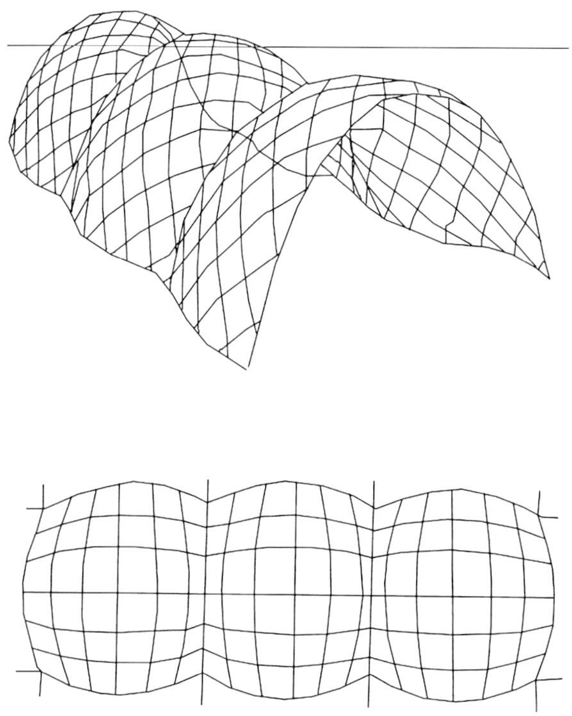

13.29 An example of grid shells

Details and construction of grid shells

There are different methods for the construction of grid shells. One of these methods is based on building the grid flat on the ground and then lifting it, applying forces to it, in order to give its form. When it is flat, the grid units are rectangular, and when the shell takes its final form, the grid units become trapezoidal. This means that all elements should be able to turn in their joints in order to achieve the expected form. Pinned joints allow rotation in all directions (Paoli, 2007). The final step of construction is to fix the form, which can also be done in different ways. Screwing of the joints can be done after achieving the final form and the action of screwing stops any further rotation at the joints (Raynon, n.d.). A second method is to use structural elements at the edges in order to fix the position of the grid shell at these edges. Figure 13.30 shows some details from grid shell joints.

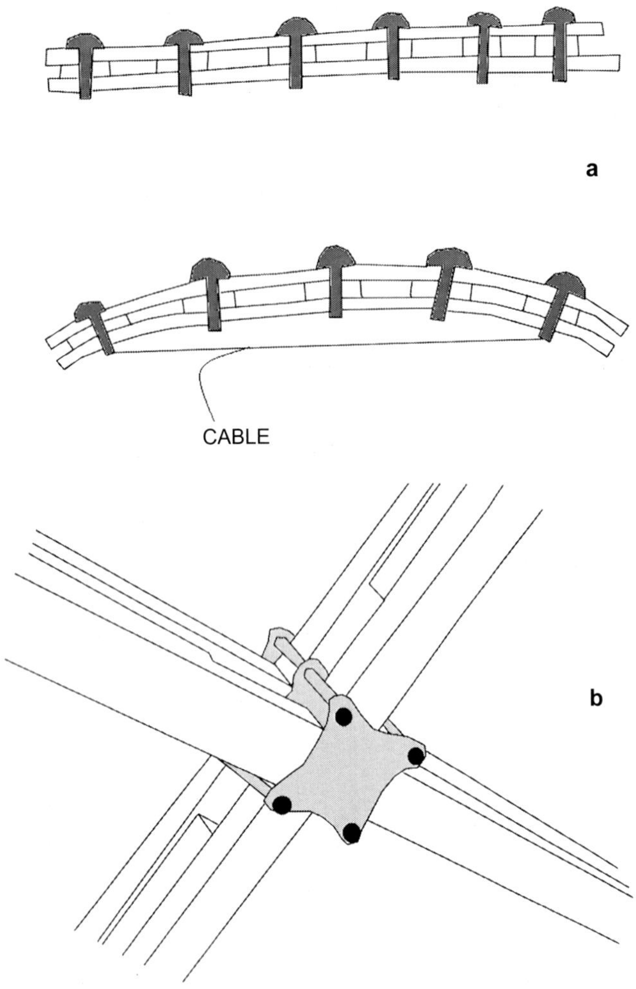

13.30 Examples of construction details for timber grid shell joints

Structural guidelines for grid shells

Structural guidelines for grid shells are listed in Table 13.5.

Table 13.5 Structural guidelines for grid shells and the associated value system

Structural guidelines	*Value system*
Curvature and height of the grid shell structure should be sufficient to work as a compression or tension structure.	Economy and aesthetics
The surface should be bendable.	Practicality
Span can be up to 90m economically.	Economy
Horizontal forces at the edges and supports should be balanced.	Safety
Buckling of the surface should be avoided.	Safety
Joints should allow rotation of members before fixing.	Practicality

13.31 A sketch of case study 29: Suan Lien Center, Taipei, Republic of China, 2009 (drawn with the help of URL18, 2011)

Case study 29: Suan Lien Center, Republic of China

J.J. Pan and Partners' Suan Lien Center in Taipei, Taiwan, is sketched in Figure 13.31; plans and section of the building are shown in Figure 13.32. Suan Lien Center is a church: the shells are in the form of two hands, which are together in prayer, and the large window between these shells is in the form of a fish, which is also a religious symbol (URL8, n.d.; URL9, 2011). The grid shell of this building is made out of aluminum alloy (URL8, n.d). The structural system of the building is the dominant physical entity. Although a sufficient curvature was given to the grid shell surfaces and rotation at the joints was provided, the form of the structure is not in the best form. Suan Lien Center has a contravening relationship with structural guidelines.

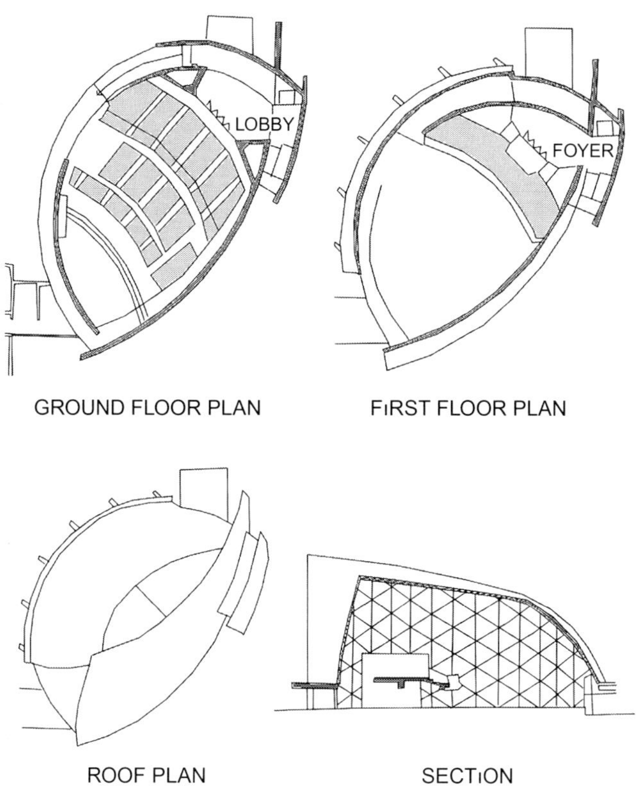

13.32 Plans and section of Suan Lien Center (drawn with the help of URL18, 2011)

CONCLUSIONS

When structural guidelines for compression structures are analysed, one can expect dominant buildings that attract the attention of people. The first expectation is to have abstract geometries. However, only two of the six case studies in this chapter (US Pavilion at Expo '67 and the Small Sports Palace) have abstract forms. Forms of the other four buildings were determined according to architectural criteria, such as the use of symbolic forms. Abstract form is usually not seen as an architectural expectation. TWA Airport Building symbolises a flying bird, and Suan Lien Center symbolises human hands in prayer. However, the form of the dome at Parliament Building in Berlin and the Kimbell Art Museum vault were not designed for symbolic purposes. Instead, they were designed to have certain effects on people. It can be stated, therefore, that architects do not prefer abstract forms and abstract articulations of forms in large-scale buildings.

Structural guidelines were mainly followed for these buildings. This shows that there was a good collaboration between the architects and structural engineers involved in these buildings.

When compared with historical compression structures, the structural elements used to solve horizontal force problem are more hidden in modern compression structures. The use of thin compression and tension rings, Y-shaped columns, and cables at foundation level are the most common solutions, and these solutions do not affect the tectonics of the building, as was the case in historical compression structures.

PROBLEMS TO SOLVE

Find the best form for the following loadings for compression structures.

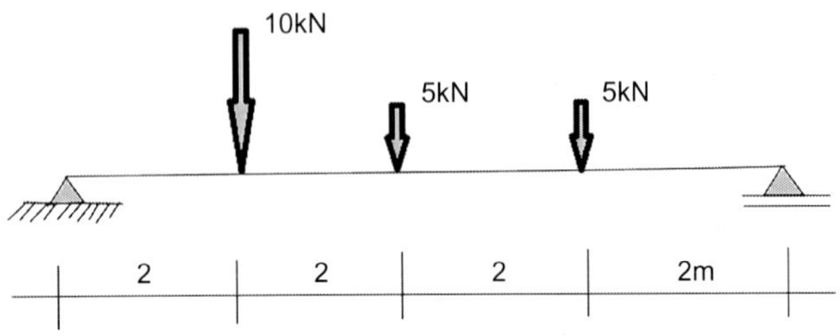

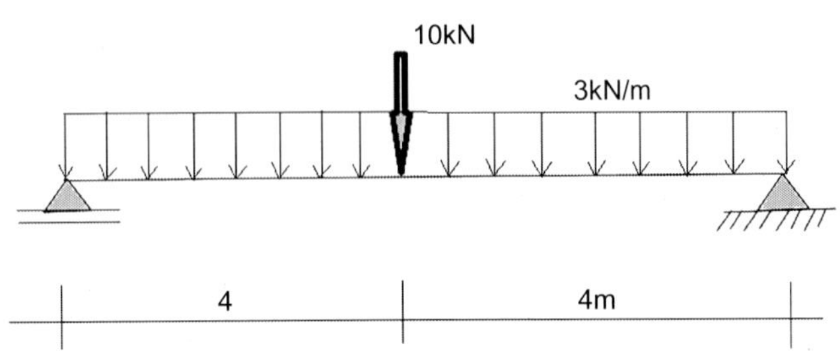

REFERENCES

ACI 318-95 (1995) *Building Code Requirements for Reinforced Concrete*, American Concrete Institute: Michigan.

Adriaenssens, S., Block, P., Veenendaal, D., Williams, C. (eds) (2014) *Shell Structures for Architecture – Form Finding and Optimization*, Routledge: London.

Altin, M. (2001) 'The structural analysis of domes: from Pantheon until Reichstag' in eds P.B. Lourenco & P. Roca *Historical Constructions*, University of Guimarães: Portugal: pp.197–208.

American Institute of Steel Construction (1976) *Modern Steel Construction* (viewed 28 August 2014: www.modernsteel.com/archives/PDFs_61-90/1976A9_16-1&2.pdf)

Bainbridge, D.A., Haggard, K. (2011) *Passive Solar Architecture*, Chelsea Green Publishing Company: Vermont.

Burger, N., Billington, D.P. (2006) 'Felix Candela, Elegance and Endurance: An Examination of the Xochimilco Shell' *Journal of the International Association for Shell and Spatial Structures*, Vol. 47, No. 3 (viewed 15 November 2014: www.iass-structures.org/index.cfm/journal.getFile/2.1.17._34_Burger___Billington_final_versionV3.pdf)

Crane, T. (1956) *Architectural Construction*, John Wiley and Sons: New York.

Encyclopedia Britannica (2014) (viewed 27 September 2014: http://kids.britannica.com/comptons/art-143312/The-steel-and-glass-dome-of-the-Reichstag-building)

Engel, H. (1997) *Structure Systems*, 3rd edition, Gerd Hatje Publishers: Ostfildern, Germany.

Foster + Partners (2014) *Reichstag, New German Parliament* (viewed 25 September 2014: www.fosterandpartners.com/projects/reichstag-new-german-parliament/)

Lin, T.Y., Stotesbury, S.D. (1981) *Structural Concepts and Systems for Architects and Engineers*, John Wiley and Sons: New York.

Lofgren, K. (2013) *Iconic Eero Saarinen JFK Airport Terminal Will Have New Life as a Standard Hotel* (viewed 28 September 2014: http://inhabitat.com/nyc/iconic-eero-saarinen-jfk-airport-terminal-will-have-new-life-as-a-standard-hotel/)

Paoli, C. (2007) *Past and Future of Grid Shell Structures*, Master thesis, Massachusets Institute of Technology: USA.

Paroissien, L. (2007) *Andrew Andersons Interviewed by Leon Paroissien* (viewed on 28 September 2014: http://recollections.nma.gov.au/issues/vol_4_no1/notes_and_comments/andrew_andersons_interviewed_by_leon_paroissien)

Raynon, J. (n.d.) *The Savill Garden Gridshell Design and Construction* (viewed 3 September 2014: www.ewpa.com/archive/2008/june/paper_297.pdf)

Savela, M. (2011) *TWA Plans* (viewed 28 September 2014: http://mikasavela.tumblr.com/post/11984130833/twa-plans)

URL1 (n.d.) *Guangzhou Sport Complex* (viewed 28 August 2014: www6.cityu.edu.hk/construction_archive/major_reference_pdf.aspx?id=169)

URL2 (2001) *Method of Constructing Geodesic Domes* (viewed 28 August 2014: www.google.it/patents/US6295785)

URL3 (2008) *Felix Candela and Restaurant Los Manantiales* (viewed 2 December 2013: http://anengineersaspect.blogspot.com/2009/06/felix-candela-and-restaurant-los.html)

URL4 (2009) *Heinz Isler Concrete Thin Shells* (viewed 2 September 2014: http://anengineersaspect.blogspot.com.tr/2009/10/27-heinz-isler-concrete-thin-shells-on.html)

URL5 (2014) *Kimbell Art Museum Kahn Building* (viewed 2 September 2014: www.kimbellart.org/architecture/kahn-building)

URL6 (2014) *Little Sports Palace* (viewed 2 September 2014: http://structurae.net/structures/little-sports-palace)

URL7 (2014) *TWA Terminal at JFK International Airport* (viewed 2 September 2014: http://structurae.net/structures/twa-terminal-at-jfk-international-airport)

URL8 (n.d.) *Religious Design Makes Chapel a Hit* (viewed 2 September 2014: http://taiwanhoy.nat.gov.tw/ct.asp?xitem=107532&ctnode=321&mp=16)

URL9 (2011) *Church of Suan-Lien Center for the Elderly/J.J.Pan and Partners* (viewed 2 September 2014: www.archdaily.com/143022/church-of-suan-lien-center-for-the-elderly-j-j-pan-partners/)

URL10 (n.d.) *Peoples Meeting Dome* (viewed 27 September 2014: www.behance.net/gallery/Peoples-Meeting-Dome/5228001?iframe=1)

URL11 (n.d.) *Shoji Sadao* (viewed 28 September 2014: www.tumblr.com/search/Shoji%20Sadao)

URL12 (2012) *Expo 67* (viewed 28 September 2014: www.tourisme-montreal.org/blog/expo-67-then-and-now/)

URL13 (2013) *Kimbell Museum* (viewed 28 September 2014: www.greatbuildings.com/buildings/Kimbell_Museum.html)

URL14 (2014) *Structures* (viewed 28 September 2014: www.studyblue.com/notes/note/n/structures/deck/10791456)

URL15 (n.d.) *Structure and Beauty* (viewed 28 September 2014: www.pinterest.com/sophievolland/structure-and-beauty/)

URL16 (2013) *Chthonic Morphosis* (viewed 28 September 2014: http://arpc167.epfl.ch/alice/WP_2013_SP/lenherr/?p=2412)

URL17 (n.d.) *Eero Saarinen's TWA Terminal Brought Back to Life* (viewed 28 September 2014: http://uk.phaidon.com/agenda/architecture/picture-galleries/2011/october/20/eero-saarinens-twa-terminal-brought-back-to-life/?view=thumbs&view=thumbs)

URL18 (2011) 'Church of Suan-Lien Center for the Elderly/J.J. Pan & Partners' *ArchDaily* (viewed 28 September 2014: www.archdaily.com/143022/church-of-suan-lien-center-for-the-elderly-j-j-pan-partners/)

URL19 (2007) *A Different Type of String Theory: Antonio Gaudi* (viewed 17 August 2014: http://memetician.livejournal.com/201202.html)

14 The Tectonics of Tension and Compression Structures

Structures that work under both tension and compression as a response to loading are generally very effective structures and their strength depends on their formal arrangement. These are trussed structures that are various compositions of triangles. Some members in these triangles work with compression and some members work with tension, but this situation might change when the loading changes.

Triangulation is the general characteristic of all trussed structures. The triangular organisation of elements is more effective than having an organisation in the form of square. If all joints are pin joints, triangular arrangement cannot be deformed by shear forces. However, a square-shaped organisation can be easily deformed by shear forces, as seen in Figure 14.1.

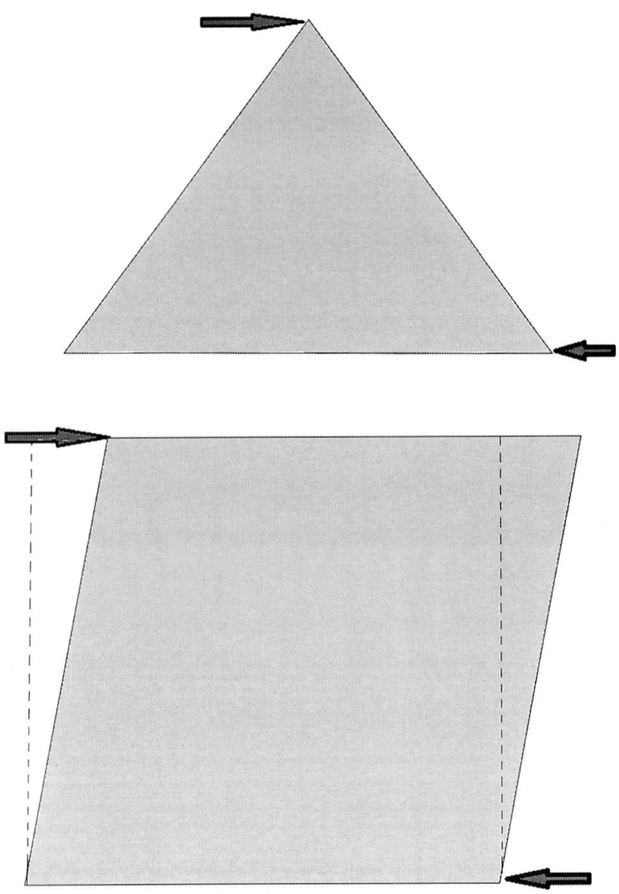

14.1 Deformation by shear forces: a triangle in comparison to a square

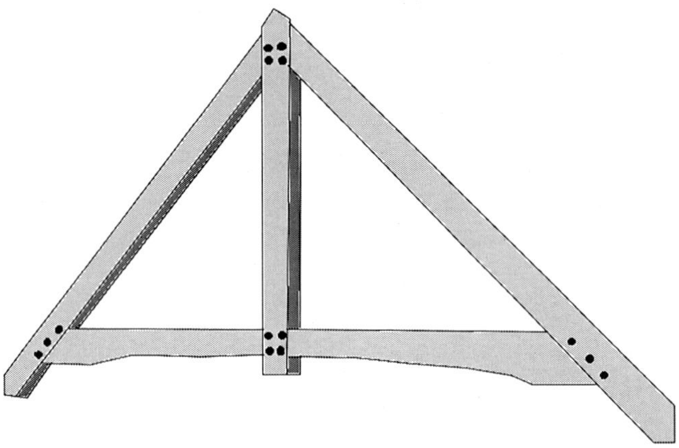

14.2 A simple truss

In the past, people used timber beams to span roofs and the size of these timber elements was limited. It was easy to find beams 4m long, but harder to find beams that were longer. This practical problem inspired the idea of the truss: a structural system that spans longer distances by combining shorter elements. Figure 14.2 shows a simple truss spanning a distance that is longer than the length of its elements.

There are three types of trussed systems, which are shown in Figure 14.3:

- 2D trusses
- 3D trusses
- Space frames.

2D TRUSSES

Two-dimensional trusses can be studied by looking at:

- Analysis of internal forces in trusses.
- Span and depth of trusses.
- Use and organisation of trusses in buildings.

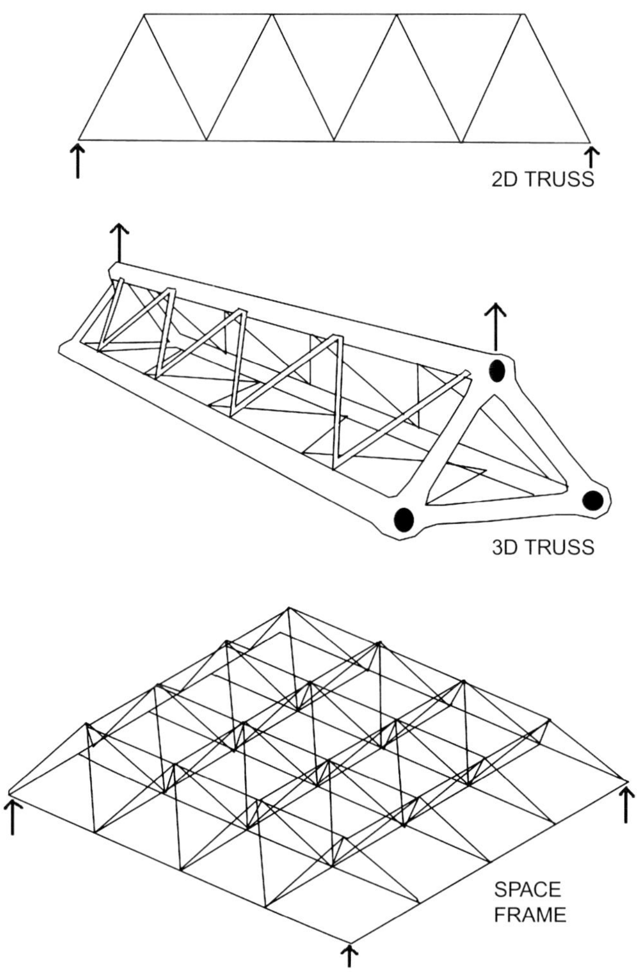

14.3 Types of trussed systems

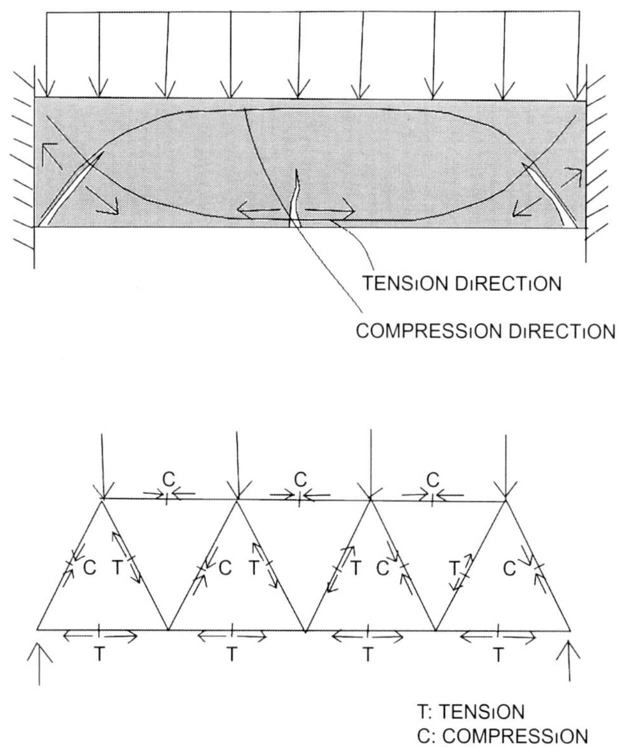

14.4 Similarity between beams and trusses

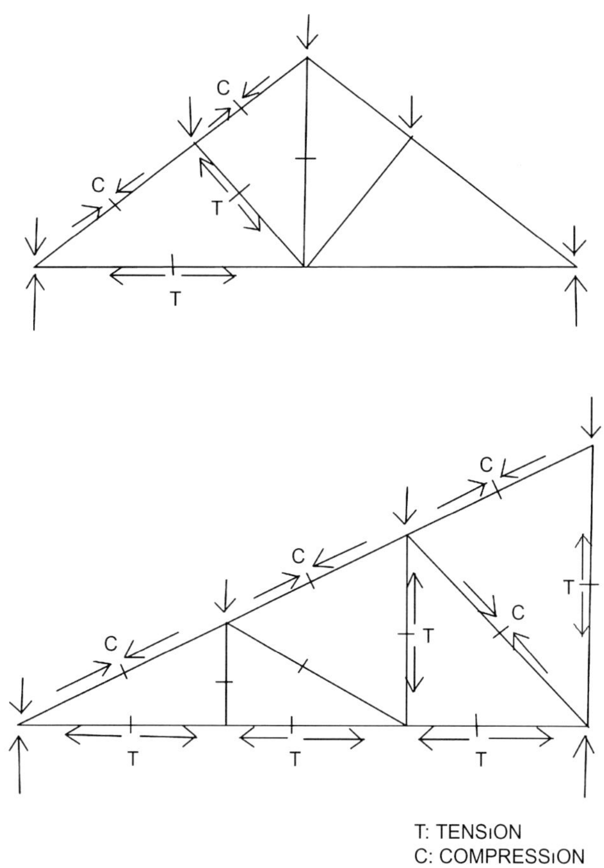

14.5 Type of internal forces in trusses

Analysis of internal forces in trusses

A truss is very similar to a beam. The only difference between them is the careful placement of structural material in the case of trusses. This similarity between a truss and a beam can be seen in Figure 14.4. If a concrete beam with fixed supports is loaded heavily, the first crack will develop at the bottom middle part and this crack will be vertical. The secondary cracks will develop at the supports and these cracks will be diagonal. The directions of these cracks show the direction of tensile forces in the beam. Tensile forces are perpendicular to the cracks. The direction of tensile forces will be parallel to the negative curve at the bottom of the beam. Similarly, the direction of compressive forces will be parallel to the positive curve at the top of the beam. The type of internal forces in trusses can be understood with the help of the same principle. The top chord of the truss will be in compression, if the loading is downwards. Similarly, the bottom chord will be in tension. The type of internal forces in the diagonal members of trusses can be understood with the help of diagonal tensile

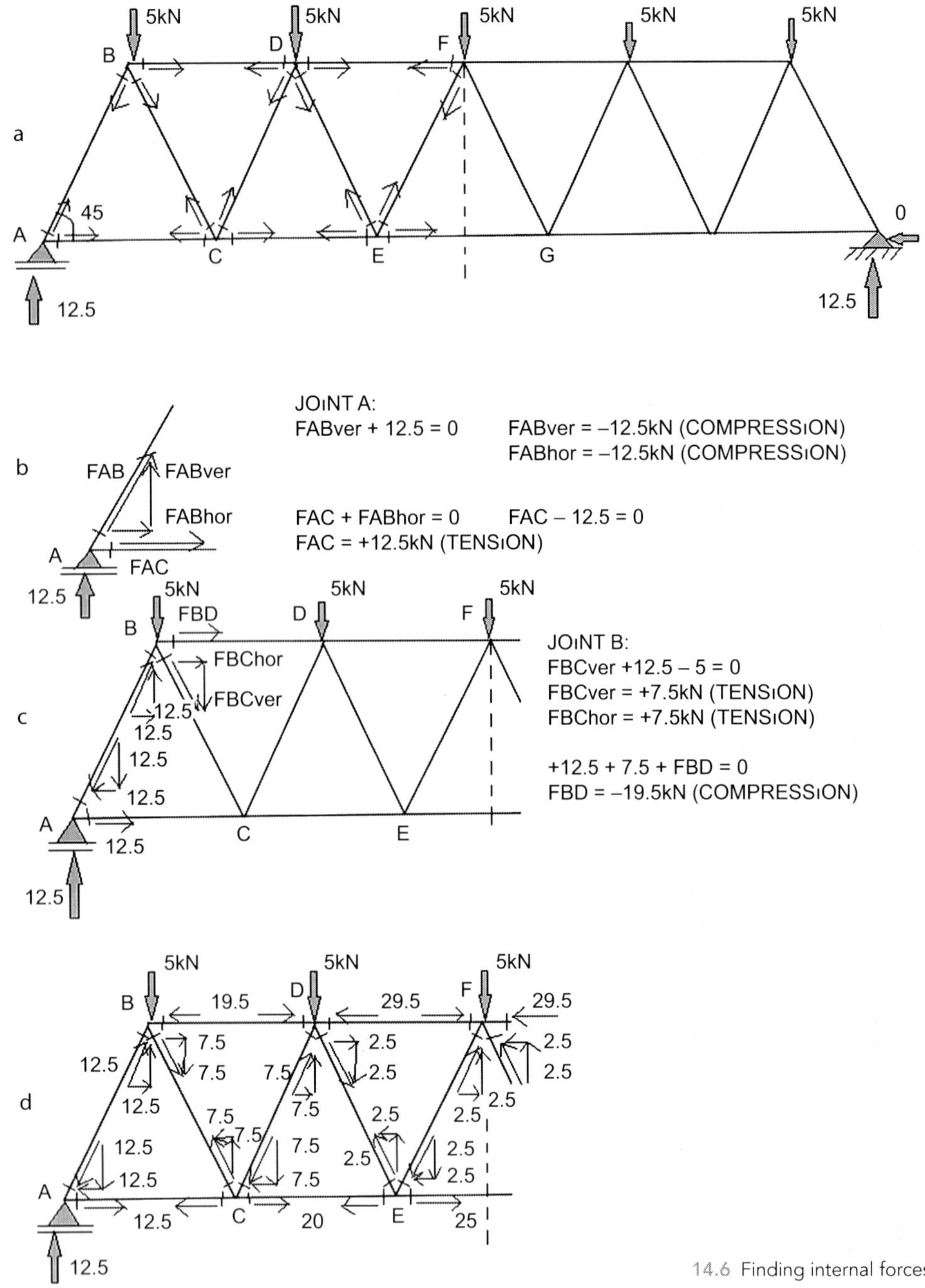

14.6 Finding internal forces in trusses 1

forces at the supports of beams. The tensile force in the diagonal members of the truss will be in the same direction as the diagonal tensile force at the supports of the beam. The same principle can be applied to different organisations of trusses, as seen in Figure 14.5.

Depending on the organisation of trusses, it might sometimes be difficult to determine the type of internal force in some of the members by using this method. Another method – the Method of Joints – can be useful in determining types of internal force in all truss members as well as their magnitudes (Dabby & Bedi, 2012).

The Method of Joints depends on the isolation of each joint in order to be able to use two equations of equilibrium to find internal forces in all members of the joint. The steps of this method are as follows:

- Find the reactions at supports.
- Name each joint with a letter.
- Assume initially that all members have tension in them (see Figure 14.6a).
- Choose a joint with maximum two unknowns, isolate that joint,

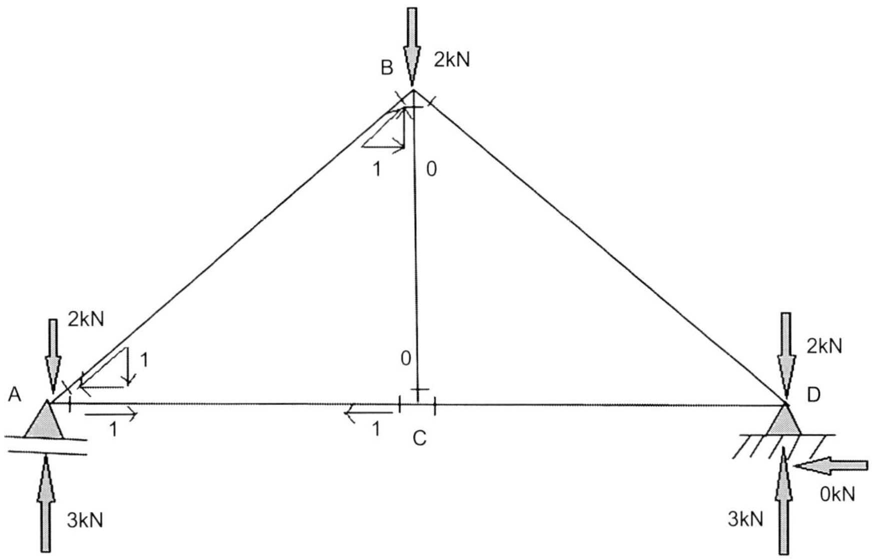

JOINT A:
+3 – 2 + FABver = 0 FABver = –1kN (COMPRESSION)
FABhor = –1kN (COMPRESSION)

–1 + FAC = 0 FAC = +1kN (TENSION)

JOINT C:
–1 + FCD = 0 FCD = +1kN (TENSION)
FCB = 0kN

JOINT B:
+1 – 2 – FCDver = 0 FCDver = –1kN (COMPRESSION)
FCDhor = –1kN (COMPRESSION)

14.7 Finding internal forces in trusses 2

and apply two equations of equilibrium (equilibrium of vertical forces and equilibrium of horizontal forces) to find internal forces in the members of this joint.

- If a result is minus, this means that there is compression in that member. If a result is plus, there is tension in that member (see Figure 14.6b). Change the direction of the arrows accordingly.
- Reflect the found results to the other side of each member.
- Choose another joint with two unknowns and apply the same procedure until all joints are analysed (see Figure 14.6c and d).

Figure 14.7 shows another truss problem that is also solved by using the Method of Joints. This system contains a member with no internal force. These types of members can be needed to decrease the length of members in the system. They can also carry loads applied by suspended ceilings.

The compression members in trusses should be thick enough to avoid buckling.

Span and depth of trusses

Trusses are depth-effective structures, similar to all other structures that develop bending stress in them as a response to loads. The axial forces in the top and bottom chords of trusses form a force couple that balances the moment due to loading on the system (see Figure 14.8).

Spans for timber trusses are between 5m and 50m, while steel truss spans are between 15m and 80m (Engel, 1997). The economic depth of a truss can be found by dividing its span by 12,

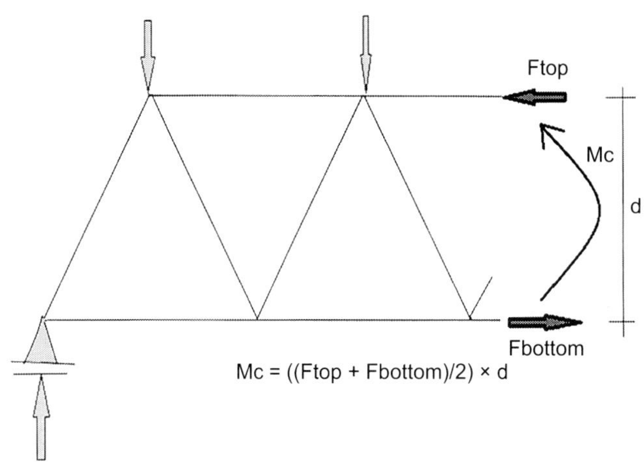

14.8 Creation of counter moment using the force couple in the top and bottom chords of a truss

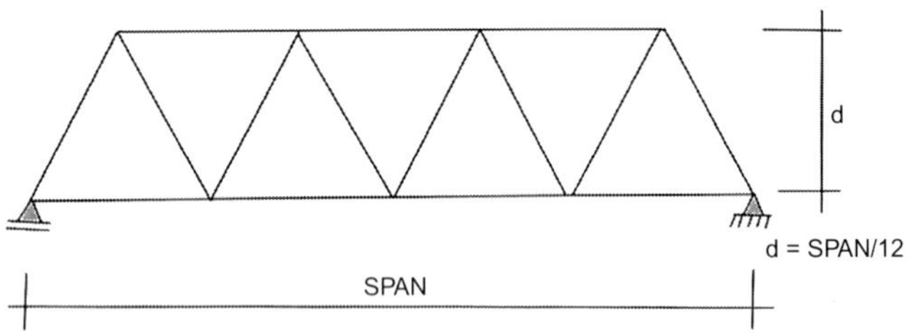

14.9 Economic depth of trusses in relation to their span

as seen in Figure 14.9 (URL2, 2000). This span to depth ratio can be increased up to 15.

This means that the economic depth of a 80m-long truss is around 5.5 metres. This value is so high that it might create some architectural complications in the determination of the height of the space. It is better to place trusses that span more than 30m close to each other. It is also necessary to make connections between them.

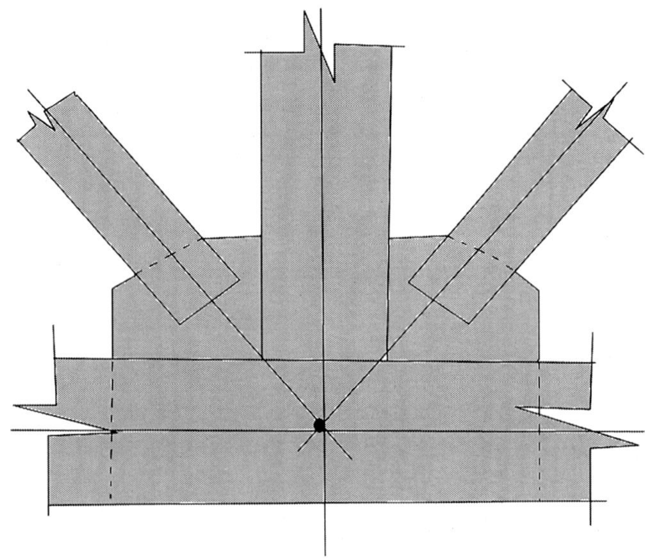

14.11 A joint detail from a steel truss

Use and organisation of trusses in buildings

Trusses can be loaded only on their joints, as seen in Figure 14.10. Thus, the rafters should be placed on them accordingly.

The axes of all members in a joint of a truss should meet at one point, as seen in Figure 14.11. If not, bending will occur. This principle is also valid for the joints at the supports, as seen in Figure 14.12.

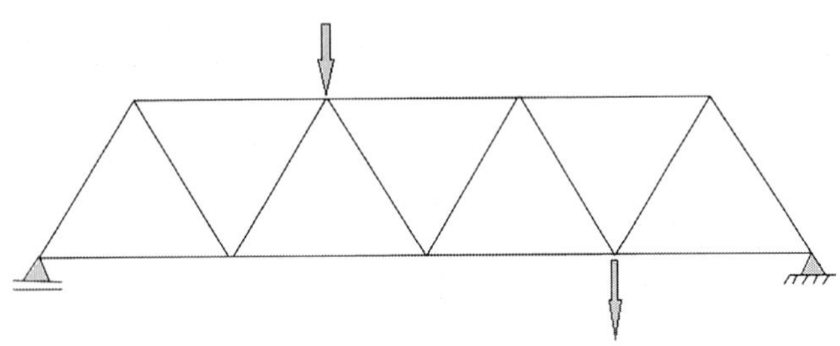

14.10 Loading on trusses

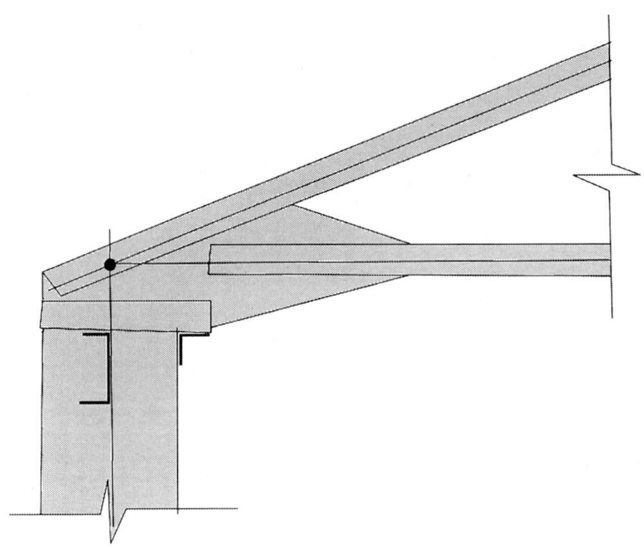

14.12 Support and truss connection

Depending on the geometry of the building, trusses can be placed within a structure in many different ways. Figure 14.13 shows a simple organisation in which truss axes are parallel to each other and the trusses are connected to each other with beams in the perpendicular direction. These beams provide resistance against horizontal loads. Cables should be used for wind-bracing purposes.

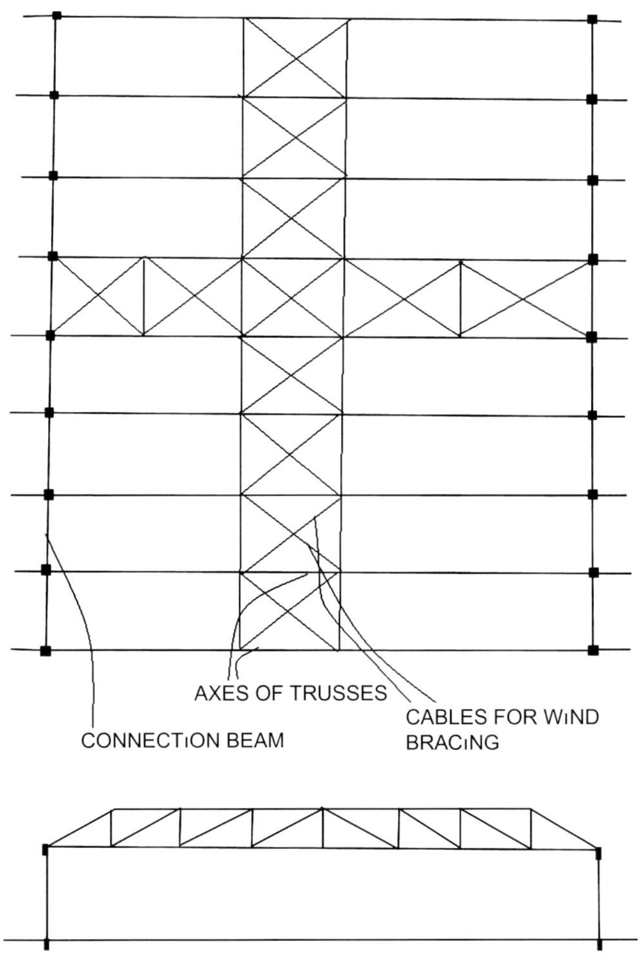

14.13 Organisation of trusses

Trusses are built at ground level and then lifted to their positions by manpower or with the help of available machinery.

Structural guidelines for trusses

Structural guidelines for trusses are listed in Table 14.1.

Table 14.1 Structural guidelines for trusses and the associated value system

Structural guidelines	*Value system*
Trusses can be formed with the help of triangulation.	By definition
Timber trusses can span up to 50m and steel trusses can span up to 80m.	Economy
Depth of trusses can be found by dividing their span with a value between 12 and 15.	Economy
Trusses can be loaded from their joints.	By definition
Axes of all members in a joint should meet at a point to avoid bending.	By definition
Axis of the supporting column should meet with the point at which axes of all members in that joint meet.	By definition
Truss axes in a building should be connected to each other with a beam in perpendicular direction in order to take the horizontal loads.	Safety
Wind bracing should be provided.	Safety

Case study 30: Cluj Arena, Romania

Dico si Tiganas' Cluj Arena was chosen to represent the tectonics of truss systems due to its architectural qualities (see Figures 14.14 and 14.15).

All structural guidelines for trusses are followed in Cluj Arena, including the large cantilevering truss. The trussed curved cantilevers turn around the arena to form its structure. The main tectonic feature of the building, which is its fluid and flowing form, was achived with the arrangement of the form of trusses. This fluid

14.14 A sketch of case study 30: Cluj Arena, Cluj-Napoca, Romania, 2011 (drawn with the help of URL3, 2013)

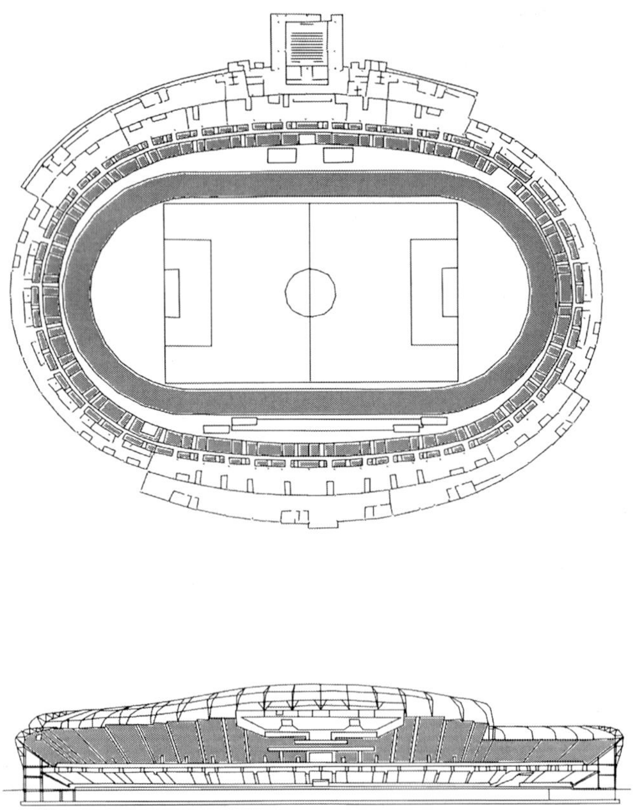

14.15 Plan and section of Cluj Arena (drawn with the help of URL4, 2012)

form is also semi-transparent due to the covering materials. This fluidity and transparancy were within the tectonic aims of the designers in order to integrate the building with the city and with the river beside it (URL1, 2012). Thus, the dominant physical entities in Cluj Arena are the structure system and the construction materials.

The structure is covered when it is seen from the outside. However, all of its details are seen from the inside. This difference between inside and outside also creates another tectonic effect. Since structural guidelines are followed in Cluj Arena, the building has an affirmative relationship with structural guidelines.

3D TRUSSES

To increase the effectiveness of trusses, they can be built as three-dimensional elements. There are various arrangements of 3D trusses, as seen in Figure 14.16. 3D trusses are usually made out of timber or steel. The characteristics of 2D trusses in relation to the formation of joints are also valid for 3D trusses.

Examples of 3D trusses are: Robin Hood Airport, Doncaster, UK; and Hamburg Airport, Germany. Hamburg Airport gets light through the space frames, and the tectonic effect is very strong because the building looks as if it does not have any structure.

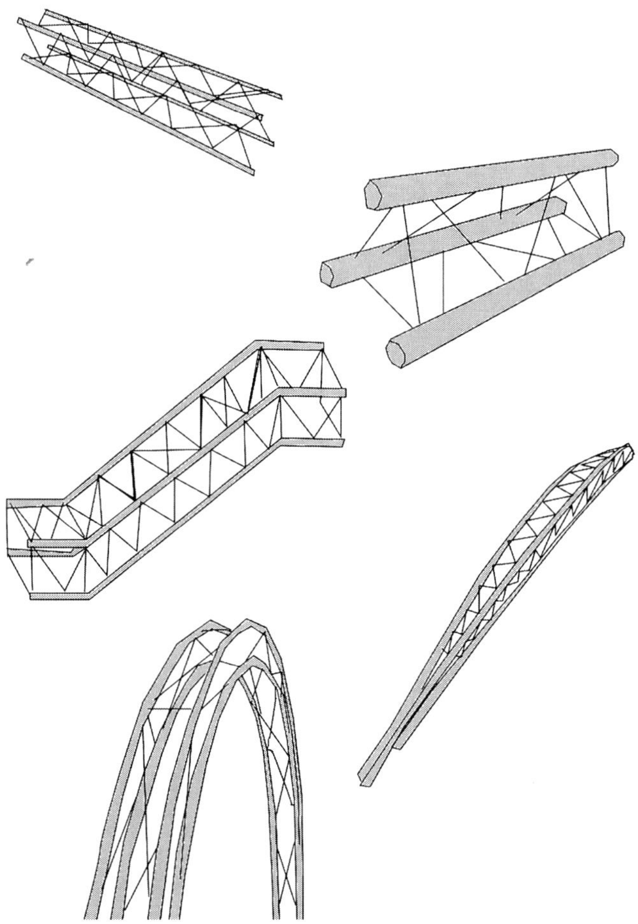

14.16 Different forms of 3D trusses

Span and depth of 3D trusses

Timber 3D trusses are used for spans between 12m and 25m. Steel 3D trusses are used for spans between 20m and 80m (Engel, 1997). However, there are 3D truss bridges spanning much longer distances. The depth of 3D trusses is similar to the depth of space frames. The span to depth ratio can change between 12 and 20, depending on the span and loading (URL2, 2000).

Structural guidelines for 3D trusses

Structural guidelines for 3D trusses are similar to structural guidelines for 2D trusses and they are listed in Table 14.2.

Table 14.2 Structural guidelines for 3D trusses and the associated value system

Structural guidelines	*Value system*
3D trusses can be formed with the help of triangulation.	By definition
Timber 3D trusses can span up to 25m and steel 3D trusses can span up to 80m.	Economy
Span to depth ratio for 3D trusses can change between 12 and 20.	Economy
3D trusses can be loaded from their joints.	By definition
Axes of all members in a joint should meet at a point to avoid bending.	By definition
Axis of the supporting column should meet with the point at which axes of all members in that joint meet.	By definition
3D truss axes in a building should be connected to each other with a beam in perpendicular direction in order to take the horizontal loads.	Safety
Wind bracing should be provided.	Safety

Case study 31: Waterloo Terminal, UK

Grimshaw Architects with Sir Alexander Gibb and Partners' Waterloo (Railway) Terminal in London was chosen as a case study because it has a very striking structural design and has been awarded a number of architectural awards. Figure 14.17 is a sketch of this building; its plans and section are shown in Figure 14.18.

Waterloo Terminal has a roof 400m long. The span of the roof varies between 35m and 50m at different parts of the building. The 3D trusses are in form of three hinged arches. However, an undulation is made to achieve the expected space. The depth of

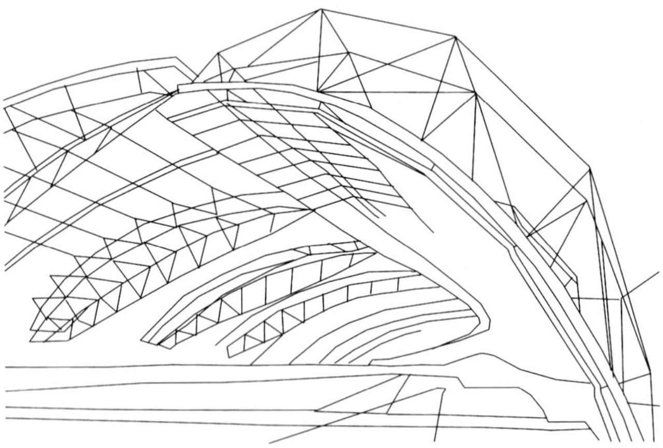

14.17 A sketch of case study 31: Waterloo Terminal, London, UK, 1993 (drawn with the help of URL5, n.d.)

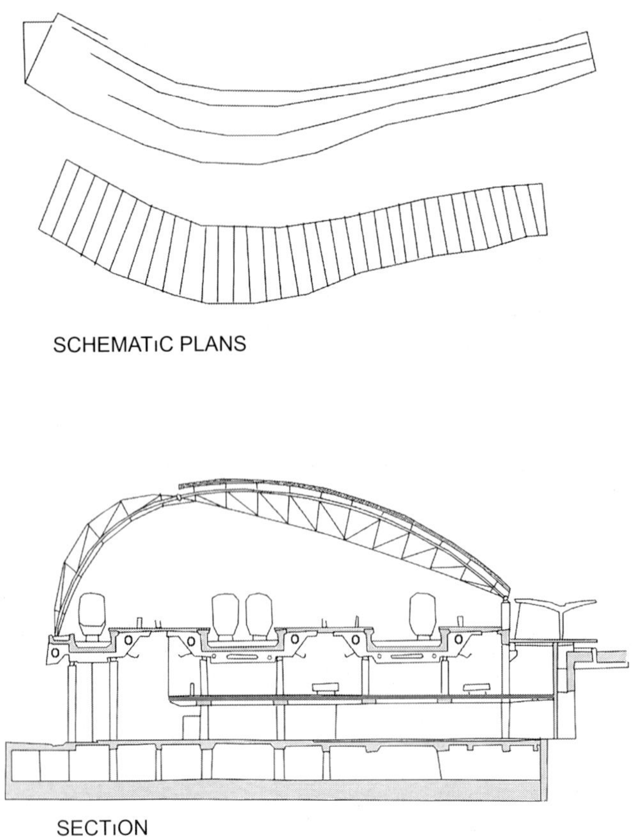

14.18 Schematic plans and section of Waterloo Terminal (drawn with the help of Archinform, 2014 and Royal Academy, n.d.)

the trusses becomes very narrow at the hinges because there is no moment at the hinges. The dominant physical entity in Waterloo Terminal is the structural system. The architectural concept of the building is also structural.

The building was designed as a monument to the railway age (Grimshaw, n.d.). It has a lightweight effect due to the use of 3D trusses and the transparency of the building on one side. The transparent side gives a view of Westminster to the rail passengers.

Although all structural guidelines are followed in this building, the design of the structure has the most important effect on the architectural value of the building. Waterloo Terminal has an affirmative relationship with structural guidelines.

SPACE FRAMES

The concept of a space frame is described in very diverse ways in structural engineering literature. Sometimes it is categorised as a three-dimensional structure. This chapter, however, describes structures that are composed of many small struts connected to each other to make a triangular geometric pattern.

Space frames can be used in different arrangements and formations of different surface geometries. They can be made out of timber, steel and aluminum. Their joints are usually a screw type, as seen in Figure 14.19. Their units and joints are usually produced by using industrialised technology. They are assembled on-site using semi-skilled labour: this makes space frames low-cost structures (Lan, 1999).

Space-frame structures are lightweight structures that provide flexibility for the inner space because they do not need columns inside the space. A square-shaped space frame structure can be

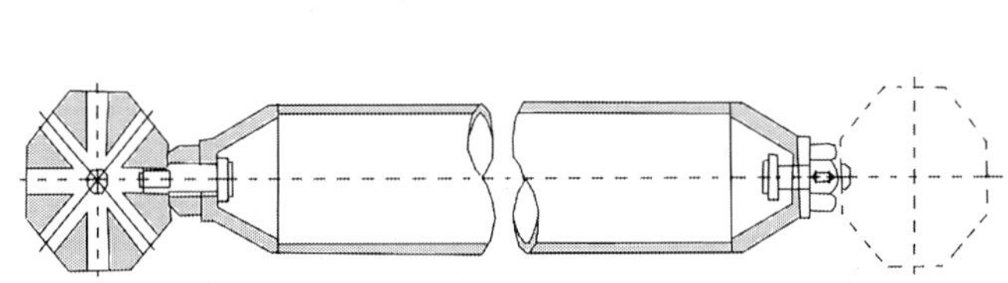

14.19 Joints in space frames

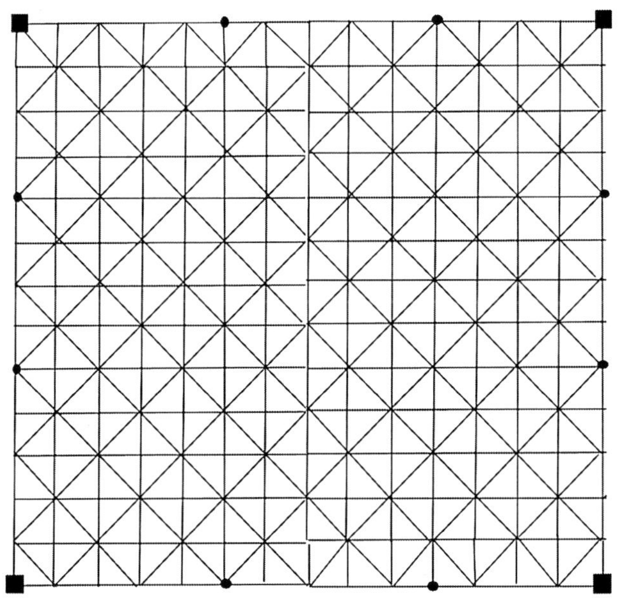

14.20 Arrangement of columns in a space frame structure

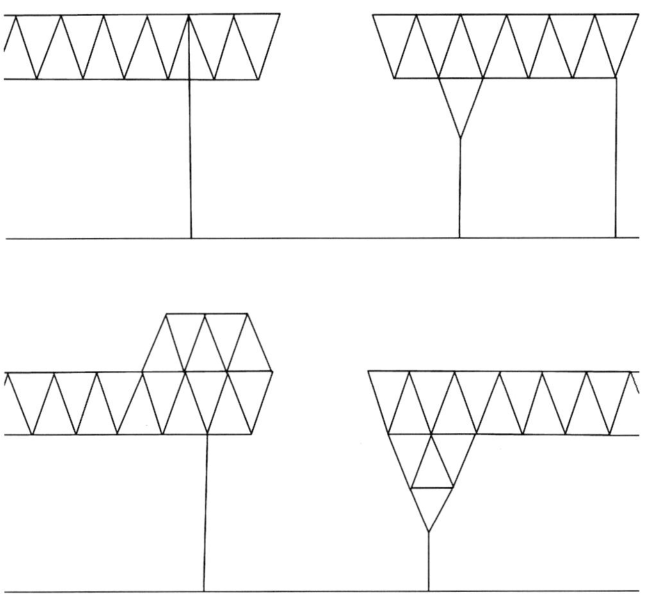

14.21 Connection of space frame to columns

supported with four columns at its four corners. However, if it is desirable to have walls to define the space, then it becomes necessary to put smaller columns within these infill walls at every 7–8m to support the wind load, as seen in Figure 14.20. The connection of space frame to columns can also be in different forms, as seen in Figure 14.21.

The construction of space frames can be realised in different ways (Lan, 1999):

- Erecting the space frame at its final location with the help of a temporary support.
- Erecting the space frame at ground level and then lifting it to its final location with the help of cranes or jacks.
- Dividing the space frame into strips or blocks: these strips and blocks are built at ground level according to a plan, and then lifted to their final locations.
- Dividing the space frame into horizontal strips: each strip is erected in the air at the same side of the building, and then slided horizontally to their final locations.

There are very different applications of space frames, such as that used in Zaha Hadid's London Aquatics Centre, UK, built in 2011. The space frame of this building is curved. Victoria Square Centre in Belfast, UK, and McCormick Place in Chicago, USA, are other examples for the application of space frames in architecture. To give an idea about the tectonic qualities of architectural applications of space frames, use an internet search to find pictures of these buildings.

Span and depth of space frames

According to some structural engineering literature, the span of timber space frames can be between 15m and 60m and the span of steel space frames can be between 25m and 100m (Engel, 1997; URL2, 2000). However, if one considers recent developments in Republic of China, for example the construction of Water Cube in Beijing that has a span of 195m, these values should be increased (Lan, n.d.).

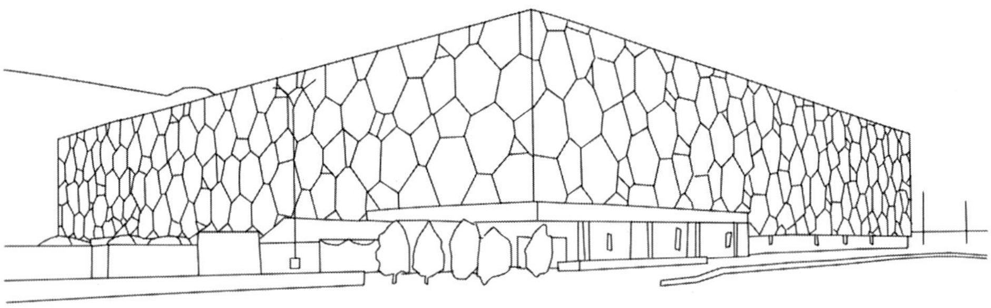

14.22 A sketch of case study 32: Water Cube, Beijing, Republic of China, 2008 (drawn with the help of URL6, 2013)

The span to depth ratio of space frames can vary from 12.5 to 25, depending on the dimensions of the elements and the loading on the system (Lan, 1999).

Structural guidelines for space frames

Structural guidelines for space frames are similar to structural guidelines for 2D and 3D trusses, and they are listed in Table 14.3.

Table 14.3 Structural guidelines for space frames and the associated value system

Structural guidelines	*Value system*
Space frames can be formed with the help of triangulation.	By definition
Timber space frames can span up to 60m and steel space frames can span up to 195m.	Economy
Span to depth ratio for space frames can change between 12.5 and 25.	Economy
Space frames can be loaded from their joints.	By definition
Axes of all members in a joint should meet at a point to avoid bending.	By definition
Axis of the supporting column should meet with the point at which axes of all members in that joint meet.	By definition

Case study 32: Water Cube, Republic of China

PTV Architects, CSCEC, CCDI and Arup's Water Cube (Beijing Olympic National Aquatics Center) was chosen as a case study due to its architectural and engineering qualities, for which the building has won several awards. Figure 14.22 is a sketch of the building; plans and a section are shown in Figure 14.23.

The facade of Water Cube is covered with the natural pattern of soap bubbles, symbolising water. 'Being a water cube' is the architectural concept of the building. The space frame used in this building is a special space frame, based on Weaire-Phelan polyhedral array. This array distributes loads equally within space (Fu et al., 2006). This special space frame and the air-inflated panels made out of ETFE (Ethylene tetrafluoroethylene) membranes give the building its unusual natural image. T.T. Lan (n.d.) says that the structure of this building is space frames with infilled 'foams'. The dominant physical entities in Water Cube are the structural system and construction materials.

The dimensions of the building are 195m × 195m × 35m (Lan, n.d.). The walls of the building are 3.6m thick and the roof is 7.2m thick due to the presence of pneumatic cushions within the space frame. The space-frame structure is an outer structure that has no relationship with the inner structure that carries the slabs.

This building represents a creative and innovative collaboration between the architectural and engineering teams. The recommendation concerning triangulation in structural guidelines was not followed, but was compensated with another geometry. Also, the span of Water Cube exceeds 100m, which is usually considered to be the span limit for space-frame structures. The

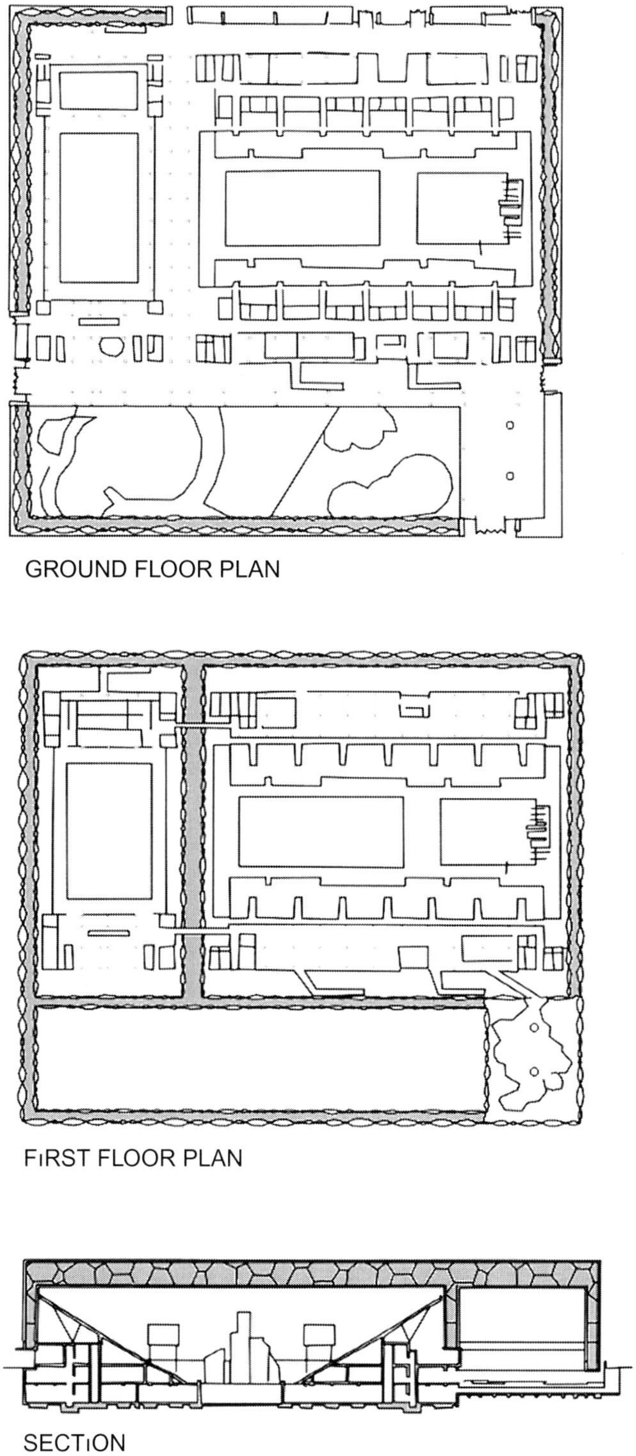

14.23 Plans and section of Water Cube (drawn with the help of URL7, n.d.)

building is located within a high-risk earthquake region: although some structural recommendations were not followed, strength against seismic loads was achieved. Thus, Water Cube has a contravening relationship with structural guidelines.

CONCLUSIONS

Two case studies in this chapter (Cluj Arena and Waterloo Terminal) followed structural guidelines and achieved strong tectonic affects. Cluj Arena achieved its tectonic effect through its fluid form, and Waterloo Terminal achieved its tectonic effect through its creative structural design to achieve a certain form. Water Cube did not follow structural guidelines about triangulation and the span limit at the time it was built. The tectonic effect of this building has been achieved through collaboration between the architectural and engineering teams. The building is very creative and very innovative. The common tectonic characteristics of these landmark buildings are their lightness and transparency.

PROBLEMS TO SOLVE

Find internal forces in the following trusses.

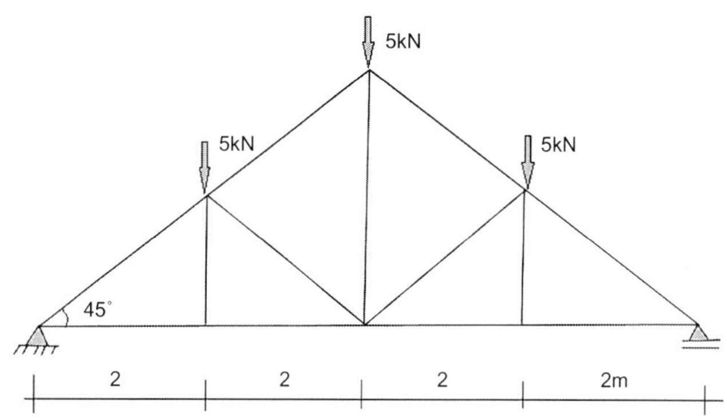

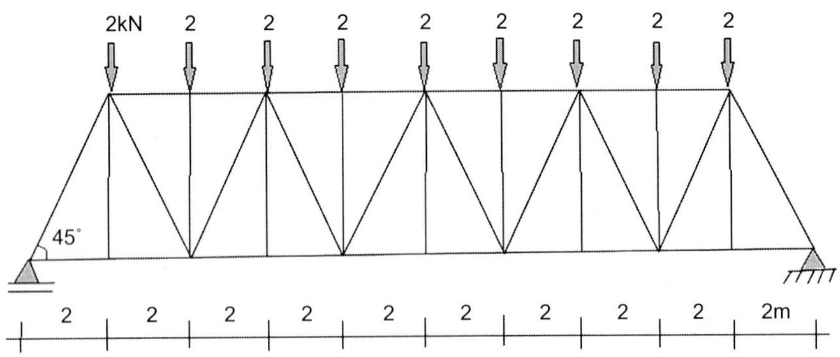

REFERENCES

Archinform (2014) *Waterloo International Terminal* (viewed 28 September 2014: http://eng.archinform.net/projekte/2325.htm)

Dabby, R., Bedi, A. (2012) *Structure for Architects*, John Wiley and Sons: New York.

Engel, H. (1997) *Structure Systems*, 3rd edition, Gerd Hatje Publishers: Ostfildern, Germany.

Fu, X., Gu, L., Yang, X., Yu, W., Chen, X. (2006) 'Design and research on structure of Beijing Olympic National Swimming Center' in eds D. Dubina & V. Ungureanu *Steel – A New and Traditional Material for Building*, Taylor and Francis: London: pp.21 –29.

Grimshaw, N. (n.d.) *International Terminal Waterloo* (viewed 7 September 2014: http://grimshaw-architects.com/project/international-terminal-waterloo/)

Lan, T.T. (1999) 'Space frame structures' in ed. Wai-Fah Chen *Structural Engineering Handbook*, 2nd edition, CRC Press LLC: Boca Raton: pp.24.1–24.49.

Lan, T.T. (n.d.) *Recent Developments of Long Span Space Structures in China* (viewed 7 September 2014: http://e-book.lib.sjtu.edu.cn/nascc2004/data/contents/PSSC%20PDF%20Files/RecentDevLongStructChina.pdf)

Royal Academy (n.d.) *Nicholas Grimshaw* (viewed 28 September 2014: www.royalacademy.org.uk/artist/nicholas-grimshaw-ppra)

URL1 (2012) *Cluj Arena – Dico si Tiganas* (viewed 7 September 2014: www.archdaily.com/210638/cluj-arena-dico-si-tiganas/)

URL2 (2000) *Rules of Thumb for Steel Design. Modern Steel Construction* (viewed 7 September 2014: www.modernsteel.com/uploads/issues/february_2000/0002_05_ruddyioannides.pdf)

URL3 (2013) *Dico si Taganas* (viewed 28 September 2014: http://en.wikipedia.org/wiki/Dico_si_Tiganas)

URL4 (2012) *Cluj Arena* (viewed 28 September 2014: www.bnab.ro/2012/proiecte/2/51/)

URL5 (n.d.) *Waterloo International Terminal* (viewed 28 September 2014: www.engineering-timelines.com/scripts/engineeringItem.asp?id=243)

URL6 (2013) *Day 3: Peking/Beijing* University (viewed 28 September 2014: http://blog.lib.umn.edu/itcomm/learningabroad/2011/05/day_3_pekingbeijing_university.php)

URL7 (n.d.) *National Swimming Center* (viewed 28 September 2014: http://openbuildings.com/buildings/national-swimming-center-water-cube-profile-3570/media)

15
The Tectonics of Folded Plates

Folded plates form the last group of form-resistant structures. Although ACI318-95 (1995) accepts that they are very similar to thin shell structures that work mainly with compression, folded plates work mainly with bending. Due to this difference in the stress type they develop, folded plates are categorised separately from thin shells in this book. Since the form is very important for folded plates, they are studied together with form-resistant structures.

Folded plates can be analysed according to their form possibilities and structural behaviour.

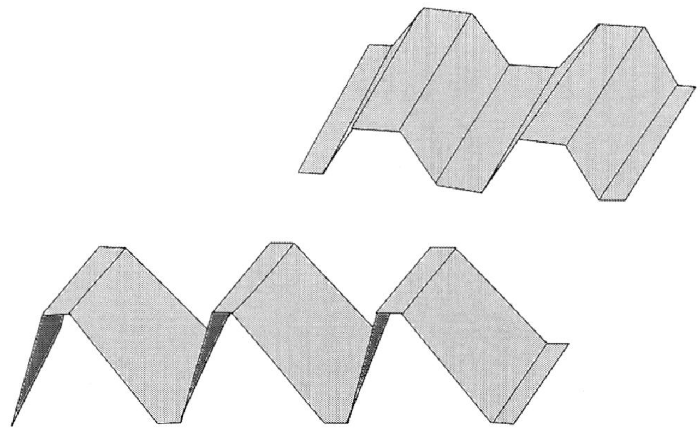

15.2 Examples of prismatic folds

FORM OF FOLDED PLATES

When we hold a paper from one side it bends downwards because it is thin and it does not have sufficient strength to keep upright. However, when we fold paper as seen in Figure 15.1, it no longer bends. Folding increases the strength of the surface by increasing its depth and moment of inertia.

Folded plates can have various different form arrangements. They can have prismatic folds, non-prismatic folds or faceted folds, as seen in Figures 15.2, 15.3 and 15.4 (ACI318-95, 1995).

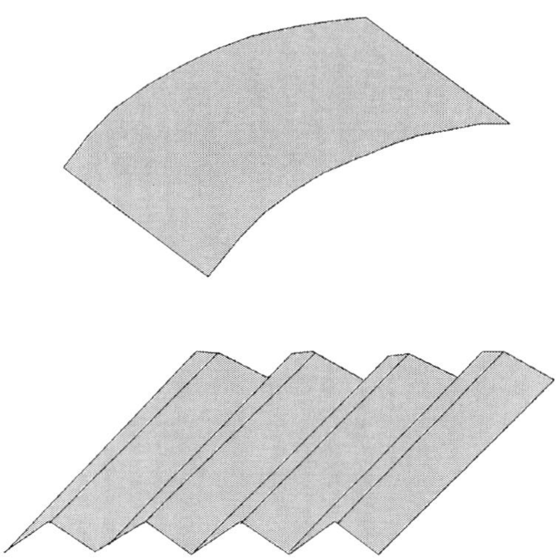

15.1 Folding a piece of paper

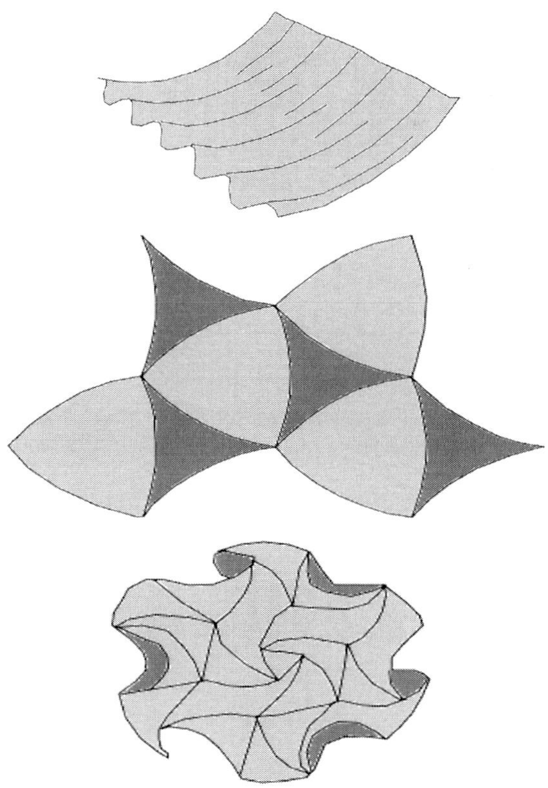

15.3 Examples of non-prismatic folds

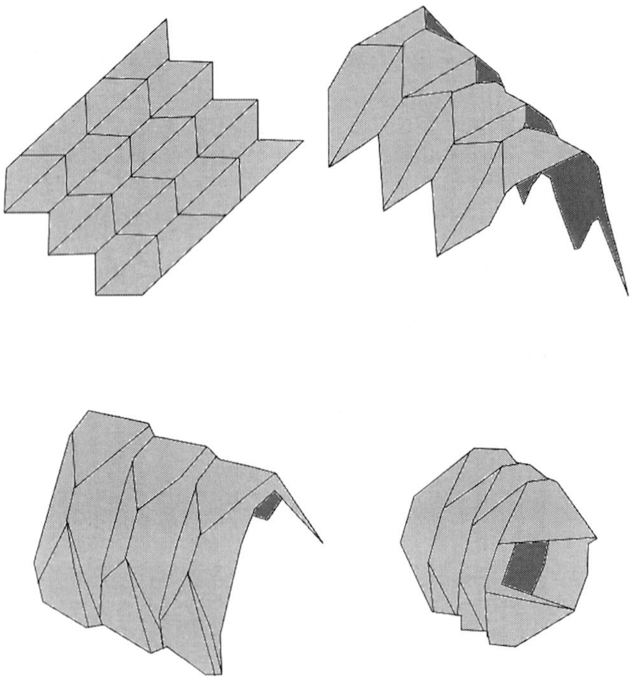

15.4 Examples of faceted folds

STRUCTURAL BEHAVIOUR AND SPAN OF FOLDED PLATES

Since the form of a folded plate is different to the form of its moment diagram, each slice in it works with bending stress, as seen in Figure 15.5. This means that each slice is similar to a beam.

Folded plates can be used for spans between 10m and 150m (Engel, 1997). Since they have more moment than thin shells, they should be thicker. The Teacher Training College in Leeds, UK, has a simple triangular prismatic folded plate roof. The span of the folded plates is around 20m; the width of each folded plate is 3.35m; and the depth of the plates is 10.2cm. This means that the span to thickness ratio in this building is around 196 (Wilby, 2005).

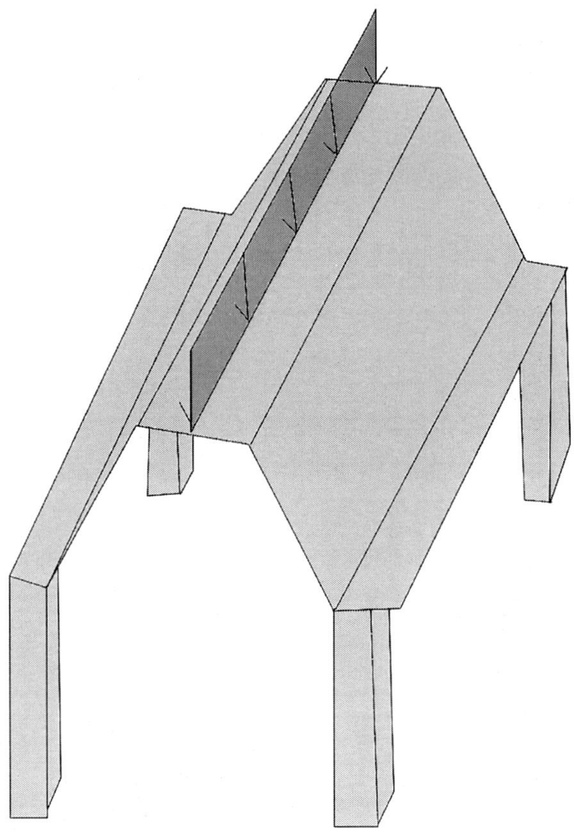

15.5 A slice of a folded plate used as a beam

The Tempodrom in Berlin, Germany, is an example of non-prismatic folded plates. Most of the folded-plate structures described in relevant literature are faceted folded plates because this type presents richer form possibilities. Examples of faceted folded-plate structures include: The United States Air Force Academy Cadet Chapel, near Colorado Springs, USA; the Commodore Munroe Stadium in Miami, USA; and the Boehringer Ingelheim company restaurant in Biberach, Germany. Amongst these examples, the Boehringer Ingelheim company restaurant (URL1, n.d.) and the Commodore Munroe Stadium (URL2, 2009) are exciting architectural examples: it would be beneficial to look at them on the internet.

STRUCTURAL GUIDELINES FOR FOLDED PLATES

Structural guidelines for folded plates are listed in Table 15.1.

Table 15.1 Structural guidelines for folded plates and the associated value system

Structural guidelines	*Value system*
Sufficient depth should be provided through folding.	Safety
Span of folded plates can be up to 150m.	Economy
Initial span to depth ratio can be accepted as 200 in horizontal folded plates.	Safety
Horizontal forces at the edges and supports can be taken by providing edge beams.	Safety
Buckling of the surfaces should be avoided.	Safety

CASE STUDY 33: YOKOHAMA INTERNATIONAL PORT TERMINAL, JAPAN

Foreign Office Architects' Yokohama International Port Terminal in Japan was chosen as a case study because the project was selected in an international architectural competition and was later awarded for its design features. Figure 15.6 shows internal and external sketches of Yokohama International Port Terminal. Plans and sections are shown in Figure 15.7.

The architects of the Yokohama International Port Terminal designed the roof of the building as an open plaza that provides continuity with the parks around the port. This is the architectural concept of the building. The curved and undulated surfaces of the roof form a natural environment. The exterior and interior spaces are continuous and they form fluid, uninterrupted and multi-directional spaces (URL3, 2007). Thus, the dominant physical entity is topography. The form of the surfaces, selected materials and design details are composed in unity to give Yokohama International Port Terminal a specific natural atmosphere.

The span of the terminal, which is located in a high-risk earthquake zone, is 42.5m. Although all structural guidelines were followed, the building presents an innovative design approach, especially in relation to the design of the folded plate and the roof above it. Yokohama International Port Terminal has an affirmative relationship with structural guidelines.

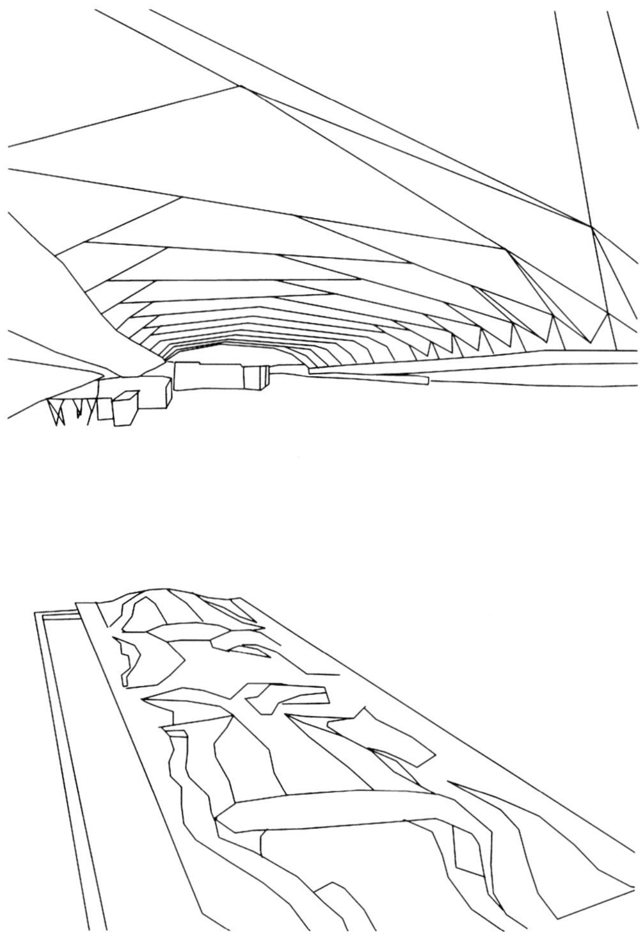

15.6 Sketches of case study 33: Yokohama International Port Terminal, Yokohama, Japan, 2002 (drawn with the help of Foreign Office Architects, n.d. and URL4, n.d.)

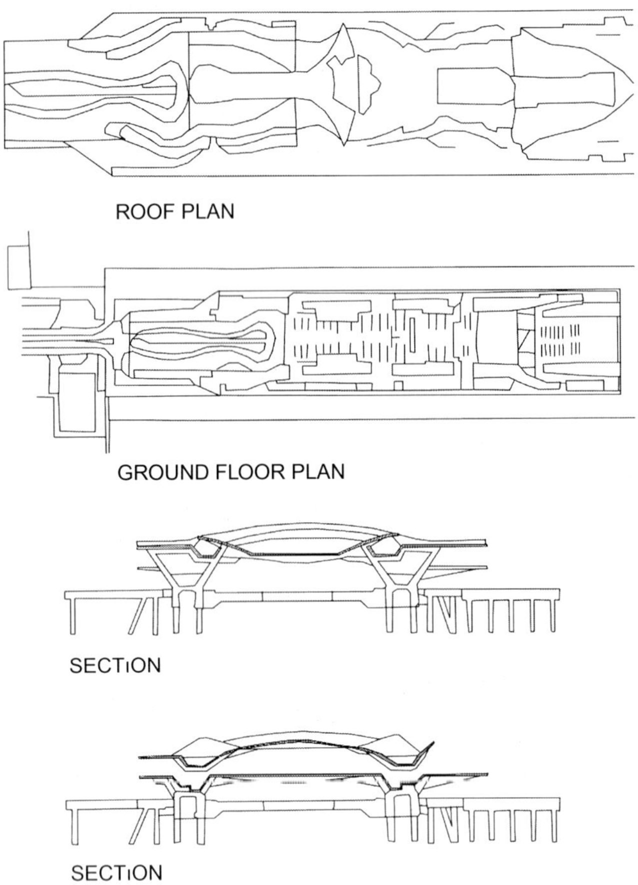

15.7 Plans and sections of Yokohama International Port Terminal (drawn with the help of URL5, 2009)

CONCLUSIONS

Folded plates can be used in landmark buildings. They usually follow structural guidelines while also presenting interesting forms and details. Yet not all buildings with folded-plate structures present innovative design approaches: Yokohama International Port Terminal is unique in being a product of a creative and innovative process.

REFERENCES

ACI 318-95 (1995) *Building Code Requirements for Reinforced Concrete*, American Concrete Institute: Michigan.

Engel, H. (1997) *Structure Systems*, 3rd edition, Gerd Hatje Publishers: Ostfildern, Germany.

FOA (Foreign Office Architects) (n.d.) *Foreign Office Architects* (viewed 28 September 2014: http://design.designmuseum.org/__entry/4867?style=design_image_popup)

Wilby, C.B. (2005) *Concrete Folded Plate Roofs*, Elsevier/Butterworth Heinemann: Oxford.

URL1 (n.d.) *Employee Restaurant Boehringer Ingelheim Biberach* (viewed 9 September 2014: www.archello.com/en/project/employee-restaurant-boehringer-ingelheim-biberach)

URL2 (2009) *Stadium Game* (viewed 9 September 2014: www.dwell.com/house-tours/article/stadium-game)

URL3 (2007) *Yokohama International Port Terminal* (viewed 9 September 2014: www.arcspace.com/features/foreign-office-architects/yokohama-international-port-terminal/)

URL4 (n.d.) *Whenislandscape* (viewed 28 September 2014: http://inlandscapegloveris.blogspot.com.tr/2009/08/yokohama-international-port-terminal.html)

URL5 (2009) *Case Study: FOA Yokohama International Port Terminal* (viewed 28 September 2014: http://classroomforall.blogspot.com.tr/2009/03/case-study-foa-foreign-office.html)

PART 5
THE TECTONICS OF OTHER STRUCTURES

Most building structures contain a combination of the structural systems that are covered in chapters 6–15. Sometimes the same system is repeated within one structure and sometimes different systems are combined side by side. There are some common methods for adding systems to each other and these approaches may be employed in many buildings. However, structural units can also be added to each other in very unexpected ways. It is also possible to integrate different structural systems to achieve a third system, and to create new structural units that have not previously existed.

These uncommon applications, which are listed below, are analysed in this book in order to define the concept of hybrid structures:

- Addition of the same structural units in a different way to form a structure.
- Addition of different structural units in an uncommon way to form a structure.
- Integration of different structural units to form a structure.
- Addition of unique structural units to form a structure.

16
The Tectonics of Hybrid Structures

BY YONCA HUROL AND BAYDU CAN AL

The term 'hybrid structure' usually refers to the integration of two different systems to form a third system. However, all unconventional additions of structural systems might cause changes in the structural behaviour of these systems and make them hybrid systems. Thus, it is worth analysing examples of these different additions in order to define the concept of hybrid structures.

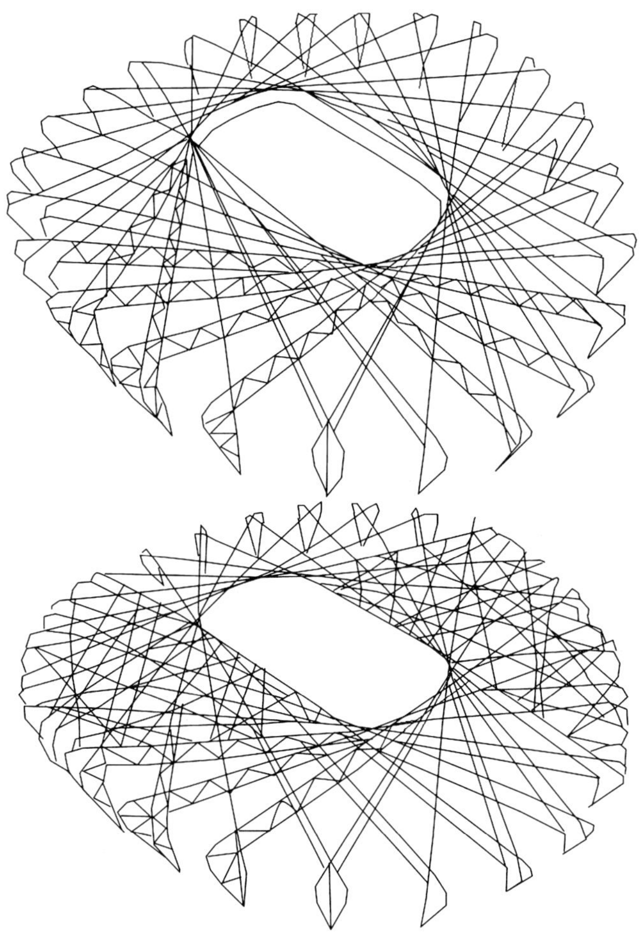

16.1 A schematic drawing of the 'Bird's Nest' (drawn with the help of URL2, 2008)

ADDITION OF THE SAME STRUCTURAL UNITS IN A DIFFERENT WAY

An example of the addition of the same structural units in a different way to form a structure is Herzog & de Meuron's Beijing National Stadium (Bird's Nest), which was built in Republic of China in 2008. A schematic drawing of the trussed frames used in this structure is given in Figure 16.1.

The structure of the 'Bird's Nest' is formed by adding a special trussed frame several times, and the gaps between these trusses are filled with additional structural elements that connect the frames to each other. This makes the structure work in a more integrated and hybrid manner. The structure becomes a hybrid of truss systems and steel shell structures. However, the hybrid nature of this building is not simply an outcome of the repetition of trusses. The addition of the elements to fill the gaps between frames also contributes to the hybrid nature of the structure.

ADDITION OF DIFFERENT STRUCTURAL UNITS IN AN UNCOMMON WAY

An example of the addition of different structural units in an uncommon way to form a structure is Santiago Calatrava's Lyon Satolas Airport Railway Station, which was built in France in 1994. Schematic drawings are given in Figure 16.2.

The structure of Lyon Satolas Airport Railway Station has a 3D truss arch along its spine, which also functions as a skylight. There are two other arches parallel to the 3D truss. These arches are connected to each other by frequently placed steel elements that form the diagonal upwards elements, which look like the opening wings of a bird. (Although the design concept of this building is actually a human eye, not a flying bird.) This structure cannot be called a hybrid structure because the elements continue to behave as expected, with a clear hierarchy between them.

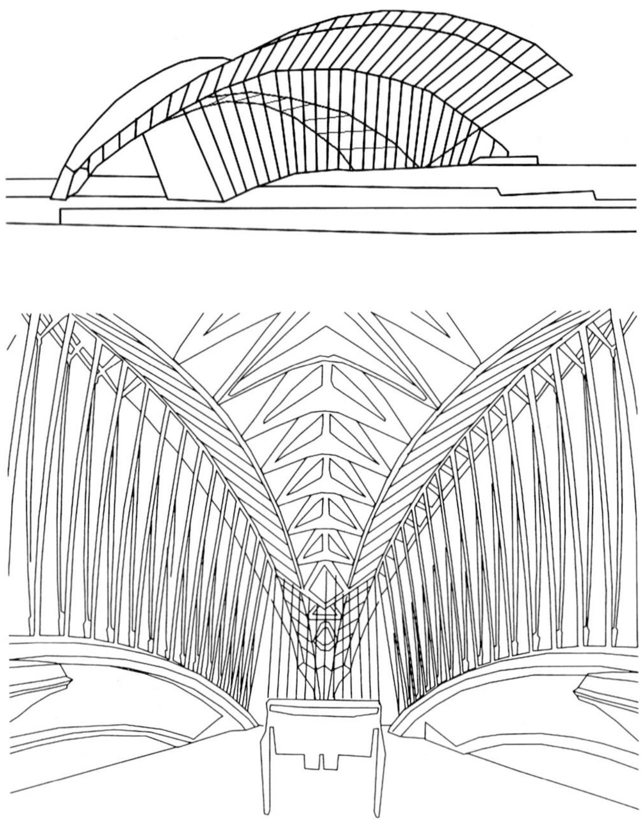

16.2 Schematic drawings of Lyon Satolas Airport Railway Station (drawn with the help of Architravel, 2013 and URL3, 2013)

INTEGRATION OF DIFFERENT STRUCTURAL UNITS TO FORM ANOTHER SYSTEM

An example of the integration of different structural units to form a third structural system is Tung Yen Lin's (structural engineer) Oklahoma State Fair Arena, USA. This structure is in the form of an inverted dome, formed by a cable-net structure. The cable-net structure is stabilised with the help of shotcrete application. Thus, the structure works as a cable structure under gravity load, but also works as a dome under the effect of the upward lift of wind, as seen in Figure 16.3 (Lin & Stotesbury, 1981). Since the structure

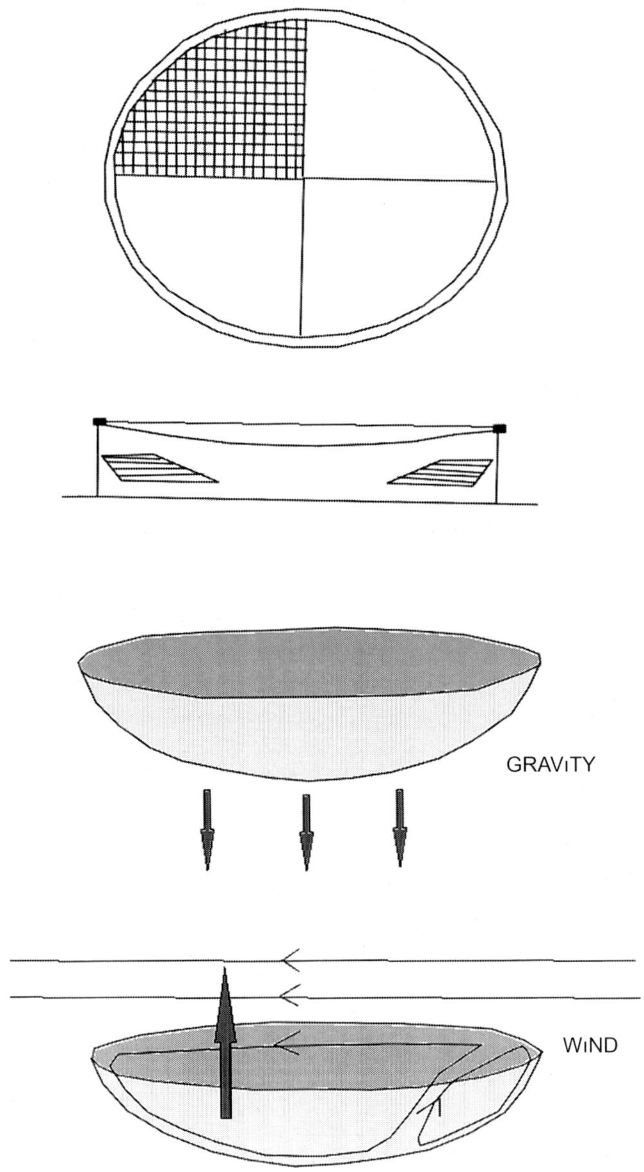

16.3 Roof of Oklahoma State Fair Arena under gravity effect and wind suction

can show two different types of structural behaviour at different times, it can be considered as a special type of hybrid structure.

Another example of the integration of different types of structures is the famous Sydney Opera House in Australia, which is a case study at the end of this chapter. The structure of this building is known as a thick shell, but it can also be seen as a hybrid of a shell structure and a waffled-slab system, as seen in Figure 16.4. The repeating pre-cast concrete ribs' geometry is extracted from a sphere. These ribs take place between a concrete pedestal and a ridge beam. Steel cables hold these ribs, which are joined together by post-tensioning (Utzon, 2002).

All ribbed shells, including Sydney Opera House and CNIT Hall in Paris, France, can be seen as hybrids of shell and ribbed-/waffled-slab structures. Ribbed shells are used in extraordinary situations, such as large spans and difficult loading conditions. The CNIT Hall has a span of 216m and the Sydney Opera House had an extraordinary form combined with an extraordinary wind loading during its construction. Its shells were like sails open to the wind before the glass surfaces in front of them were placed (URL1, 2003).

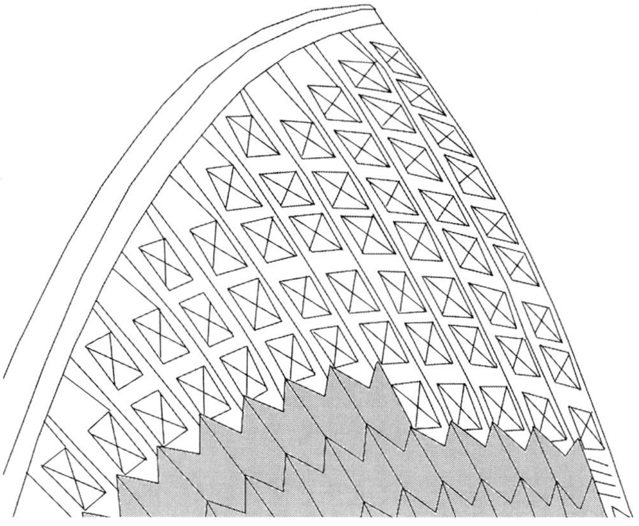

16.4 Ribs of Sydney Opera House (drawn with the help of Lin & Danziger, 2007)

ADDITION OF UNIQUE STRUCTURAL UNITS TO FORM A HYBRID STRUCTURE

An example of the addition of unique structural units to form a structure is Norman Foster's (architect) and Peter Rice's (structural engineer) Stansted Airport, which was built near London, UK, in 1991. This building has a specially designed structural unit that repeats in a very simple way within a rectangular building. This repeating unit is shown in Figure 16.5.

The structural unit in Stansted Airport contains steel inclined elements in the form of a tree. The utility pillar under this tree contains the service systems of the building. There are cross-bracing cables inside the tree structure, which take the horizontal forces. There is a steel grid shell roof at the top of the unit.

Each structural unit is surrounded with a simplified version of the same structural unit that contains only the grid shell roof. The plans and section of Stansted Airport are shown in Figure 16.6.

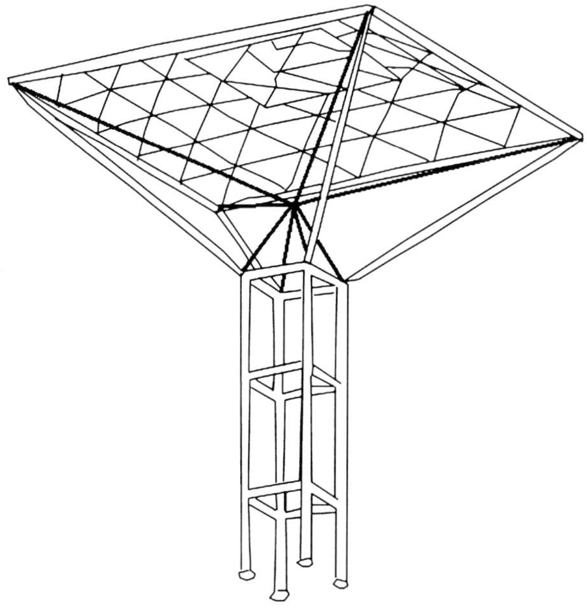

16.5 The unique structural unit in Stansted Airport (drawn with the help of URL5, 2012)

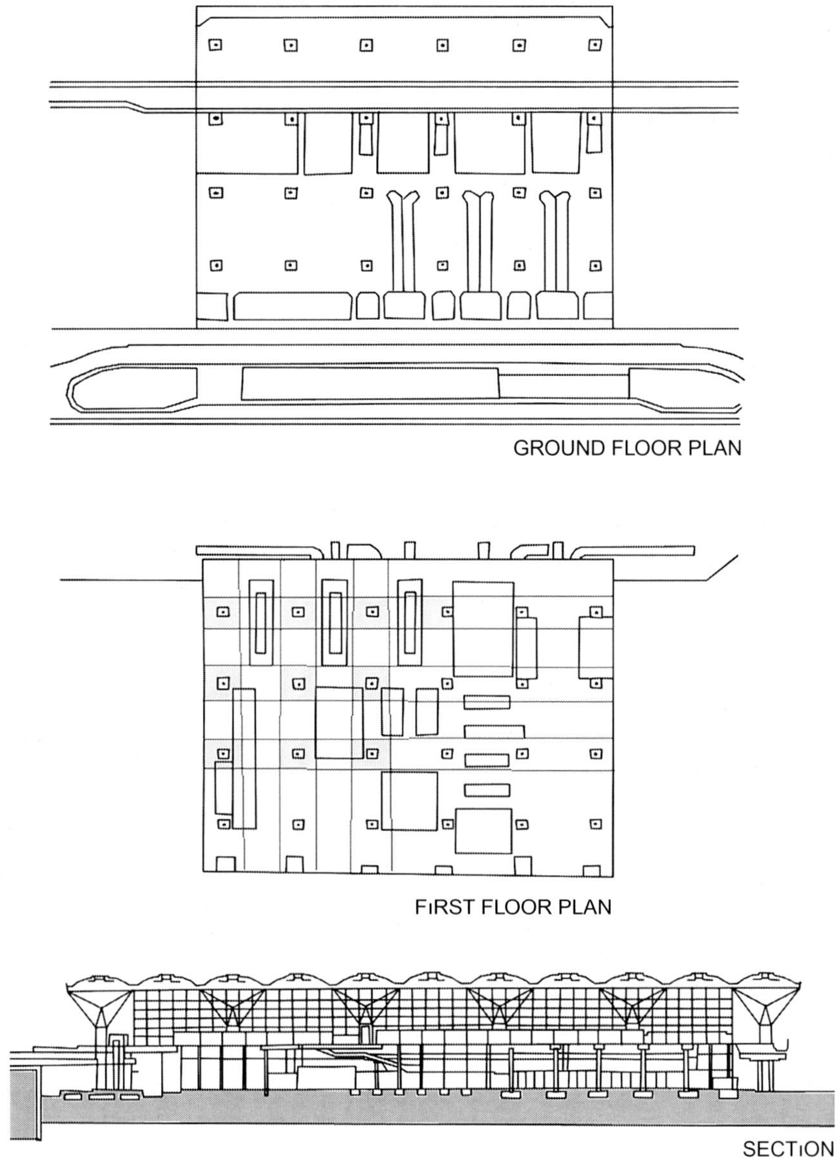

16.6 Plans and section of Stansted Airport (drawn with the help of URL4, n.d.)

The structure of Stansted Airport is a very creative combination of its elements. However, since the stress type in these elements are as expected, it is not a hybrid structure. The author of this book does not know of any hybrid example achieved through the addition of unique structural units.

ON HYBRIDITY OF STRUCTURES

According to the above explanations, the 'Bird's Nest', the Oklahoma State Fair Arena and Sydney Opera House can be seen as examples of hybrid structures.

Hybridity is not due to the simple addition of structures, and nor is it due to the addition of structures in uncommon ways. Hybridity is due to the articulation of at least two different types of structures with each other: at the end of this process, the structural behaviour of the system changes.

CASE STUDY 34: SYDNEY OPERA HOUSE, AUSTRALIA

Jorn Utzon (architect) and Ove Arup and Partners' (structural engineers) Sydney Opera House was chosen as a case study not only due to its architectural qualities, but also due to the very peculiar relationship it presents between structure and aesthetics. Figure 16.7 is a sketch of the building; the plan and section are shown in Figure 16.8. Referring to Sydney Opera House, architect Louis Kahn said: 'The sun did not know how beautiful its light was, until it was reflected off this building' (Utzon, 2002).

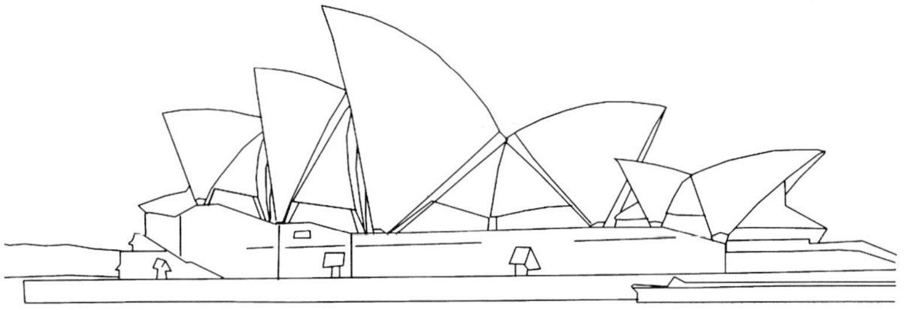

16.7 A sketch of case study 34: Sydney Opera House, Sydney, Australia, 1957–1973 (drawn with the help of URL6, 2014)

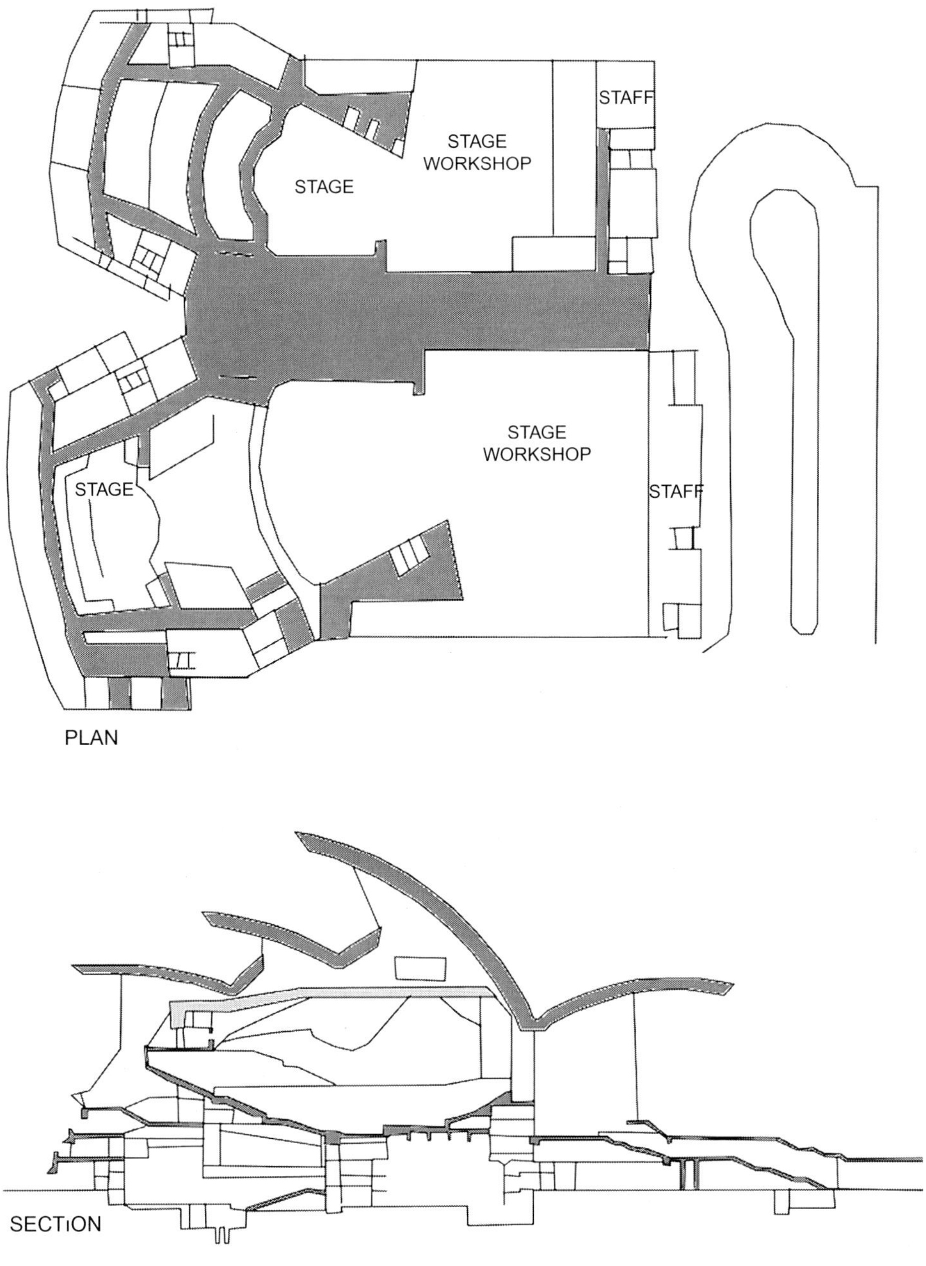

16.8 Plan and section of Sydney Opera House (drawn with the help of URL7, 2014 and URL8, 2014)

Sydney Opera House, which was awarded The Pritzker Architecture Prize, was designed to be located on a 12m high platform extending into Sydney Harbour. Large stairs take people to the two large auditoriums that are located over this platform. The shells of this building were designed as a white light sculptural roof, which follows the profile of the spaces underneath (Utzon, 2002). The platform under the shells and the sculptural form of the shells form the architectural concept of the building.

The geometry of the shells was not defined in the original project. The complication in the geometry of the building delayed the construction of it between 1959 and 1973. Structural guidelines concerning the form definition of shell structures, and having an appropriate curvature to decrease the bending moment in shells, were not followed. After winning the architectural competition, architects and structural engineers working on this project first had to define the geometry of the shells as sections from a sphere. The decision to use pre-cast ribs reduced the cost of this very expensive building.

The tectonics of the initial project were not successful due to the problems with the structural system. However, the project later passed through an innovative transformation to become one of the most important masterpieces of tectonics. The dominant physical entity is the structural system and Sydney Opera House has a contravening relationship with structural guidelines.

CONCLUSIONS

It is difficult to categorise structural systems of hybrid structures and to determine their structural guidelines. Thus, they usually have a natural appearance like the 'Bird's Nest' and Sydney Opera House. Most hybrid structures are products of innovative design procedures that make them landmark buildings with particular tectonic qualities.

REFERENCES

Architravel (2013) *Lyon-Satolas Airport Railway Station* (viewed 29 September 2014: www.architravel.com/architravel/building/lyon-satolas-airport-railway-station/)

Lin, L., Danziger, B. (2007) 'The Imaginative Engineer' *Structure* (viewed 29 September 2014: www.structurearchives.org/article.aspx?articleID=476)

Lin, T.Y., Stotesbury, S.D. (1981) *Structural Concepts and Systems for Architects and Engineers*, John Wiley and Sons: New York.

Utzon, J. (2002) *Sydney Opera House Utzon Design Principles* (viewed 13 September 2014: www.environment.gov.au/system/files/pages/59ca36d1-4581-4d7d-83d7-04b124d801b1/files/soh-utzondesignprinciples.pdf)

URL1 (2003) *Model of Sydney Opera House 1960* (viewed 13 September 2014: www.powerhousemuseum.com/mob/collection/database/?irn=12041&img=7593)

URL2 (2008) *Beijing Bird's Nest* (viewed 29 September 2014: www.architectureweek.com/2008/0827/building_1-2.html)

URL3 (2013) *Le Confident* (viewed 29 September 2014: http://leconfident.wordpress.com/2012/06/07/cage-bird/)

URL4 (n.d.) *Architecture of Stansted Airport* (viewed 29 September 2014: http://new-architecture-archi.blogspot.com.tr/2012/02/architecture-of-stansted-airport.html)

URL5 (2012) *Stansted Airport* (viewed 29 September 2014: www.greatbuildings.com/cgi-bin/gbc-drawing.cgi/Stansted_Airport.html/Stansted_Structure.gbd)

URL6 (2014) *Sydney Opera House Arts Center in Sydney* (viewed 29 September 2014: http://travelinnate.com/sydney-opera-house-arts-centre-in-sydney/)

URL7 (2014) *Sydney Opera House Floor Plans* (viewed 29 September 2014: http://furnituredir.blogspot.com.tr/2014/04/sydney-opera-house-floor-plans.html)

URL8 (2014) *Sydney Living Museums* (viewed 29 September 2014: http://sydneylivingmuseums.com.au/public-sydney-drawing-city)

17 Evaluation of Case Studies and Conclusions

The Tectonics of Structural Systems questions the relationship between the knowledge of structures, which is represented by structural guidelines for specific structural systems, and aesthetics in architecture, which is represented by successful architectural examples employing different types of structural systems. The 34 case studies are analysed according to their relationship with the recommendations in structural guidelines and their aesthetic characteristics. Table 17.1 shows the relationship of each case study to the recommendations in structural guidelines, the way in which those not following the recommendations were compensated, and the implications for the architectural concept of the building. Table 17.2 shows the relationship between structural guidelines, earthquake risk, physical entities and innovation.

Knowledge should be in the service of human freedom and there is no freedom without knowledge. Architectural design requires freedom.

The concept of structural guidelines introduced in this book, and the reasons behind each recommendation, represent a

Table 17.1 Attitude towards structural guidelines and its relation to architectural concept in case studies

No	*Building*	*Material/structural system*	*Non-adherence to structural recommendations*	*Method of compensation*	*Reason for non-adherence to structural recommendations*	*Architectural concept's relationship with non-adherence to structural recommendations*
1	Cologne Cathedral	Stone masonry	Height of the walls	Thick exterior walls with ribs	Economy	Related. Height of the space and large openings were needed to gain spiritual character.
			Having many openings	Increasing thickness of walls	Safety	
2	Stone tower at Norman Castle	Stone masonry	Height of the walls	Thick walls	Economy	Related. Need for a tower.
			Size of spaces		Safety	
3	Summer House	Brick masonry	Not symmetrical	–	Economy	Related. Being a part of nature.
			Size of spaces		Safety	
			Having many openings		Safety	
4	Harran Houses	Adobe masonry	Size of spaces	–	Safety	Related. Need for practicality.
			Not using a light roof system		Safety	
			Height of the dome		Economy	
5	Great Mosque of Djenne	Adobe masonry	Size of spaces	Thick walls with cross-walls and buttresses	Safety	Related. Bringing people together.
			Height of the walls		Economy	
6	İsmail Hacı Çakır House	Timber masonry	Large openings around sofa	–	Safety	Related. Having open space during summer.
7	Arches at Cordoba Mosque	Stone arch	–	–	–	Repetition of arches and change in colours.

Table 17.1 *continued*

No	*Building*	*Material/structural system*	*Non-adherence to structural recommendations*	*Method of compensation*	*Reason for non-adherence to structural recommendations*	*Architectural concept's relationship with non-adherence to structural recommendations*
8	Vaults at Cologne Cathedral	Stone ribbed groined vault	–	–	–	Repetition of groined vaults and continuity of ribs.
9	Dome at Pantheon	Unreinforced concrete dome	Span	Thickness of dome, coffers, large oculus	Economy	Related. Need for large space for all pagan gods.
10	Villa Müller	Reinforced brick masonry	–	–	–	Achieving wall-free space inside.
11	Falling Water	Stone masonry with RC frame	Dimensions of cantilevers	–	Safety	Related. Horizontality versus verticality.
12	Villa Savoye	RC frame	–	–	–	Continuity with nature.
13	National Assembly in Dacca	RC frame and shear walls	Large galleries	Solved with expansion joints	Safety	Related. Giving outdoor effect to indoor spaces.
14	Church of Light and Sunday School	RC shear walls	One shear wall is cancelled Unity of structure is problematic	–	Safety	Related. Creating light effects.
15	Barcelona Pavilion	Steel frame	Slender columns	Not compensated	Safety	Related. Expression of steel frame.
16	Photographer's Penthouse	Steel frame	–	–	–	Lightness, large opening to river view.
17	Suspended Bedroom	Steel frame	–	–	–	Having a lightweight white cube.
18	World Trade Center	Steel framed tube	No decrease in wind load	Tubular structure	Economy	Verticality.
19	Shanghai World Financial Center	Steel and RC frame with outrigger trusses	–	–	–	Aesthetics of aerodynamic form.
20	Zagreb Arena	Cable	Form of columns	–	Economy	Related. Symbolises fingers.
21	Science and Technology Museum	Suspended glass system	Impossible wish of achieving a dematerialised surface	Invention of suspended glass systems	Impossibility	Related. Dematerialisation.
22	Olympic Stadium in Munich	Membrane	–	–	–	Like a cloud over the area.
23	Swarovski Pavilion	Pneumatic	–	–	–	Gemstone geometry.
24	Dome at Parliament Building in Berlin	Steel dome	–	–	–	Expression of a society in which people are above the government.
25	US Pavilion at Expo '67	Steel geodesic dome	–	–	–	Having a building inside the structure.

No	*Building*	*Material/structural system*	*Non-adherence to structural recommendations*	*Method of compensation*	*Reason for non-adherence to structural recommendations*	*Architectural concept's relationship with non-adherence to structural recommendations*
26	Kimbell Art Museum	RC vault	Not best form for the vaults Vaults are divided into two	Acceptable curvature Connected with struts	Economy Safety	Related. A good and modest room for exhibition purposes.
27	Small Sports Palace	RC dome	–	–	–	Abstract geometric perfection.
28	TWA Airport Building	RC shell	Not in the best form	More material used	Economy	Related. Symbolises a bird in flight.
29	Suan Lien Center	Aluminum alloy grid shell	Not in the best form	–	Economy	Related. Symbolises two hands in prayer.
30	Cluj Arena	Steel truss	–	–	–	Fluidity of form.
31	Waterloo Terminal	Steel 3D truss	–	–	–	Undulated form.
32	Water Cube	Steel space frame	Span limit No triangulation	Innovation Another geometry	Economy Definition of structure	Related. Symbolises water.
33	Yokohama International Port Terminal	RC folded plate	–	–	–	Roof as open plaza.
34	Sydney Opera House	RC ribbed thick shell	Form not defined Not using the correct curvature	Innovation	Practicality Economy	Related. Sculptural roof structure.

Table 17.2 Attitude towards structural guidelines, earthquake risk, dominant physical entities and innovation in case studies

No	*Building*	*Material/structural system*	*Earthquake risk*	*Dominant physical entity*	*Tectonics achieved through*		*Innovation*
					Adherence to structural recommendations	*Non-adherence to some structural recommendations*	
1	Cologne Cathedral, Germany	Stone masonry	–	Structural system		Height of the walls Having many openings	–
2	Stone tower at Norman Castle, UK	Stone masonry	–	Structural system		Height of the walls Size of spaces	–
3	Summer House, Finland	Brick masonry	–	Structural material		Not symmetrical Size of spaces Having many openings	–

Table 17.2 *continued*

No	*Building*	*Material/structural system*	*Earthquake risk*	*Dominant physical entity*	*Tectonics achieved through*		*Innovation*
					Adherence to structural recommendations	*Non-adherence to some structural recommendations*	
4	Harran Houses, Turkey	Adobe masonry	High	Structural system		Size of spaces Not using a lightweight roof system Height of the dome	–
5	Great Mosque of Djenne, Mali	Adobe masonry	–	Structural system Structural material		Size of spaces Height of the walls	–
6	İsmail Hacı Çakır House, Turkey	Timber masonry	–	Structural system Structural material		Large openings around sofa	–
7	Arches at Cordoba Mosque, Spain	Stone arch	High	Structural system	+	–	–
8	Vaults at Cologne Cathedral, Germany	Stone ribbed groined vault	–	Structural system	+	–	–
9	Dome at Pantheon, Itlay	Unreinforced concrete dome	High	Structural system		Span	Substantially larger than previous domes.
10	Villa Müller, Czech Republic	Reinforced brick masonry	–	Structural system	+	–	–
11	Falling Water, USA	Stone masonry with RC frame	–	Topography		Dimensions of cantilevers	–
12	Villa Savoye, France	RC frame	–	Structural system	+	–	–
13	National Assembly in Dacca, Bangladesh	RC frame and shear walls	High	Materials Details		Large galleries	–
14	Church of Light and Sunday School, Japan	RC shear walls	High	Materials Details		One shear wall is cancelled Unity of structure is problematic	–
15	Barcelona Pavilion, Spain	Steel frame	High	Structural system Details		Slender columns	–
16	Photographer's Penthouse, USA	Steel frame	–	Structural system	+	–	–
17	Suspended Bedroom, France	Steel frame	–	Structural system Construction methods	+	–	–
18	World Trade Center, USA	Steel framed tube	–	Structural system		No decrease in wind load	–

No	Building	Material/structural system	Earthquake risk	Dominant physical entity	Tectonics achieved through		Innovation
					Adherence to structural recommendations	Non-adherence to some structural recommendations	
19	Shanghai World Financial Center, Republic of China	Steel and RC frame with outrigger trusses	–	Wind Structural system	+	–	Aerodynamic form.
20	Zagreb Arena, Croatia	Cable	–	Structural system Construction materials		Form of columns	–
21	Science and Technology Museum, France	Suspended glass system	–	Structural system Details		Impossible wish of achieving a dematerialised surface	Invention of suspended glass systems.
22	Olympic Stadium in Munich, Germany	Membrane	–	Structural system	+	–	–
23	Swarovski Pavilion, Switzerland	Pneumatic	–	Structural system	+	–	–
24	Dome at Parliament Building in Berlin, Germany	Steel dome	–	Environmental control systems	+	–	–
25	US Pavilion at Expo '67, Canada	Steel geodesic dome	–	Structural system	+	–	Development of the mathematics of geodesic domes.
26	Kimbell Art Museum, USA	RC vault	–	Natural light		Not best form for the vaults Vaults are divided into two	–
27	Small Sports Palace, Italy	RC dome	High	Structural system	+	–	–
28	TWA Airport Building, USA	RC shell	–	Structural system		Not in the best form	–
29	Suan Lien Center, Republic of China	Aluminum alloy grid shell	High	Structural system		Not in the best form	–
30	Cluj Arena, Romania	Steel truss	High	Structural system Construction materials	+	–	–
31	Waterloo Terminal, UK	Steel 3D truss	–	Structural system	+	–	–
32	Water Cube, Republic of China	Steel space frame	High	Structural system Construction materials		Span limit No triangulation	A geometry other than triangulation. Use of pneumatic elements.

Table 17.2 *continued*

No	Building	Material/structural system	Earthquake risk	Dominant physical entity	Tectonics achieved through		Innovation
					Adherence to structural recommendations	Non-adherence to some structural recommendations	
33	Yokohama International Port Terminal, Japan	RC folded plate	High	Topography	+	–	Use of top of folded plate as open plaza.
34	Sydney Opera House, Australia	RC ribbed thick shell	–	Structural system		Form not defined Not using the correct curvature	New methods of analysis.

body of knowledge of structures for architects. It is expected that architects should follow these guidelines, but not become slaves to them. It is possible to achieve tectonics, which is the balance between technology and aesthetics, by following structural guidelines or by not following them, and solving the ensuing problems by some other means. Whether one follows structural guidelines or not, it is necessary to have the knowledge of structures. Structural knowledge brings freedom to structural design.

New technologies can be developed to meet the demands of architects. Building codes and structural guidelines also need to be changed to reflect new developments in building technology.

Table 17.2 shows that structural system is usually the dominant physical entity in most of the case studies. Since this book is about structures, it seems obvious to select case studies in which the structural system is dominant. However, structural system does play an important role in architectural design.

The recommendations set out in structural guidelines influence architectural design to varying degrees. The influence of structural recommendations increases if the designer does not know the reasons behind them. Yet if the architect does knows the reasons behind each recommendation, it becomes possible either to follow it knowingly or to solve the problems caused by non-adherence in unexpected ways. Structural recommendations represent common solutions to structural problems, but there are always alternatives.

This book brings together structural guidelines for different structural systems and architects' responses to these guidelines. The theory parts of each chapter develop structural guidelines and the case studies present architects' responses. When all the case studies are revised together, it is seen that more than half of the cases (19 of 34) did not follow structural guidelines and four of these compensated for this non-adherence with innovative solutions. These are the dome at Pantheon (example of a masonry dome), the Science and Technology Museum in Paris (example of a suspended glass system), the Water Cube in Beijing (example of a space frame) and Sydney Opera House (example of a hybrid structure).

The historical examples compensated for the non-adherence to recommendations by increasing the thickness of walls and by adding other structural elements, such as ribs. Whereas contemporary examples either developed creative compensation methods (such as using expansion joints effectively), or used more structural material in order to solve problems.

The author of this book was expecting to discover that non-adherence to structural recommendations is based on economic value. It is true that recommendations based on economic value might not be followed, depending on the economic power and expectations of the client. However, the analysis of the 34 case studies showed that structural recommendations with all types of values (including safety, economy and practicality) were not followed, and the problems arising due to non-adherence were compensated.

The author was also expecting that buildings in high-risk earthquake zones would follow structural guidelines and buildings in risk-free areas would not. However, although structural guidelines differ according to earthquake risk, this proved not to be the case. Many buildings in high-risk earthquake zones, such as the Church of Light and Sunday School in Japan and the Water Cube in Republic of China, did not follow some structural recommendations, but solved the problems that arose through non-adherence. There are also many buildings that follow all structural guidelines in risk-free areas.

There are usually architectural reasons behind not following structural recommendations. Architects for the majority of the case studies (18 of 19) did not follow guidelines in order to achieve the design concept behind their architecture. For example, Cologne Cathedral and Mosque of Djenne had to have high walls to be able to give a spiritual effect to the interior space. Kimbell Art Museum had to have modest vaults that are not in the best structural form. The TWA Airport Building had to combine forms other than the best form in order to acheive the flying bird symbolism.

Considering the case studies in relation to their structural systems, the following statements can be made:

- Vertical elements (walls) of historical structures did not follow structural guidelines.
- Except in the case of Pantheon, horizontal structural elements (such as vaults and arches) of historical structures did follow structural guidelines.
- Half of the case studies with framed structures did not follow structural guidelines.
- Most of the case studies with tensile structures did follow structural guidelines concerning the horizontal elements in the system. One of them did not follow structural guidelines concerning the vertical elements (the columns of Zagreb Arena).
- Half of the case studies with compression structures did not follow structural guidelines.
- Two-thirds of the case studies with trussed structures did not follow structural guidelines.
- One case study with folded plates followed structural guidelines.
- One case study with a hybrid structure did not follow structural guidelines.

Fewer than half of the case studies (15 of 34) followed structural guidelines. However, this does not mean that they are less creative than the ones in which the structural guidelines were not followed. These buildings followed structural guidelines and still achieved high tectonic value: for example, Waterloo Terminal and the dome at Parliament Building in Berlin. Although these 15 buildings followed structural guidelines, three of them were contributing to knowledge with their use of innovation. These are Shanghai World Financial Center, US Pavilion Expo '67 and Yokohama International Port Terminal.

Architects are free to follow structural guidelines or not, subject of course to local building codes. Architects should, however, be aware of the implications of their decision. Tectonics can be achieved through the balance between technology and aesthetics.

Index

Aalto, A. 64–5
across wind 157
adobe masonry 65–70
aerodynamic form 157–9
Akashi Kaikyo Bridge 176–7
along wind 157
Ando, T. 140–1
arch 77–81

Bachelard, G. 14–15
Banque Populaire de l' Ouest et de l' Armorique 183–4
Barcelona Pavilion 142
beam 104–5
bending 30–2
bicycle wheel structure 182–3
Bird's Nest 235
Botticher, K. 11
bracing 118–19
brick masonry 62–5
buckling 29–30, 34–5

cables and suspension structures 45, 48, 176–80
cable-truss 181–2
Calatrava, S. 2–3, 235–6
Candela, F. 20, 205
Cantilever Method 155–6
Caribbean hut 11–12
centre of gravity 33–4
Church of Light and Sunday School 140–1
Circus Lane House 2–3
Cite la Muette 21
Cluj Arena 221–2
Cologne Cathedral 60–1, 83–4
Columbus'92 'Bigo' 187–8
column 105–6
column axes 130–1
compression 28–30
compression structures 44–5, 195–215
confined masonry 90–1
construction load 26–7
continuous foundation 57
Cordoba Mosque 80–1
corrugation 206–7
curvature 196, 202–3

damping systems 163–4
dead load 26
deflected shape 103, 120–6, 128–9
deformation limit 39–40
design of the optimum structure 21
dome 45, 48, 84–8, 197–202
Doric Temple 11
Dorton Arena 178
dynamic load 27

earthquake load 27
earthquake resistant design of frame systems 131–6
economy of structures 19–24
equations of equilibrium, three 36–9
equilibrium 36–9
evolutionary structural optimisation 20–1
expansion joint 116–17

faceted folds 229–30
Falling Water 97–8
Federal Reserve Bank Building 177–8
flat arch 47, 78
flat slab 107–8
folded plate 46, 48, 229–32
folding 206–7
force 25–6
Foreign Office Architects 231–2
form of structural systems 44–9
Foster, N. 198–9, 237–8
foundations 57, 114–16
Fox, K. P. 165–7
frame system 46, 48–9, 103–44, 147–53, 155–6, 159–63
Frampton, K. 12–13
Fuller, B. 202–3

Gaudi, A. 175
Gehry, F. 22–3
geodesic dome 48, 200–2
Great Mosque of Djenne 69–70
grid shell 211–13
Grimshaw, N. 223–4
Guggenheim Museum 22–3

Harran Houses 68–9
Hartoonian, G. 10, 12
height limits of structural systems 49, 161
Herzog & de Meuron 235
high dome 84–5
high-rise building structures 154–67
horizontal force problem 78–9, 174, 196
hybrid structures 235–40
hyperbolic paraboloid 205

individual footings 114
interior architecture 147–53
İsmail Hacı Çakır House 73–4

Johnson, P. 22

Kahn, L. 139–40, 207–8
Kimbell Art Museum 207–8

Lanificio Gatti 20
Le Corbusier 138–9
Lipstick Building 22
live load 26
load 26–7
Loos, A. 95–6

Los Manantiales Restaurant 20
Lyon Satolas Airport Railway Station 235–6

Masjid-i Jami 11–12
masonry arch 77–81
masonry dome 84–8
masonry structures 51–99
masonry vault 81–4
membrane 186–9
Method of Joints 218–19
Millennium Dome 47
moment of inertia 31–2
Munich Olympic Stadium 188–9
Murphy, R. 2–3

National Assembly Building in Dacca 139–40
Nervi, P. L. 20, 208–9
non-prismatic folds 229
N, V, M diagrams 120–9

Oklahoma State Fair Arena 236
one-way slab 106–7
optimisation of the designed structure 21–3
optimum 19
Otto, F. 188–9
overturning 33–4

Pan J.J. and Partners 112–13
Pantheon 87–8
Parliament Building in Berlin 198–9
partition walls 109–10, 134–6
Photographer's Penthouse 150–1
pile foundation 115–16
plan irregularities 132–4
pneumatic structures 189–91
Portal Method 126–9
post and lintel system 103
pre-stressing 40
pre-tensioned cables 173–4, 179–81
prismatic folds 229
PTV Architects 226–7

raft foundation 115
reinforced masonry 90–9
ribbed slab 106–8
Rice, P. 21, 183, 185–6, 237
Rogers, R. 47

Saarinen, E. 179, 209–10
San Lorenzo Church in Turin 86
Science and Technology Museum in Paris 21, 23, 183, 185–6
Sekler, E. F. 11–12
Semper, G. 11–12
shallow dome 84–5
Shanghai World Financial Center 165–7
shear 29–30
shear wall 49, 118–19, 159–62
shell structures 202–10
short-column problem 135–6
slab 106–10
slab-on-ground foundation 114–15
sliding 34–5
Slovak Radio Building 34
Small Sports Palace 208–9
soft-storey problem 134–5
space frame 48, 224–7
span limits of structural systems 48
stability 33–6
stairs 110–13
Stansted Airport 237–8
State Hermitage Museum 2–3
stone masonry 53–62
stone tower in Norman Castle 61–2
strength 25–33
stress 27–33
structural efficiency 19
structural guidelines: for adobe masonry 67–8; for bicycle wheels and suspended glass systems 184; for brick masonry 63–4; for cables and suspension structures 179; for different approaches to economy 23–4; for folded plates 231; for frame and shear wall systems 137–8; for geodesic domes 202; for grid shells 212; for high-rise building structures 163; for interior architecture 150; for masonry arches 80; for masonry domes 87; for masonry vaults 83; for membranes 188; for pneumatic structures 190; for reinforced masonry 92–3; for shells 207; for space frames 226; for steel vaults and domes 198; for stone masonry 59–60; for strength, stability, equilibrium and deformation limit requirements 41; for 3D trusses 223; for timber masonry 73; for trusses 221; definition 5-6; due to building form and size 48–9
Suan Lien Center 212–13
Summer House 64–5
Suspended Bedroom 151–2
suspended glass systems 183–6
Swarovski Pavilion in Basel 190–1
Sydney Opera House 238–40

tectonics 2–4, 10–15
temperature load 27
tensile structures 173–92
tension 27–9
3D truss 46–8, 222–4
three equations of equilibrium 36–9
Tiganas, si D. 221–2
timber masonry 70–4
torsion 32–3

truss 216–22
Tschumi, B. 21, 183, 185–6
tubular structures 159–62
TWA Airport Building 209–10
twisting instability 133–4
two-way slab 106–7

UHPFRC 105
uneven settlement 35
UPI-2M Ltd 179–81
US Pavilion at Expo'67 202–3
Utzon, J. 204, 237, 238–40

van der Rohe, M. 142–3
vault 45, 48, 81–4, 197–8
Veech Media Architecture 190–1
vertical irregularities 134–6
Villa Muller 95–6
Villa Savoye 138–9
vortex shedding 157–8

waffled slab 108–9
Water Cube 226–7
Waterloo Terminal 223–4
wind bracing 221
wind instability 36, 173–4
wind load 26–7
World Trade Center 164–5
Wright, F. L. 97–8

Yale Hockey Rink 179
Yamasaki, M. 164–5
Yokohama International Port Terminal 231–2

Zagreb Arena 179–81